ENCYCLOPAEDIA OF
GEOGRAPHY

ENCYCLOPAEDIA OF GEOGRAPHY

Vivek Sharma

ANMOL PUBLICATIONS PVT. LTD.
NEW DELHI - 110 002 (INDIA)

ANMOL PUBLICATIONS PVT. LTD.
H.O.: 4374/4B, Ansari Road, Darya Ganj,
New Delhi-110 002 (India)
Ph.: 23278000, 23261597

B.O.: No. 1015, Ist Main Road, BSK IIIrd Stage
IIIrd Phase, IIIrd Block
Bangalore - 560 085 (India)
Visit us at: www.anmolpublications.com

Encyclopaedia of Geography

First Published, 2006

ISBN 81-261-3023-7

PRINTED IN INDIA

Printed at Mehra Offset Press, Delhi.

Preface

This book is intended for all who have an interest in Geography, be it a student, a teacher and professionals, who must understand, interpret, and communicate information to others. Many of the terms used are technical, and few people can understand all of them. This Encyclopeadia of Geography is intended to help such understanding. It provides clear, authoritative definitions of terms associated with all aspects of Geography.

This Encyclopaedia of Geography broadly covers various aspects of Geography with illustrations. A very well written Encyclopaedia of geographical terms covering all aspects of theoretical and practical Geography, including Foundations of Geography, Natural Geography, Physical Geography, Human Geography, Economic Geography, Maps and Mapping, Applied Technologies, Earth and marine Geography. Terms are easy to find and the definitions are exact and to the point and where needed are explained in detail. The description of minerals is particularly helpful. Whether a complete beginner, a college student, or a seasoned professional, this book will fill the gaps in background knowledge and make Geography easy to understand for everyone.

We hope that professionals and students will find this reference volume useful. Although we could not include every concept in Geography, we have endeavored to include those that our colleagues and we deemed most important. The aim of this volume is to provide a concise repository of number of major concepts in this field. In preparation of this book, the author has freely consulted large number of books and journals so no authenticity is claimed. Author is especially thankful to Anmol Publications Pvt. Ltd., New Delhi for shaping this book in its final form. Suggestions for further improvement of this book are not only welcome but also greatly appreciated.

Author

A

A horizon Soil horizon normally found below the O horizon and above the B horizon. This layer is characterized by the following two features:

1. A layer in which humus and other organic materials are mixed with mineral particles.

2. A zone of translocation from which eluviation has removed finer particles and soluble substances.

Abiotic Non-living. Ecosystems consist of abiotic or non-living components and biotic or living components. Abiotic components include physical factors such as soil, water and climate. The mix of these elements will influence the nature and number of living organisms in a particular environment. Human activities have altered abiotic components in many ecological systems; positively through soil improvement and irrigation in agriculture, for example, or negatively through pollution and other forms of environmental degradation.

Ablation Surface removal of ice or snow from a glacier or snowfield by melting, sublimation, and/or calving.

Ablation zone Region in a glacier where there is a surface net removal of snow and/or ice by melting, sublimation, and/or calving.

Abrasion Physical wearing and grinding of a surface through friction and impact by material carried in air, water, or ice.

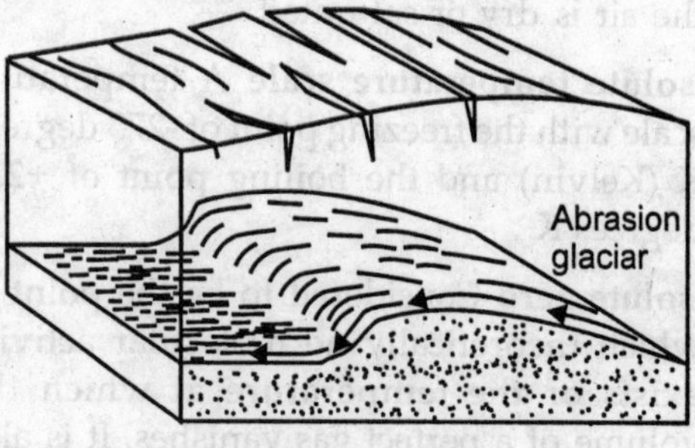

Fig. Abrasion

Absolute humidity A measure of the amount of water vapour in the atmosphere. It is

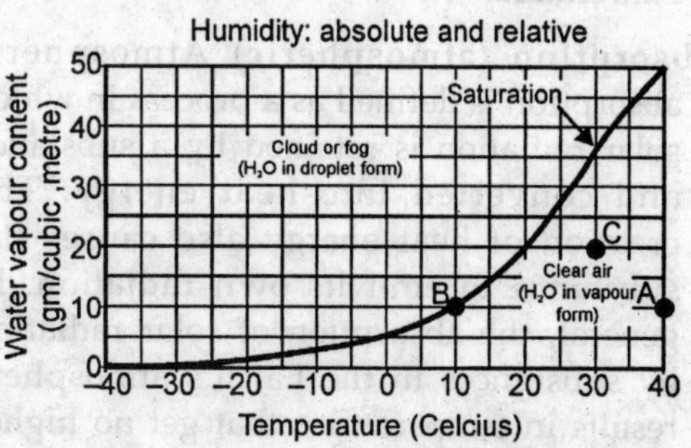

Fig. Absolute humidity

usually expressed as the mass of water vapour per unit volume of air – grams per cubic metre. Since it is based on volume, which responds to temperature and pressure changes, absolute humidity values (in a moving air mass, for example) will change, even when no water vapour is added or lost. To overcome this, meteorologists often use mixing ratio or specific humidity values as indicators of the amount of water vapour in the air. Since both involve mass rather than volume, they are not affected by changing temperature or pressure. Absolute humidity values are not easily determined through direct sampling, but can be calculated through wet-bulb temperature readings.

Absolute instability When the lapse rate of a column of air is greater than the dry adiabatic lapse rate. The term absolute is used because this applies whether or not the air is dry or saturated.

Absolute temperature scale A temperature scale with the freezing point of -273 degrees K (Kelvin) and the boiling point of +273 degrees K.

Absolute zero Considered to be the point at which theoretically no molecular activity exists or the temperature at which the volume of a perfect gas vanishes. It is also the zero point of the Kelvin Temperature Scale. The value is -273.16 degrees Kelvin, -273 degrees Celsius and -459.6 degrees Fahrenheit.

Absorption (atmospheric) Atmospheric absorption is defined as a process in which solar radiation is retained by a substance and converted into heat energy. The creation of heat energy also causes the substance to emit its own radiation. In general, the absorption of solar radiation by substances in the Earth's atmosphere results in temperatures that get no higher than 1800° Celsius. According to Wien's Law, bodies with temperatures at this level or lower would emit their radiation in the long wave band.

Abstract space Geographic model or representation of the real world. For example, maps and globes are abstractions of the real world or concrete space.

Abyssal fan Fan shaped accumulation of sediment from rivers that is deposited at the base of a submarine canyon within a ocean basin.

Abyssal plain The flat, gently sloping or nearly level region of the sea floor.

Abyssal zone The deepest part of the oceans lying more than 2000 m beneath the surface. The main physical feature of this zone is the abyssal plain, a region of low relief beyond the continental margins. Its surface is broken in places by volcanic hills or seamounts rising out of the plain, and, particularly in the Pacific Ocean, by linear trenches that reach depths in excess of 12,000 m beneath the sea surface. Sediments deposited in the upper levels of the oceans ultimately reach the abyssal zone, but little light penetrates, energy levels are low and plant or animal life is sparse. The water in the abyssal zone is considered to be in long-term storage, and its participation in the hydrological cycle has a time-scale of centuries.

Accessibility A locational characteristic that permits a place to be reached by the efforts of those at other places.

Accessibility resource A naturally occurring landscape feature that facilitates interaction between places.

Acclimation Slow adjustment of an organism to new conditions in its environment.

Accretion The growth of the continental masses over geologic time via the addition of marine sediments. These sediments are added on to the edges of the continents through tectonic collision with other oceanic or continental plates.

Accumulation Surface addition of snow to a glacier or snowfield.

Accumulation zone 1. Region in a glacier where there is a surface net addition of snow.

2. Part of a hill slope that has a net gain of material leading to a progressive raising of the slope's surface.

Accuracy Is the degree to which information on a map or in a digital database matches true or accepted values. Accuracy is an issue pertaining to the quality of data and the number of errors contained in a dataset or map. In discussing a GIS database, it is possible to consider horizontal and vertical accuracy with respect to geographic position, as well as attribute, conceptual, and logical accuracy. The effect of inaccuracy and error on a GIS solution is the subject of sensitivity analysis . Accuracy, or error, is distinguished from precision, which concerns the level of measurement or detail of data in a database.

Acid 1. Substance having a pH less than 7.

2. Substance that releases hydrogen ions (H^+).

Acid deposition Atmospheric deposition of acids in solid or liquid form on the Earth's surface.

Acid mine drainage The seepage of sulphuric acid (H_2SO_4) from mining operations into adjacent waterways. It is most common in coal-mining areas, but is also associated with nickel (Ni) and copper (Cu) mining where the ores contain sulphur (S) compounds. The most common source of sulphur is iron sulphide (pyrites), which on exposure to oxygen (O) and water, and with the help of bacteria such as Thiobacillus thioxidans, is converted into sulphuric acid. In the process ferric hydroxide is precipitated, giving acid mine drainage its characteristic yellow-brown colour. The precipitate also builds up as a brown coating on rocks and sediments in stream beds. Acid mine drainage is highly acidic and corrosive, and contains toxic metals such as aluminium (Al), copper (Cu), zinc (Zn), manganese (Mn) and beryllium (Be) leached from the local bedrock by the acid. This combination of acid and minerals completely disrupts the aquatic ecosystem, and may render the water unsuitable for municipal, industrial or recreational use. In the United States, 2.7 million tonnes of acids are estimated to drain every year from existing and abandoned mine workings, polluting some 12,000 km of streams. Methods developed to reduce the acidity of mine drainage include the addition of an alkaline buffer such as limestone to the backfill when strip-mining areas are being rehabilitated and the storage of water from working mines in holding ponds, where it can be neutralized before release. Such methods can be incorporated relatively easily into current operations, but to apply them to abandoned mine workings, which make a major contribution to acid mine drainage, would take many years and would be costly.

Acid precipitation Atmospheric precipitation with a pH less than 5.6. Normal pH of precipitation is 5.6.

Acid rain Acid rain is normally considered to be a by-product of modern atmospheric pollution. Even in a pure, uncontaminated world, however, it is likely that the rainfall would be acidic. Carbonic acid (H_2CO_3) is formed when carbon dioxide (CO_2) is absorbed by atmospheric water. Nitric acid (HNO_3 is created during thunderstorms and sulphuric acid (H_2SO_4) is formed from the sulphur dioxide (SO_2) released during volcanic eruptions and forest fires or from the sulphur (S) emitted by phytoplankton during their seasonal bloom period. All of these are natural processes which contribute to make normal rain acid. Current concern over acid rain is not with the naturally produced variety, but rather with that which results from modern industrial activity and the acid gases it produces.

Considerable amounts of sulphur dioxide are released into the atmosphere as a by-product of metal smelting, particularly when non-ferrous ores are involved. The burning of coal and oil to provide energy for space heating or to fuel thermal electric power stations also produces sulphur dioxide. The continuing growth of transportation systems using the internal combustion engine contributes to acid rain through the release of oxides of nitrogen (NO_x) into the atmosphere.

Acid precipitation produced by human activities differs from natural acid precipitation not only in its origins, but also in its quality. Anthropogenically produced acid rain tends to be many times more acidic than the natural variety. The difference is commonly of the order of 1.0 to 1.5 points on the pH scale used to measure acidity. In North America, for example, naturally acid rain has a pH of about 5.6, whereas measurements of rain falling in southern Ontario, Canada frequently provide values in the range of 4.5 to 4.0. To put these values in perspective, vinegar has a pH of 2.7 and milk a pH of 6.6. Similar values for background levels of acidic rain are indicated by studies in Europe with the average annual pH of rain over Britain between 1978 and 1980 being within the range of 4.5 to 4.2.

The quality of the rain is determined by a series of chemical processes set in motion when acidic materials are released into the atmosphere. Some of the sulphur dioxide and oxides of nitrogen emitted will return to the surface quite quickly, and close to their source, as dry deposition. The remainder will be carried up into the atmosphere, to be converted into sulphuric and nitric acid, which will eventually return to earth as acid rain. The processes involved are fundamentally simple.

Oxidation converts the gases into acids, in either a gas or liquid phase reaction, at a rate which depends upon such variables as the concentration of heavy metals in airborne particulate matter, the presence of ammonia and the intensity of sunlight. The metals and ammonia appear to act as catalysts, while the sunlight provides the energy required by the chemical reactions. Acid rain remained mainly a local problem in the past. The introduction of the tall stacks policy in an attempt to reduce local pollution, however, increased its geographical extent. The release of pollutants at the greater altitudes possible with the taller smokestacks placed them outside the boundary layer circulation and into the larger scale atmospheric circulation system with its potential for much greater dispersal through the mechanisms involved in the Long Range Transportation of Air Pollution (LRTAP). Local problems were mitigated at the expense of the larger environment. The main sources of acid rain are to be found in the industrialized areas of the northern hemisphere. North-eastern North America, Britain and Western Europe have received most attention, although their output of sulphur dioxide has been declining. Levels of oxides of nitrogen have not yet experienced significant decline and continue to rise in some areas. The emission of acid gases from Eastern Europe and the republics of the former USSR – Russia, Ukraine and Kazakhstan – remains high. In Asia, Japanese industries emit large quantities of sulphur dioxide, while the industrial areas of China are also major contributors. Acid emissions remain limited outside of the major industrial nations, but concern has been expressed over growing levels of air pollution, often associated with urban automobile exhaust emissions, which may already have provided a base for acid rain in some Third World countries.

Once the acid gases have been released into the atmosphere, they are at the mercy of the prevailing circulation patterns. With almost all of the areas currently producing large amounts of acidic pollution located within the mid-latitude westerly wind belt, emissions are normally carried eastwards, or perhaps north-eastwards, often for several hundred kilometres before being redeposited. In this way, pollutants originating in the US Midwest cause acid rain in Ontario, Quebec and the New England states. Emissions from the smelters and power stations of the English Midlands and the Ruhr contribute to the acidity of precipitation in Scandinavia, and acidity in the Arctic originates as far as 8000 km away to the south in North America and Eurasia. Thus the problem of acid rain transcends national boundaries, introducing political overtones to the problem and creating the need for international cooperation if a solution is to be found. The impact of acid rain on the environment depends not only on the level of acidity in the rain, but also on the nature of the environment itself. Areas underlain by acidic rocks such as granitic or quartzitic bedrock, for example, are particularly susceptible to damage, lacking as they do the ability to 'buffer' or neutralize additional acidity from the precipitation. Acid levels therefore rise, the environmental balance is disturbed and serious ecological damage is the inevitable result. In contrast, areas which are geologically basic – underlain by limestone or chalk, for example – are much less sensitive, and may even benefit from the additional acidity. The highly alkaline soils and water of areas underlain by chalk or limestone ensure that the acid added to the environment by the rain is very effectively neutralized. The impact of acid rain on the environment was first recognized in the lakes and rivers of these areas. Reduced pH values which indicated the rising acidity were accompanied by low levels of calcium (Ca) and magnesium (Mg), elevated sulphate concentrations and an increase in the amounts of potentially toxic metals such as aluminium (Al). When aquatic communities were examined in areas as far apart as New York State, Nova Scotia, Norway and Sweden there was clear evidence that increased surface water acidity had adverse effects on fish.

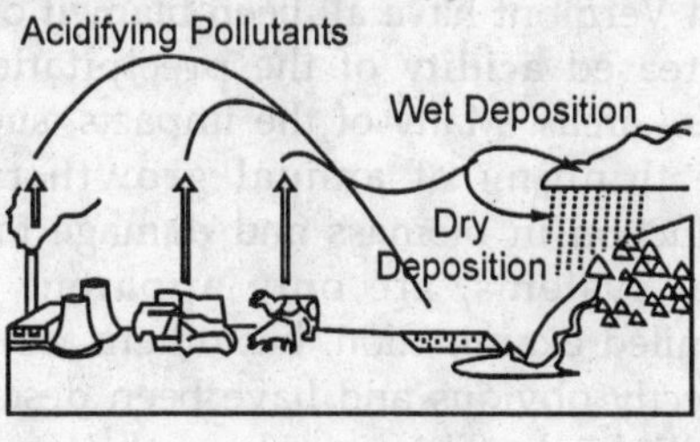

Fig. Acid rain

In some cases the acidity was sufficiently high that mature fish died, but, more commonly, fish populations began to decline because of the effects of the increasing acidity on reproduction. Damage to the eggs during spawning and the inability of the young fry to survive the higher acidity, particularly during the spring flush, ensured that the older fish were not replaced as they died. As a result, fish populations in many rivers and lakes in eastern North America, Britain and Scandinavia have declined noticeably in the last two to three decades and hundreds of lakes are now completely devoid of fish. Although the impact on the fish is most obvious, all aquatic organisms will decline in number and variety during progressive acidification. Waterbodies that have lost or are in the process of losing their flora and fauna are often described as 'dead' or 'dying'. This is not strictly correct, however, for some organisms are remarkably acid-tolerant, and even the most acid lakes have some life in them. Species of protozoans are found at pH levels as low as 2.0, for example. There is growing evidence that those areas in which the waterbodies have already succumbed to acidification must also face the effects of increasing acid stress on their forests and soils. The threat is not universally recognized, however, and there remains a great deal of controversy over the amount of damage directly attributable to acid rain. Reduction in forest growth in Sweden, physical damage to trees in West Germany, and the death of sugar maples in Quebec and Vermont have all been blamed on the increased acidity of the precipitation in these areas. Many of the impacts, such as the thinning of annual growth rings, reduction in biomass and damage to fine root systems, are only apparent after detailed examination, but others are more directly obvious and have been described as dieback. This involves the gradual wasting of the tree inwards from the outermost tips of its branches. The process is cumulative over several years until the tree dies. The symptoms of dieback have been recognized in the maple groves and red spruce forests of north-eastern North America, and across fifteen countries and some 70,000 km^2 of forest in Europe. In Germany, where it was first linked with acid rain, dieback, or Waldsterben, was particularly extensive and in the 1980s was seen as a major threat to the survival of the German forests. More recent research has indicated that forest decline in Europe is a multifaceted process in which acid precipitation may be only one of a series of contributors along with tree harvesting practices, drought and fungal attacks, although its exact role remains controversial. Present concern over acid rain is concentrated mainly on its effect on the natural environment, but acid rain also contributes to deterioration in the built environment: it attacks limestone and marble used as building stone, for example.

The faceless statues and crumbling cornices of the world's famous palaces, castles, abbeys and cathedrals, from the Parthenon in Greece to the Taj Mahal in India, attest to its power. By attacking the fabric of these buildings it not only causes physical and economic damage, but threatens the world's cultural heritage. Acid rain in the form of aerosols or attached to smoke particles can cause respiratory problems in humans, as was the case with the infamous London Smog of 1952. There is no evidence that wet deposition is directly damaging to human health, although in its ability to mobilize metals, from lead (Pb) or copper (Cu) pipes, for example, it may have important indirect effects. Solutions to the problem of acid rain are deceptively simple. In theory, a reduction in the emission rate of acid forming gases is all that is required to slow down and eventually stop the damage being caused by the acidification of the environment. Translating that concept into reality has proved difficult, however. One of the first approaches involved adaptation, such as the addition of lime to land or lakes to counteract the increased acidity, and to allow the recovery mechanisms to work more effectively. Experiments involving the liming of lakes in Canada and Sweden, for example, have provided encouraging results. Artificial buffering of lakes in this way is only a temporary measure which may be likened to the use of antacid to reduce acid indigestion. The neutralizing effects of the lime may last longer than those of the antacid, but they do wear off in three to five years and re-liming is necessary as long as acid loading continues. Most of the current proposals for dealing with acid rain tackle the problem at its source. They attempt to prevent, or at least reduce, emissions of acid gases into the atmosphere, and since sulphur dioxide makes the greatest contribution to acid rain in North America and Europe, it has received most attention in the development of abatement procedures.

Sulphur dioxide is formed when coal and oil are burned to release energy, and the technology to control it may be applied before, during or after combustion. The exact timing will depend upon such factors as the amount of acid reduction required, the type and age of the system and the cost-effectiveness of the particular process. Flue gas desulphurization (FGD) is the most common approach to sulphur dioxide reduction. It uses scrubbers to neutralize the acidity of the flue gases before they are released into the atmosphere. With sulphur dioxide removal rates of between 80 and 95 per cent, FGD is more effective than methods such as fuel switching or fuel desulphurization applied prior to combustion and it is technically simpler than fluidized bed combustion (FBC) and lime injection multi-stage burning (LIMB) incorporated in the combustion process itself. Although FGD systems do little to reduce emissions of oxides of nitrogen they are the preferred method of dealing with acid rain, and future requirements to reduce sulphur dioxide levels will be met in large part by the installation of FGD equipment.

The study of acid rain has declined remarkably since the 1980s, when it was viewed by many as the major environmental problem facing the northern hemisphere. It is one environmental issue in which abatement programmes have met with some success. Sulphur dioxide levels continue to decline, the rain in many areas is considerably less acid than it was a decade ago, and the transboundary disputes which absorbed large amounts of time, energy and money in the 1980s have been resolved. There are indications that lakes and forests are showing some signs of recovery in the most vulnerable areas of North America and Europe, although this is still a matter of dispute

Developments such as these have created the perception that the acid rain problem is being solved. It would be more accurate, however, to say that it has changed in nature and in geographical extent. In the developed nations, oxides of nitrogen are gradually making a larger contribution to acidity as levels of sulphur dioxide decline. In less developed areas, from Eastern Europe to China, the full impact of acid rain may still be in the future, and the research activities currently winding down in Western Europe and North America may have to be revived to deal with it.

Acid shock A sudden acidification of runoff waters from the spring melting of accumulated snow in the middle latitudes because of the winter deposition of acidic precipitation.

Acidic Any substance with a pH below 7.

Acidic solution Any water solution that is acidic (pH less than 7) or has more hydrogen ions (H^+) than hydroxide ions (OH^-).

Acquisition of signal The time you begin receiving a signal from a spacecraft. For polar-orbiting satellites, radio reception of the APT signal can begin only when the polar-orbiting satellite is above the horizon of a particular location. This is determined by both the satellite and its particular path during orbit across the reception range of a ground station.

Actinomycetes Group of micro organisms that are intermediate between bacteria and fungi.

Active layer Upper zone of soil in higher latitude locations that experiences daily and seasonal freeze-thaw cycles.

Active remote sensing Form of remote sensing where the sensor provides its own source of electromagnetic radiation to illuminate the object understudy. Radar is an example of an active remote sensing device.

Active system (active sensor) A remote-sensing system that transmits its own radiation to detect an object or area for observation and receives the reflected or transmitted radiation. Radar is an example of an active system. Compare with passive system.

Actual evapotranspiration Is the amount of water that is actually removed from a surface due to the processes of evaporation and transpiration.

Actual mixing ratio Another term used to describe mixing ratio.

Adaptation 1. Evolutionary adaptation - a genetically based characteristic expressed by a living organism. Particular adaptations found in populations become frequent and dominant if they enhance an individual's ability to survive in the environment.

2. Physiological adaptation - change in an organism's physiology as a result of exposure to some environmental condition.

Adaptive radiation The evolution of a number of new species from one or a few ancestor species over many thousands or millions of years. Normally occurs after a mass extinction creates a number of vacant ecological niches or when a radical change in the environment produces new ecological niches.

Adiabatic A process in which heat does not enter or leave a system. In the atmospheric sciences, adiabatic processes are often used to model internal energy changes in rising and descending parcels of air in the atmosphere. When a parcel of air rises in expands because of a reduction in pressure. If no other non-adiabatic processes occur (like condensation, evaporation and radiation), expansion causes the parcel of air to cool at a set rate of 0.98° per 100 metres. The opposite occurs when a parcel of air descends in the atmosphere. The air in a descending parcel becomes compressed. Compression causes the temperature within the parcel to increase at a rate of 0.98° per 100 metres.

Adiabatic cooling The cooling of a rising parcel of air due to adiabatic processes.

Adiabatic process A process that involves thermodynamic change in a system with no transfer of heat or mass between the system and its surroundings. It applies in the operation of such devices as air compressors, internal combustion engines and aerosol spray cans, but the impact of adiabatic change is particularly prominent as air masses travel through the atmosphere. In the absence of heat exchange, the temperature of an air mass is governed by changing atmospheric pressure. When an air mass rises it moves into an area of reduced pressure, and therefore expands. That expansion in turn leads to a fall in temperature which is proportional to the pressure change. A descending air mass will be compressed as the atmospheric pressure increases, and its temperature will rise. If the air mass is dry, the rate of change will be constant at the dry adiabatic lapse rate (DALR). If the air mass is saturated or becomes saturated, it will cool (or warm up) at the saturated adiabatic lapse rate (SALR), which is not constant. The relationship between the adiabatic lapse rates and the environmental lapse rate (ELR) determines the stability of the atmosphere, which has implications for the accumulation or dispersal of air pollutants.

Adjustment Process designed to remove inconsistencies in measured or computed quantities by applying derived corrections to compensate for random or accidental errors.

Adjustment, land- line Positioning land lines on a map to indicate their true, theoretical, or approximate location relative to the adjacent terrain and culture, by reconciling the information shown on Bureau of Land Management plats and field records with the ground evidence of the location of the lines.

Adjustment, standard accuracy Adjustment of a survey resulting in values for positions and (or) elevations that comply with the National Map Accuracy Standards.

Advanced very high resolution radiometre (AVHRR) A five-channel scanning instrument that quantitatively measures electromagnetic radiation, flown on NOAA environmental satellites. AVHRR remotely determines cloud cover and surface temperature. Visible and infrared detectors observe vegetation, clouds, lakes, shorelines, snow, and ice. TIROS Automatic Picture Transmissions (APT) are derived from this instrument.

Advection Advection involves the transfer of heat energy by means of horizontal mass motions through a medium.

Advection fog Fog generated by winds that contrast in temperature with the Earth's surface. Warm air advection can produce fog through contact cooling with a cold surface.

Advisory Statements that are issued by the National Weather Service for probable weather situations of inconvenience that do not carry the danger of warning criteria, but, if not observed, could lead to hazardous situations. Some examples include snow advisories stating possible slick streets, or fog advisories for patchy fog condition causing temporary restrictions to visibility.

Aeolian Geomorphic process involving wind.

Aeolian landform A landform formed from the erosion or deposition of weathered surface materials by wind. This includes landforms with some of the following geomorphic features: sand dunes, deflation hollows, and desert pavement.

Aerial photography Form of remote sensing that captures images of objects using photographic cameras and film from platforms in the atmosphere.

Aerobic 1. Presence of molecular oxygen.

2. Occurring only in the presence of molecular oxygen.

3. Growing in the presence of molecular oxygen.

Aerosol Particles of liquid or solid dispersed as a suspension in gas.

Aerotriangulation The process of developing a network of horizontal and or vertical positions from a group of known positions using direct or indirect measurements from aerial photographs and mathematical computations.

Afforestation The act or process of establishing a forest, especially on land not previously forested.

Aftershock Smaller earth tremors that occur seconds to weeks after a major earthquake event.

Aggradation Readjustment of the stream profile where the stream channel is raised by the deposition of bed load.

Agronomy Field of science that studies phenomena related to agriculture.

Air This is considered the mixture of gases that make up the earth's atmosphere. The principal gases that compose dry air are Nitrogen (N_2) at 78.09%, Oxygen (O_2) at 20.946%, Argon (A) at 0.93%, and Carbon Dioxide (CO_2) at 0.033%. One of the most important gases in meteorology is water vapour (H_2O).

Air mass A body of air whose temperature and humidity characteristics remain relatively constant over a horizontal distance of hundreds to thousands of kilometres. Air masses develop their climatic characteristics by remaining stationary over a source region for a number of days. Air masses are classified according to their temperature and humidity characteristics.

Air mass thunderstorm A thunderstorm that is produced by convection within an unstable air mass through an instability mechanism. Such thunderstorms normally occur within a tropical or warm, moist air mass during the summer afternoon as the result of afternoon heating and dissipate soon after sunset. Although such thunderstorms are not generally associated with fronts, they are less likely to become severe than other types of thunderstorms. However, that does not preclude them from having brief heavy down pours.

Air pollution Toxification of the atmosphere through the addition of one or more harmful substances in the air. Substance must be in concentrations high enough to be hazardous to humans, other animals, vegetation, or materials. Air pollution is a broad term applied to any chemical, physical (e.g. particulate matter), or biological agent that modifies the natural characteristics of the atmosphere. The atmosphere is a complex, dynamic natural system that is essential to support life on planet earth. Stratospheric ozone depletion due to air pollution has long been recognized as a threat to human health as well as to the earth's ecosystems.

Worldwide air pollution is responsible for large numbers of deaths and cases of respiratory disease. Setting of air quality standards, has begun to improve air quality for some pollutants. While major stationary sources are often identified with air pollution, the greatest source of emissions are actually mobile sources, principally the automobile. There are many available air pollution control technologies and urban planning strategies available to reduce air pollution; however, worldwide costs of addressing this issue are high. The most immediately available method of improving air quality is adoption of greater use of bio ethanol fuel, bio diesel, solar energy and increased reliance upon hybrid vehicle technology. It is estimated that three million people indirectly die of respiratory and cardiovascular disease worldwide, many of which cases are linked to air quality and smog. Many of these mortalities are largely attributable to indoor air pollution. Worldwide more deaths per year are linked to air pollution than the number deaths caused by automobile accidents. Research published in 2005 suggests that 310,000 Europeans die from air pollution annually. Direct causes of air pollution related deaths include aggravated asthma, bronchitis, emphysema, lung and heart diseases, and other respiratory allergies. The US EPA estimates that a proposed set of changes in diesel engine technology could result in 12,000 fewer premature mortalities, 15,000 fewer heart attacks, 6000 fewer emergency room visits by children with asthma, and 8900 fewer respiratory-related hospital admissions each year in the United States. Air pollutants are classified as either directly released or formed by subsequent chemical reactions. A direct release air pollutant is one that is emitted directly from a given source, such as the carbon monoxide or sulphur dioxide, all of which are by-products of combustion; whereas, a subsequent air pollutant is formed in the atmosphere through chemical reactions involving direct release pollutants. The formation of ozone in photochemical smog is the most important example of a subsequent air pollutant.

Air pressure The weight of the atmosphere over a particular point, also called barometric pressure. Average air exerts approximately 14.7 pounds (6.8 kg) of force on every square inch (or 101,325 Newton on every square metre) at sea level.

Air quality standards The maximum level which will be permitted for a given pollutant. Primary standards are to be sufficiently stringent to protect the public health. Secondary standards must protect the public welfare, including property and aesthetics.

Alaskan winds The down slope air flow that blows through the Alaskan valleys. It is usually given local names, such as Knik, Matanuska, Pruga, Stikine, Taku, Take, Turnagain, or Williwaw.

Albedo The percent of reflectivity of wavelengths of an object's surface. This varies according to the texture, colour, and expanse of the object's surface. Surfaces with high albedo include sand and snow, while low albedo rates include forests and freshly turned earth.

Alberta clipper A fast moving, snow-producing weather system that originates in the lee of the Canadian Rockies. It moves quickly across the northern United States, often bring gusty winds and cold Arctic air.

Aleutian low A semi-permanent, sub polar area of low pressure located in the Gulf of Alaska near the Aleutian Islands. It is a generating area for storms and migratory lows often reach maximum intensity in this area. It is most active during the late fall to late spring. During the summer, it is weaker, retreating towards the North Pole and becoming almost nonexistent. During this time, the North Pacific High pressure system dominates.

Alfisols Soil associated with forest vegetation. Upper layers of this soil are relatively rich in organic matter. Whitish layer found in the A horizon because of eluviation. Illuvial layer forms in the B horizon.

Algae A simple photosynthetic plant that usually lives in moist or aquatic environments. The bodies of algae can be unicellular or multicellular is design.

Algorithm A mathematical relation between an observed quantity and a variable used in a step-by-step mathematical process to calculate a quantity.

In the context of remote sensing, algorithms generally specify how to determine higher-level data products from lower-level source data. For example, algorithms prescribe how atmospheric temperature and moisture profiles are determined from a set of radiation observations originally sensed by satellite sounding instruments.

Alidade Instrument, or part of an instrument, for determining direction , either horizontal or vertical . In its simplest form, a peepsight or telescope mounted on a straightedge and used for plotting directions graphically. In such instruments as transits and theodolites, the alidade is the part containing the telescope and its attachments.

Alien species Species that is not naturally found in a region.

Alkaline Substance capable of neutralizing acid, with a pH greater than 7.0.

Allele Alternative forms of a gene. Each form produces a unique inheritable characteristic.

Allelopathy A particular form of amensalism found in plants. In this interaction, one species produces and releases of chemical substances that inhibit the growth of another species.

Allogenic succession A succession caused by a change in environmental conditions that is unrelated to the activities of the developing plant community.

Allopatric speciation The evolution of a new species because of the isolation of a small group of individuals from the other members of a population.

Allophane Nondescript hydrated alumino-silicate substance associated with clay minerals.

Alluvia Soil or sediments deposited by a river or other running water. The structure formed is called an alluvial deposit. Flowing water associated with glaciers may also deposit alluvium, but deposits directly from ice are not alluvium. The word comes from Latin alluvius, from alluere, "to wash against." .

A river is continually picking up and dropping solid particles of rock and dirt from its bed throughout its length. Where the river flow is fast, more particles are picked up than dropped. Where the river flow is slow, more particles are dropped than picked up. Areas where more particles are dropped are called alluvial or flood plains, and the dropped particles are called alluvium. Even small streams make alluvial deposits, but it is in the flood plains and deltas of large rivers that large, geologically-significant alluvial deposits are found. Alluvium is typically made up of a variety of materials. There are fine particles of silt and clay. There are also larger particles of sand and gravel, and these are also typically present in a wide range of sizes.

The amount of solid matter carried by a large river is enormous. The names of many rivers derive from the colour that the transported matter gives the water. For example, the Huang He in China is, literally translated, Yellow river and the Missouri River in the United States is also called Big Muddy. It has been estimated that the Mississippi River annually carries 406 million tons of sediment to the sea, the Huang He 796 million tons, and the Po River in Italy 67 million tons. Alluvial deposits often contain valuable ores such as gold and platinum and a wide variety of gemstones. Throughout history many shallow lakes have been filled in with alluvium, leaving fertile plains. Alluvial soils are almost always very fertile. The alluvial mud annually deposited by the Nile has enabled the Egyptians to grow crops since at least the 4th millennium BC without artificial fertilization.

Alluvial fan Large fan shaped terrestrial deposit of alluvial sediment on which a braided stream flows over. Form as stream load is deposited because of a reduction in the velocity of stream flow.

Alluvial soils Soils deposited through the action of moving water. These soils lack horizons and are usually highly fertile.

Alluvial terraces Flat elevated benches composed of unconsolidated alluvium found either side of a stream channel. Formed when a stream down cuts into its floodplain.

Alluvium Sediment that originates from a stream.

Alpha particle Particle of matter that is positively charged. This particle consists of two neutrons and two protons and is emitted as a form of radioactivity from the nuclei of some radioisotopes.

Alpine glacier Small glacier that occupies a U-shaped valley on a mountain. Also called a mountain glacier.

Alpine permafrost Form of permafrost that exists at high altitudes in mountainous environments.

Alpine tundra High altitude biome dominated by a few species of dwarf shrubs, a few grasses, sedges, lichens, and mosses. Productivity is low in this biome because of the extremes of climate. Alpine tundra occurs at high enough altitude at any latitude on Earth. Alpine tundra also lacks trees, but does not usually have permafrost, and alpine soils are generally better drained than permafrost soils. Alpine tundra transitions to subalpine forests below the tree-line; stunted forests occurring at the forest-tundra ecotone are known as Krummholz. Notable animals in the alpine tundra include, Kea parrots, marmots, Mountain goats, and pika. Alpine tundra does not map directly to specific WWF ecoregions. Portions of Montane grasslands and shrublands ecoregions include alpine tundra. The alpine tundra also has elks,marmots, mountain goats, pikas, and sheep.

Alternative hypothesis (H1) Is a hypothesis that has been suggested because it is believed to be false or because it is to be used as a starting point for scientific argument. Used in statistical testing to organize arguments.

Altimetre Instrument for measuring altitudes or elevations with respect to a reference level, usually mean sea level. The most common type is an aneroid barometre. A radar altimetre determines the height of an aircraft above the terrain by measuring the time required for an electromagnetic pulse to travel from aircraft to the ground and back.

Altimetre setting The pressure value to which an aircraft altimetre scale is set so that it will indicate the altitude above mean sea level of an aircraft on the ground at the location for which the value was determined.

Altitude In meteorology, the measure of a height of an airborne object in respect to a constant pressure surface or above mean sea level.

Altocumulus clouds Composed of flattened thick, gray, globular masses, this middle cloud genus is primarily made of water droplets. In the mid-latitudes, cloud bases are usually found between 8,000 and 18,000 feet. A defining characteristic is its layer effect, often with a wavy billowy appearance, giving it the nickname of "sheep" or "woolpack" clouds. Sometimes confused with cirrocumulus clouds, its elements (individual clouds) have a larger mass and cast a shadow on other elements. It may form several sub-types, such as altocumulus castellanus or altocumulus lenticularis Virga may also fall from these clouds.

Altostratus clouds Gray-looking middle altitude cloud that is composed of water droplets and ice crystals. Appears in the atmosphere as dense sheet like layer. Can be recognized from stratus clouds by the fact that you can see the sun through it. Found in an altitude range from 2,000 to 8,000 metres.

AM/FM Is an acronym for automated mapping and facilities management. AM/FM applications are a specialized GIS. The systems are used by public and private utilities, city and county governments, and other organizations that require careful management of a widely distributed collection of capital equipment, stock, or "facilities." AM/FM most commonly describes databases designed to manage networks of roads, pipes, and wires. These GIS applications are the particular concern of AM/FM international, a professional organization headquartered in Aurora, Colourado.

Amensalism Interspecific interaction where one species suffers in terms of fitness, while the fitness of the other species does not change.

Amino acid Organic nitrogen containing acids which are used to construct proteins.

Ammonia Chemical compound composed of nitrogen and hydrogen (NH_3). Component of the nitrogen cycle. Immediately released from organic matter upon decomposition.

Ammonium Chemical compound composed of nitrogen and hydrogen (NH_4). Component of the nitrogen cycle. Product of organic matter decomposition. Can be fixed to clay minerals and later exchanged.

Ampere (AMP) Standard unit to measure the strength of an electric current. One amp is the amount of current produced by an electromotive force of one volt acting through the resistance of one ohm. The ampere is 10^{-1} of the theoretical electromagnetic unit of current. Named for the French physicist Andre Marie Ampere.

Amphibian Group of vertebrate animals that can inhabit both terrestrial and aquatic habitats. This group of animals consists of frogs, newts, and salamanders. These organisms live at the land/water interface and spend most of their life in water. Descended from fish and ancestors to reptiles.

Amphibole A group of double chained inosilicate minerals whose basic chemical unit is the tetrahedron (SiO_4). They are common rock forming minerals and are found in most igneous and metamorphic rocks. They form at low temperatures with the presence of water in the crystallization environment. There are about 60 recognized mineral types in this group.

Amplitude The magnitude of the displacement of a wave from a mean value.

For a simple harmonic wave, it is the maximum displacement from the mean. For more complex wave motion, amplitude is usually taken as one-half of the mean distance (or difference) between maxima and minima.

Amplitude Modulation (AM) One of three ways to modify a sine wave signal in order to make it "carry" information. The strength (amplitude) of a signal varies (modulates) to correspond to the transmitted information. As applied to APT an audible tone of 2400 Hz is amplitude modulated, with the maximum signal corresponding to light areas of a photograph, the minimum levels black, and the intermediate strengths various shades of grey.

Anabatic wind A wind that is created by air flowing uphill. Valley breezes, produced by local daytime heating, are an example of these winds. An anabatic wind, from the Greek anabatos, meaning moving upward, is a wind which blows up a steep slope or mountain side. It is also known as an upslope flow. These winds typically occur during the daytime in calm sunny weather. A hill or mountain top will be radiatively warmed by the Sun which in turn heats the air just above it. Air at a similar altitude over an adjacent valley or plain does not get warmed so much because of the greater distance to the ground below it. The effect may be enhanced if the lower lying ground is shaded by the mountain and so receives less heat.

Fig. Anabatic wind

The air over the hill top is now warmer than the air at a similar altitude around it and will rise through convection. This creates a lower pressure region into which the air at the bottom of the slope flows, causing the wind. It is common for the air rising from the tops of large mountains to reach a height where it cools adiabatically to below its dew point and forms cumulus clouds. These can then produce rain or even thunderstorms. Anabatic winds are particularly useful to soaring glider pilots who can use them to increase the aircraft's altitude. Katabatic winds are down-slope winds, frequently produced at night by the opposite effect, the air near to the ground losing heat to it faster than air at a similar altitude over adjacent low-lying land.

Anaerobic 1. Absence of molecular oxygen.

2. Occurring only in the absence of molecular oxygen.

3. Growing in the absence of molecular oxygen.

Analemma Analemma is a term in astronomy used to describe the plot of the positions of the Sun on the celestial sphere at the same time of day (at approximately 24 hour intervals) and from the same location on Earth on successive days through the calendar year. This appears as a figure eight. If the course of the Earth around the sun were a precise circle and if the Earth's polar axis stood perpendicular to the plane of its orbit, the Sun would always appear at the same point in the sky at the same time of day throughout the year and the analemma would be a dot. If the orbit were circular, but the polar axis tilted as it is, the northern and southern lobes of the figure of eight would be equal in size; if the polar axis were not tilted but the orbit had its current eccentricity then the analemma would be a straight east-to-west line. It is difficult but possible to construct an analemma by taking a photograph of the sun each day of the year at the same time and place and with a camera positioned in exactly the same way each time. Due to the earth's tilt on its axis (23.45°) and its elliptical orbit around the sun, the relative location of the sun above the horizon is not constant from day to day when observed at the same time on each day. Depending on one's geographical latitude, this loop will be inclined at different angles.

The vertical coordinate of a point corresponding to a date corresponds to the declination of the sun on that date, while the horizontal coordinate indicates whether the sun is "fast" or "slow" compared to mean time as measured by clocks. The deviation between solar time and mean time is known as the equation of time and is due to the variation in the length of the synodic day, which is due to variation in the distance between the earth and the sun. An analemma is a visual representation or illustration of the equation of time. Plotting the analemma with the width exaggerated shows that it is slightly asymmetrical due to the misalignment of apsides and solstices.

Analog Transmission of a continuously variable signal as opposed to a discretely variable signal. Compare with digital. A system of transmitting and receiving information in which one value (i.e., voltage, current, resistance, or, in the APT system, the volume level of the video tone) can be compared directly to the information (in the APT system, the white, black, and grey values) in the image.

Ancillary data Data other than instrument data required to perform an instrument's data processing. Ancillary data includes such information as orbit and/or attitude data time information spacecraft engineering data and calibration information.

Andesite An extrusive igneous rock that develops from a magma that is chemically between felsic and mafic and whose mineral crystals are fine.

Andisols Soil order (type) of the United States Department of Agriculture Comprehensive Soil Classification System. These soils develop from parent materials that are volcanic in origin.

Anemometre Mechanical instrument used to measure wind speed. These instruments commonly employee three methods to measure this phenomenon: 1. A device with three or four open cups attached to a rotating spinal. The speed of rotation is then converted into a measurement of wind speed;

2. A pressure plate that measures the force exerted by the moving wind at right angles;

3. A device consisting of a heated-wire where electrical resistance (temperature of the wire) is adjusted to account for heat lost by air flow. The faster the wind the greater the heat loss and thus the more energy that is required to keep the wire at a constant temperature. As a result, wind speed is measured through the drain of electrical current.

Aneroid barometre An instrument for measuring the atmospheric pressure. It registers the change in the shape of an evacuated metal cell to measure variations on the atmospheric pressure. The aneroid is a thin-walled metal capsule or cell, usually made of phosphor bronze or beryllium copper. The scales on the glass cover measure pressure in both inches and millibars.

Angiosperms Group of vascular plants who encase their seeds in a mature ovary or fruit.

Angle of incidence Angle at which the sun's rays or insolation strike the Earth's surface. If the sun is positioned directly over head or 90° from the horizon, the incoming insolation strikes the surface of the Earth at right angles and is most intense.

Angle of repose Measurement commonly used in civil engineering. It is the maximum angle at which a material can be inclined without failing. Geomorpologist use this measurement for determining the stability of slope to mass movements.

Animal Organisms that belong to the kingdom Animalia. General characteristics of these organisms include: eukaryotic cell type, mitochondria, and a complex nervous system. This group of life includes organisms like sponges, jellyfishes, arthropods (insects, shrimp, and lobsters), mollusks (snails, clams, oysters, and octopuses), fish, amphibians (frogs, toads, and salamanders), reptiles (turtles, lizards, alligators, crocodiles, snakes), birds, and mammals (kangaroos, bats, cats, rabbits, elephants, whales, porpoises, monkeys, apes, and humans).

Animalia Multicellular organisms that have a eukaryotic cell type, mitochondria, and a complex nervous system.

Anion An ion carrying a negative atomic charge.

Annual plant Plant species that completes its life in one growing season.

Anomalous propagation This refers to the non-standard propagation of a beam of energy, radio or radar, under certain atmospheric conditions, appearing as false (non-precipitation) echoes. Often referred to as A.P.

Anomaly 1. The deviation of (usually) temperature or precipitation in a given region over a specified period from the normal value for the same region.

2. The angular distance of an Earth satellite (or planet) from its perigee (or perihelion) as seen from the centre of the Earth (sun).

Antarctica The area around the geographic South Pole, from 90 degrees South to the Antarctic Circle at approximately 66 1/2 degrees South latitude, including the continent of Antarctica. Antarctica is the southernmost continent and encompasses the South Pole. Antarctica is surrounded by the Southern Ocean and divided by the Transantarctic Mountains. On average, it is the coldest, driest, and windiest continent and has the highest average elevation of all the continents. At 14 million km², Antarctica is the third-smallest continent after Europe and Australia and 98% of it is covered in ice. Because there is little precipitation, except at the coasts, the interior of the continent is technically the largest desert in the world. There are no permanent human residents and Antarctica has never had an indigenous population. Only cold-adapted plants and animals survive there, including penguins, fur seals, mosses, lichens, and many types of algae.

Although myths and speculation about a Terra Australis ("Southern Land") date back to antiquity, the first confirmed sighting of the continent is commonly accepted to have occurred in 1820 by the Russian expedition of Mikhail Lazarev and Fabian Gottlieb von Bellingshausen. The continent remained largely neglected for the rest of the 19th century however, because of its hostile environment, lack of resources, and isolated location. The Antarctic Treaty was signed in 1959 by 12 countries. The treaty prohibits military activities, supports scientific research, and protects the continent's ecozone. Ongoing experiments are conducted by more than 4,000 scientists from many nationalities and with different research interests.

Belief in the existence of a Terra Australis — a vast continent located in the far south of the globe to "balance" the northern lands of Europe, Asia and north Africa — had existed since Ptolemy suggested the idea in order to preserve symmetry of landmass in the world. Depictions of a large southern landmass were common in maps such as the early 16th century Turkish Piri Reis map. Even in the late 17th century, after explorers had found that South America and Australia were not part of

"Antarctica," geographers believed that the continent was much larger than its actual size. European maps continued to show this land until Captain James Cook's ships, Resolution and Adventure, crossed the Antarctic Circle on January 17, 1773 and again in 1774. The first confirmed sighting of Antarctica can be narrowed down to the crews of ships captained by two individuals. According to various organizations, ships captained by three men sighted Antarctica in 1820: Fabian von Bellingshausen (a captain in the Russian Imperial Navy), Edward Bransfield (a captain in the British Navy), and Nathaniel Palmer (an American sealer out of Stonington, Connecticut). Von Bellingshausen supposedly saw Antarctica on January 27, 1820, three days before Bransfield sighted land, and ten months before Palmer did so in November 1820. On that day the two ship expedition led by Von Bellingshausen and Mikhail Petrovich Lazarev reached a point within 32 km (20 miles) of the Antarctic mainland and saw ice fields there.

In 1841, explorer James Clark Ross passed through what is now known as the Ross Sea and discovered Ross Island. He sailed along a huge wall of ice that was later named the Ross Ice Shelf. Mount Erebus and Mount Terror are named after two ships from his expedition: HMS Erebus and HMS Terror. During an expedition led by Ernest Shackleton in 1907, parties led by T. W. Edgeworth David became the first to climb Mount Erebus and to reach the South Magnetic Pole. On December 14, 1911, a party led by Norwegian polar explorer Roald Amundsen from the ship Fram became the first to reach the geographic South Pole, using a route from the Bay of Whales and up the Axel Heiberg Glacier. This area had been previously visited by the famous "Claus Expedition".

Richard Evelyn Byrd led several voyages to the Antarctic by plane in the 1930s and 1940s. He is credited with implementing mechanized land transport and conducting extensive geological and biological research. However, it was not until October 31, 1956 that anyone set foot on the South Pole again; on that day a U.S. Navy group led by Rear Admiral George Dufek successfully landed an aircraft there.

Antarctic circle The Antarctic Circle is one of the five major circles of latitude that mark maps of the Earth. It is at latitude 66° 33' 39" south of the equator (in 2000; like for its northern counterpart, the Arctic Circle, the value is currently slowly decreasing over time, pushing the Antarctic Circle southwards with about 15 m per year). For everywhere within the Antarctic Circle, there is at least twenty-four hours of continuous daylight on the Summer Solstice in December, and at least twenty-four hours of continuous nighttimes on the Winter Solstice in June. That is to say, one whole day during which the sun does not set, and one whole day during which the sun does not rise. This is because the earth is tilted at a 23.5 degree angle, and during the winter solstice, the southern hemisphere is tilted away from the Sun, meaning that the Antarctic circle is completely tilted away from the Sun, hence it experiences 24 hour nighttimes, and vice versa. In practice these numbers are somewhat relaxed for various factors, the most important being: the atmospheric refraction, the real altitude of the observer above sea level, and mirages. Mirages on the Antarctic continent tend to be even more spectacular than in arctic regions, creating, for example, a series of sunsets and sunrises while in reality during that period the sun was under the horizon all the time. The area south of this circle is known as the Antarctic, and the zone to the north is the Southern Temperate Zone. The continent of Antarctica forms a land mass covering much of the area within the Antarctic Circle.

Antarctic high A region of high pressure that occupies central Antarctic throughout the year. This pressure system is responsible for very cold temperatures and extremely low humidity.

Antarctic ocean Although not officially recognized as a separate ocean body, it is commonly applied to those portions of the Atlantic, Pacific, and Indian Oceans that reach the Antarctic continent on their southern extremes.

Antebellum Before the war; in the United States, belonging to the period immediately prior to the Civil War (1861-1865).

Antenna A wire or set of wires used to send and receive electromagnetic waves. Two primary features must be considered when selecting antennas: beamwidth, or the "width" of the antenna pattern (wide beamwidth suggests the ability to receive signals from a number of different directions), and gain, or the increase in signal level. Generally beamwidth or gain can be increased only at the expense of the other. Gain can be increased by multiplying the number of antenna elements, although this adds "directionality" that reduces beamwidth.

Important antenna considerations include the following: 1. The physical size of antenna components is determined by the frequency of the transmissions it will receive—the higher the frequency the shorter the elements. At high frequencies, use of a satellite dish will compensate for the reduced amount of energy intercepted by shortened components.

2. The antenna design should fit the type of radio frequency (RF) signal polarization it will receive. The orientation of radio waves in space is a function of the orientation of the elements of the transmitting antenna. A circularly polarized wave rotates as it propagates through space. Antennas can be designed for either right or left-handed circular polarization. Earth-based communication antennas are either vertical or horizontal in polarization, and not suited for space communication. Police and cellular phone transmissions use vertical polarization because a simple vertical whip antenna is the easiest sort of omnidirectional antenna to mount on a vehicle.

3. The antenna needs to produce sufficient signal gain to produce noise-free reception.

4. The antenna should be clear of conductive objects such as power lines, phone wires, etc., so height above the ground becomes important.

Basic antenna components are: 1. Driven element—the parts connected to and receiving power from the receiver/transmitter;

2. Parasitic elements—the parts dependent upon resonance rather than connection to a power source;

A director or parasitic element that rein forces radiation on a line pointing to it from the driven element;

A reflector or parasitic element that rein forces radiation on a line pointing from it to the driven element.

A fundamental form of antenna is a single wire whose length approximately equals half the transmitting wavelength. Known as a dipole antenna, it is the unit from which many more complex forms of antennas are constructed.

One of the most common forms of VHF antenna is the Yagi/beam, named for the Japanese scientist who first described the principles of combining a basic dipole (driven element) and parasitic elements. A common TV antenna is an example of this type. A Yagi/beam antenna is directional and therefore includes a rotator to aim (direct) the antenna.

An omnidirectional antenna has a wide beamwidth and consequently does not require "tracking" (aiming the antenna toward the signal source). An example of an omnidirectional antenna is the turnstile antenna, a variation of the standard dipole antenna well suited for space communications. The quadrifilar helix antenna is omnidirectional and an inherently excellent antenna for ground station use. Quadrifilars are also used on NOAA's polar-orbiting environmental satellites.

The parabolic reflector or satellite dish antenna collects RF signals on a passive dish-shaped surface. A feedhorn antenna—a simple dipole antenna mounted in a resonant tube structure (cylinder with one open end)—transfers the RF energy to a transmission line. The bigger the dish, the greater the amount of RF energy intercepted, and therefore the greater the gain from the signal.

Antenna array An ordered assembly of elementary antennae spaced apart and fed in such a manner that the resulting radiation is concentrated in one or more directions.

Antenna beam The focused pattern of electromagnetic radiation that is either received or transmitted by an antenna.

Anthracite A hard coal containing little volatile matter.

Anticline A fold in rock layers that forms an arch.

Anticyclone A relative pressure maximum. An area of pressure that has diverging winds and a rotation opposite to the earth's rotation. This is clockwise the Northern Hemisphere and counter-clockwise in the Southern Hemisphere. Also known as an area of high pressure, it is the opposite of an area of low pressure, or a cyclone.

Antipodes Antipodes are a pair of points that are on opposite sides of a planet (like the North Pole and the South Pole).

Anvil The upper portion of a cumulonimbus cloud that becomes flat and spread-out, sometimes for hundreds of miles downstream from the parent cloud. It may look smooth or fibrous, but in shape, it resembles a blacksmith's anvil. It indicates the mature or decaying stage of a thunderstorm.

Aphelion The point on the earth's orbit that is farthest from the sun. Although the position is part of a 21,000 year cycle, currently it occurs around July, when the earth is about 3 million miles farther from the sun than at perihelion. This term can be applied to any other celestial body in orbit around the sun. It is the opposite of perihelion.

Apogee The point farthest from the earth on the moon's orbit. This term can be applied, to any other body orbiting the earth, such as satellites. It is the opposite of perigee.

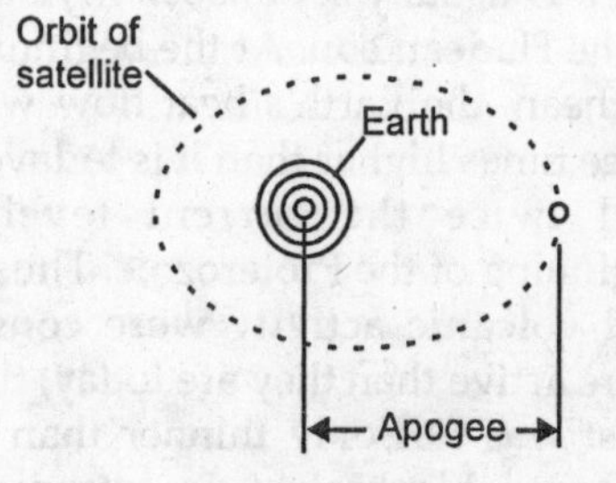

Fig. Apogee

Applied physical geography The field of Applied Physical Geography uses theoretical information from the various fields of Physical Geography to manage and solve problems related to natural phenomena found in the real world.

Aquatic With reference to water.

Aquiclude Rock formations that are impermeable to groundwater water.

Aquifer Layer of water-bearing permeable rock, sand, or gravel capable of providing significant amounts of water.

Aquifer recharge area Surface area that provides water for an aquifer.

Archaea A group of recently discovered organisms that resemble bacteria. However, these organisms are biochemically and genetically very different from bacteria. Some species of the domain Archaea live in the most extreme environments found on the Earth.

Archaebacteria Term used to describe organisms that belong to the biological domain Archaea.

Archean The Archean (Archaean; previously known as the Archaeozoic or Archeozoic) is a geologic eon that refers to the time before the Proterozoic, 2500 Ma (million years ago). Instead of being based on stratigraphy, this date is defined chronometrically. The lower boundary has not been officially recognized by the International Commission on Stratigraphy, but it is usually set to 3800 Mya at the end of the Hadean eon. At the beginning of the Archean, the Earth's heat flow was nearly three times higher than it is today, and was still twice the current level by the beginning of the Proterozoic. Thus, tectonic and volcanic activity were considerably more active than they are today; the Earth's crust was not only thinner than is today, but probably broken up into many more plates, with numerous hot spots, rift valleys, and transform faults.

There were no large continents until late in the Archean; small protocontinents were the norm, prevented from coalescing into larger units by the high rate of geologic activity. These felsic protocontinents probably formed at hot spots rather than subduction zones, from a variety of sources: mafic magma melting more felsic rocks, partial melting of mafic rock, and from the metamorphic alteration of felsic sedimentary rocks.

The Archean atmosphere apparently lacked free oxygen. Temperatures appear to have been near modern levels, although astronomers think that the sun was about one-third dimmer. This is thought to reflect larger amounts of greenhouse gases than later in the Earth's history. Although a few stones are known that are older, the oldest rock formations exposed on the surface of the Earth are Archean or slightly older. Archean rocks are known from Greenland, the Canadian Shield, Western Australia, and southern Africa. Although the first continents formed during this eon, rock this age makes up only 7% of the world's current cratons; even allowing for erosion and destruction of past formations, evidence suggests that only 5-40% of the present continental crust formed during the Archean.

In contrast to the Proterozoic, Archean rocks are often heavily metamorphized deep-water sediments, such as graywackes, mudstones, volcanic sediments, and banded iron formations. Greenstone belts are typical Archean formations, consisting of alternating high and low-grade metamorphic rocks. The high-grade rocks were derived from volcanic island arcs, while the low-grade metamorphic rocks represent deep-sea sediments eroded from the neighbouring island arcs and deposited in a forearc basin. In short, greenstone belts represent sutured protocontinents.

Fossils of cyanobacterial mats (stromatolites) are found throughout the Archean—becoming especially common late in the eon—while a few probable bacterial fossils are known from chert beds. In addition to the domain Bacteria (once known as Eubacteria), microfossils of the extremophilic domain Archaea have also been identified. Life was probably present throughout the Archean, but may have been limited to simple non-nucleated single-celled organisms, called Prokaryota (and formerly known as Monera); there are no known eurkaryotic fossils, though they might have evolved during the Archean and simply not left any fossils. However, no fossil evidence yet exists for ultramicroscopic intracellular organisms such as viruses.

Archipelago A group of islands that have an arc shaped distribution. These islands are usually of volcanic origin and are associated with subduction zones.

Arctic The Arctic is the area around the Earth's North Pole. The Arctic includes parts of Russia, Alaska (United States), Canada, Greenland (a territory of Denmark), Iceland, Norway, Sweden and Finland, as well as the Arctic Ocean. There are numerous definitions of the Arctic region.

The boundary is generally considered to be north of the Arctic Circle (66° 33'N), which is the limit of the midnight sun and the polar night. Other definitions are based on climate and ecology, such as the 10°C (50°F) July isotherm, which also roughly corresponds to the tree line in most of the Arctic. Socially and politically, the Arctic region includes the northern territories of the eight Arctic states, including Lapland, although by natural science definitions much of this territory is considered sub arctic.

The Arctic is mostly a vast, ice-covered ocean, surrounded by treeless, frozen ground. It teems with life, including organisms living in the ice, fish and marine mammals, birds, land animals and human societies. The Arctic region is by its nature a unique area. The cultures in the region and the Arctic indigenous peoples have adapted to its cold and extreme conditions. From the perspective of the physical, chemical and biological balance in the world, the Arctic region is in a key position. It reacts sensitively particularly to changes in the climate, which reflect extensively back on the global state of the environment.

From the perspective of research into climatic change, the Arctic region is considered an early warning system. The Arctic is also known as the Land of the Midnight Sun as it is within the Arctic Circle. The name Arctic comes from the ancient Greek, meaning 'bear', and is a reference to the constellations of the Great Bear and Little Bear, which are located near the North Star (which is actually part of the Little Bear).

Arctic air mass An air mass that develops around the Arctic and is characterized by being cold aloft and extending to great heights. The boundary of this air mass is often defined by the Arctic front, a semi-permanent, semi-continuous feature.

Arctic circle Latitude of 66.5° North. The southern limit of the area of the Earth that experiences 24 hours of darkness or 24 hours of day at least one day during the year.

Arctic jet The jet stream that is situated high in the stratosphere in and around the Arctic or Antarctic Circles. It marks the boundary of polar and arctic air masses.

Arctic sea smoke A type of advection fog that forms primarily over water when cold air passes across warmer waters.

Area studies tradition Academic tradition in modern Geography that investigates an area on the Earth from a geographic perspective at either the local, regional, or global scale.

Arete A sharp, narrow mountain ridge. It often results from the erosive activity of alpine glaciers flowing in adjacent valleys.

Arête Sharp topographic ridge that separates cirques on a mountain that is or has been glaciated.

Argon (a) A colourless, odourless inert gas that is the third most abundant constituent of dry air. It comprises 0.93%.

Argument of perigee One of the six Keplerian elements, it gives the rotation of the satellite on the orbit. The argument (argument meaning angle) of perigee—perigee is the point on an orbital path when the satellite is closest to the Earth—is the angle (measured from the centre of the Earth) from the ascending node to perigee. Example: When ARGP = 0 degrees, the perigee occurs at the same place as the ascending node. That means that the satellite would be closest to Earth just as it rises up over the equator. When ARGP = 180 degrees, apogee would occur at the same place as the descending node. This means that the satellite would be farthest from Earth just as it rises over the equator.

Arid A term used for an extremely dry climate. The degree to which a climate lacks effective, life-promoting moisture. It is considered the opposite of humid when speaking of climates.

Aridisols Aridisols are commonly found in dry environments that are low in organic matter and rich in deposited salts. Aridisols (from the Latin aridus, for "dry") form in an arid or semi-arid climate. Aridisols dominate the deserts and xeric shrublands which occupy about one third of the Earth's land surface. Aridisols have a very poor concentration of organic matter (In this case decomposed material, humus). Water deficiency is the major defining characteristic of Aridisols. Also required is sufficient age to exhibit sub-soil weathering and development. Imperfect leaching in Aridisols often results in one or more subsurface soil horizons in which suspended or dissolved minerals have been deposited: silicate clays, sodium, calcium carbonate, gypsum or soluble salts. These subsoil horizons can also be cemented by carbonates, gypsum or silica. Accumulation of salts on the surface can result in salinization.

Arkose A type of sedimentary sandstone that contains a large quantity of weathered feldspar grains. This type of sedimentary rock forms in arid conditions.

Arroy A deep gully cut by a stream that flows only part of the year; a dry gulch. A term normally used only in desert areas.

Artesian water Groundwater that is confined by two impermeable layers beneath the Earth's surface.

Artificial intelligence Neural networks. The branch of computer science that attempts to program computers to respond as if they were thinking—capable of reasoning, adapting to new situations, and learning new skills. Examples of artificial intelligence programs include those that can locate minerals underground and understand human speech.

Ascending node The point in an orbit (longitude) at which a satellite crosses the equatorial plane from south to north.

Asexual reproduction Any process of reproduction that does not involve the fusion of gametes.

Aspect ratio The ratio of image width to image height. Weather Facsimile (WEFAX) images have a 1:1 aspect ratio (square); a conventional TV aspect ratio is 4:3 (rectangle).

Assimilation 1. Absorption and creation of food resources.

2. Organic metabolic products of food digestion. Usually the various organic constituents of the organism.

Asthenosphere The asthenosphere is the region of the Earth between 100-200 km below the surface — but perhaps extending as deep as 400 km — that is the weak or "soft" zone in the upper mantle. It lies just below the lithosphere, which is involved in plate movements and isostatic adjustments. In spite of its heat, pressures keep it plastic, and it has a relatively low density. Seismic waves, the speed of which decrease with the softness of a medium, pass relatively slowly though the asthenosphere, the cue that originally alerted seismologists to its presence; thus it has been given the name *low-velocity zone.* Under the thin oceanic plates the asthenosphere is usually much nearer the seafloor surface, and at mid-ocean ridges it rises to within a few kilometres of the ocean floor.

The upper part of the asthenosphere is believed to be the zone upon which the great rigid and brittle lithospheric plates of the Earth's crust move about. Due to the temperature and pressure conditions in the asthenosphere, rock becomes ductile, moving at rates of deformation measured in cm/yr over lineal distances eventually measuring thousands of kilometres. In this way, it flows like a convection current, radiating heat outward from the Earth's interior. Above the asthenosphere, at the same rate of deformation, rock behaves elastically and, being brittle, can break, causing faults. The rigid lithosphere is thought to "float" or move about on the slowly flowing asthenosphere, creating the movement of crustal plates described by Plate tectonics theory. Although its presence was suspected as early as 1926, the worldwide occurrence of the asthenosphere was confirmed by analyses of earthquake waves from the Chilean Great earthquake of May 22, 1960.

Astronomical twilight The time after nautical twilight has commenced and when the sky is dark enough, away from the sun's location, to allow astronomical work to proceed. It ends when the centre of the sun is 18 degrees below the horizon.

Astronomical unit (AU) The distance from the Earth to the sun. On average, the sun is 149,599,000 kilometres from Earth.

Astronomy Field of knowledge that studies the nature, motion, origin, and constitution of celestial bodies.

ATLAS (atmospheric laboratory for applications and science) mission The focus of ATLAS is to study the chemistry of the Earth's upper atmosphere (mainly the stratosphere/mesosphere) and the solar radiation incident on the Earth system (both total solar irradiance and spectrally resolved radiance, especially ultraviolet). Science operations onboard ATLAS 1 (March 1992) and ATLAS 2 (March-April, 1993) began a comprehensive and systematic collection of data that will help establish benchmarks for atmospheric conditions and the sun's stability.

Atlas A bound collection of maps.

Atmosphere The air surrounding the Earth, described as a series of shells or layers of different characteristics. The atmosphere, composed mainly of nitrogen and oxygen with traces of carbon dioxide, water vapour, and other gases, acts as a buffer between Earth and the sun. The layers, troposphere, stratosphere, mesosphere, thermosphere, and the exosphere, vary around the globe and in response to seasonal changes. Troposphere stems from the Greek word tropos, which means turning or mixing. The troposphere is the lowest layer of the Earth's atmosphere, extending to a height of 8-15 km, depending on latitude. This region, constantly in motion, is the most dense layer of the atmosphere and the region that essentially contains all of Earth's weather. Molecules of nitrogen and oxygen compose the bulk of the troposphere.

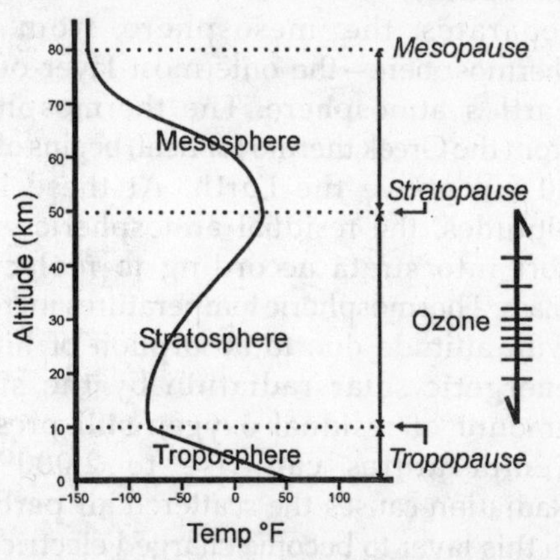

Fig. Structure of the Atmosphere

The tropopause marks the limit of the troposphere and the beginning of the stratosphere. The temperature above the tropopause increases slowly with height up to about 50 km. The stratosphere and stratopause stretch above the troposphere to a height of 50 km. It is a region of intense interactions among radiative, dynamical, and chemical processes, in which horizontal mixing of gaseous components proceeds much more rapidly than vertical mixing. The stratosphere is warmer than the upper troposphere, primarily because of a stratospheric ozone layer that absorbs solar ultraviolet energy.

The mesosphere, 50 to 80 km above the Earth, has diminished ozone concentration and radiative cooling becomes relatively more important. The temperature begins to decline again (as it does in the troposphere) with altitude. Temperatures in the upper mesosphere fall to -70° to -140° Celsius, depending upon latitude and season. Millions of meteors burn up daily in the mesosphere as a result of collisions with some of the billions of gas particles contained in that layer The collisions create enough heat to burn the falling objects long before they reach the ground. The stratosphere and mesosphere are referred to as the middle atmosphere. The mesopause, at an altitude of about 80 km, separates the mesosphere from the thermosphere—the outermost layer of the Earth's atmosphere. The thermosphere, from the Greek thermo for heat, begins about 80 km above the Earth. At these high altitudes, the residual atmospheric gases sort into strata according to molecular mass. Thermospheric temperatures increase with altitude due to absorption of highly energetic solar radiation by the small amount of residual oxygen still present. Temperatures can rise to 2,000° C. Radiation causes the scattered air particles in this layer to become charged electrically, enabling radio waves to bounce off and be received beyond the horizon. At the exosphere, beginning at 500 to 1,000 km above the Earth's surface, the atmosphere blends into space. The few particles of gas here can reach 4,500°F (2,500° C) during the day.

Atmospheric circulation The large-scale movement of air around and above the earth, associated with complex but distinct patterns of pressure systems and wind belts. It is driven by latitudinal variations in the earth's energy budget, and through the redistribution of energy helps to reduce these variations. George Hadley first proposed the classic model of the general circulation of the atmosphere in 1735 as a simple convective system, based on the concept of a non-rotating earth with a uniform surface, that was warm at the equator and cold at the poles.

Warm buoyant air rising at the equator spread north and south in the upper atmosphere, eventually returning to the surface in high or polar latitudes. From there, it flowed back across the surface towards the equator to close the circulation. Hadley's original model, with its single convection cell in each hemisphere, was eventually replaced by a three-cell model as technology advanced and additional information became available, but his contribution was recognized in the naming of the tropical cell. The three-cell model continued to assume a uniform surface, but the rotation of the earth was introduced, and with it the Coriolis effect, which causes moving objects to swing to the right in the northern hemisphere and to the left in the southern. Thus, the winds became westerly or easterly in this new model, rather than blowing north or south as in the one-cell version. The three cells and the Coriolis effect, in combination, produced alternating bands of high and low pressure, separated by wind belts which were easterly in equatorial and polar regions and westerly in mid-latitudes. Although only theoretical, elements of this model can be recognized in existing global wind and pressure patterns, particularly in the southern hemisphere, where the greater expanse of ocean more closely resembles the uniform surface of the model. In the late 1940s and 1950s, as knowledge of the atmosphere improved, it became increasingly evident that the three-cell model oversimplified the general circulation.

The main problem arose with the mid-latitude cell. Observations indicated that most energy transfer in mid-latitudes was accomplished by horizontal cells – such as travelling high and low pressure systems – rather than the vertical cell indicated by the model. Modern interpretations of the general circulation of the atmosphere retain the tropical Hadley cell, but horizontal eddies have come to dominate mid-latitudes, and have even replaced the simple thermal cell of polar latitudes. Conditions in the upper atmosphere are an integral part of modern studies of the atmospheric circulation. The upper atmospheric circulation is quite complex in detail, but in general terms it is characterized by an easterly flow in the tropics and a westerly flow in mid- to high latitudes. The upper westerlies include a pattern of waves, called Rossby waves, which vary in amplitude in a quasi-regular sequence represented by the index cycle.

During the run of any one cycle there is significant latitudinal energy transfer. Within these broad airflows, at the tropopause, there are relatively narrow bands of rapidly moving air called jet streams, in which wind speeds may average 125–130 km per hour, although much higher speeds may occur. Modern representations of the general circulation of the atmosphere take into account the non-uniform nature of the earth's surface, with its mixture of land and water, and include consideration of seasonal variations in energy flow. Differences in their physical properties ensure that land and sea warm up and cool down at different rates. This creates significant temperature differences between land and water, and produces a series of pressure cells rather than the simple belts of the original models. By altering the regional airflow, such pressure differences cause disruption of the theoretical wind patterns. The changing location of the zone of maximum insolation with the seasons also causes variations in the location, extent and intensity of the pressure cells and wind belts. Although such changes are repeated year after year they are not completely reliable, and this adds an additional element of variability to the representation of the atmospheric circulation.

Atmospheric infrared sounder Advanced sounding instrument selected to fly on the EOS-PM 1 mission (intermediate-sized, sun-synchronous, morning satellite) in the year 2000. It will retrieve vertical temperature and moisture profiles in the troposphere and stratosphere. Designed to achieve temperature retrieval accuracy of 1°C with a 1 km vertical resolution, it will fly with two operational microwave sounders. The three instruments will constitute an advanced operational sounding system, relative to the TIROS Operational Vertical Sounder (TOVS) currently flying on NOAA polar-orbiting satellites.

Atmospheric pressure The amount of force exerted over a surface area, caused by the weight of air molecules above it. As elevation increases, fewer air molecules are present. Therefore, atmospheric pressure always decreases with increasing height. A column of air; 1 square inch in cross section, measured from sea level to the top of the atmosphere would weigh approximately 14.7 lb/in^2. The standard value for atmospheric pressure at sea level is: 29.92 inches or 760 mm of mercury 1013.25 millibars(mb) or 101,325 pascals (Pa).

Atmospheric radiation measurements program (ARM) U.S. Department of Energy program for the continual, ground-based measurements of atmospheric and meteorological parametres over approximately a ten-year period. The program will study radiative forcing and feedbacks, particularly the role of clouds. The general program goal is to improve the performance of climate models, particularly general circulation models of the atmosphere.

Atmospheric response variables Variables that reflect the response of the atmosphere to external forcing (e.g., temperature, pressure, circulation, and precipitation).

Atmospheric stability/instability The stability or instability of the atmosphere is determined by its response to change. A stable atmosphere will resist change, whereas an unstable one will allow the change to develop. The terms are most commonly applied to vertical motion in the atmosphere. If a parcel of air is forced to rise in a stable atmosphere, for example, it will return to its original position once the force which caused it to rise has been removed. In an unstable atmosphere, the parcel would continue to rise away from its original position. The stability or instability of the atmosphere can be determined by comparing the environmental lapse rate (ELR) with the dry or saturated adiabatic lapse rates (DALR or SALR). If the DALR is greater than the ELR, then a dry parcel of air rising in the atmosphere will always be cooler and therefore denser than the surrounding air and will therefore tend to fall once the force causing it to rise ceases. The atmosphere is therefore stable. Conversely, if the DALR is less than the ELR the atmosphere will be unstable. A similar relationship between the ELR and the SALR will produce stability or instability also. As air rises and cools, however, its relative humidity increases and it may ultimately become saturated. In that case it will cool at the SALR, which is always less than the DALR. It is possible therefore that a parcel of air will begin to ascend cooling at a DALR greater than the ELR, but following saturation will cool at a rate less than the ELR, which leads to instability. Such instability which depends upon saturation is called conditional instability. The stability or instability of the atmosphere is important air pollution climatology. Pollutants tend to accumulate under stable conditions and are more easily dispersed when the atmosphere is unstable.

Atmospheric turbidity A measure of the atmosphere's aerosol content as indicated by the attenuation of solar radiation passing through it. For most purposes, it can be considered as an indication of the dustiness or dirtiness of the atmosphere. Volcanic activity, dust storms and a variety of physical and organic processes provide aerosols which are incorporated into the atmosphere. Once there, they are redistributed by way of the wind and pressure patterns, remaining in suspension for periods ranging from several hours to several years depending upon particle size and the altitude attained. Human industrial and agricultural activities also help to increase turbidity levels, but the anthropogenic production of aerosols cannot match the volume of material produced naturally. The presence of aerosols in the atmosphere disrupts the inward and outward flow of energy through the earth/ atmosphere system. Studies of periods of intense volcanic activity suggest that the net effect of increased atmospheric turbidity is cooling, and some of the coldest years of the Little Ice Age have been correlated with major volcanic eruptions. The eruption of Mount Pinatubo in 1991 was blamed for the cool summer of 1992 in North America and was ultimately linked with a global temperature reduction of 0.5°C. Present opinion sees increased atmospheric turbidity actually producing a slight warming.

Atmospheric windows The range of wavelengths at which water vapour;

carbon dioxide, or other atmospheric gases only slightly absorb radiation. Atmospheric windows allow the Earth's radiation to escape into space unless clouds absorb the radiation.

Atoll A coral island consisting of a ring of coral surrounding a central lagoon. Atolls are common in the Indian and Pacific Oceans.

Atom Smallest unit of an element that still maintains its chemical characteristics.

Atomic energy Energy released from an atomic nucleus at the expense of its mass.

Atomic mass number Combined number of an atom's protons and neutrons.

Atomic number Number of protons in the nucleus of an atom.

Atomic weight Combined weight of an atom's electrons, protons, and neutrons.

Attenuation The decrease in the magnitude of current, voltage, or power of a signal in transmission between points. Attenuation may be expressed in decibels, and can be caused by interferences such as rain, clouds, or radio frequency signals.

Attribute Data is descriptive information about features or elements of a database. For a database feature like census tract, attributes might include many demographic facts including total population, average income, and age. In statistical parlance, an attribute is a "variable," whereas the database feature represents an "observation" of the variable.

Audio frequencies Frequencies that the human ear can hear (usually 30 to 20,000 cycles per second).

Aurora The aurora is a glow observed in the night sky, usually in the polar zone. For this reason some scientists call it a "polar aurora" (or "aurora polaris"). In northern latitudes, it is known as "aurora borealis" which is named after the Roman goddess of the dawn, Aurora and the greek name for north wind, Boreas since in Europe especially, it often appears as a reddish glow on the northern horizon as if the sun were rising from an unusual direction. The aurora borealis is also called the "northern lights". The *aurora borealis* most often occurs from September to October and March to April. Its southern counterpart, "aurora australis", has similar properties. The aurora is now known to be caused by electrons of typical energy of 1-15 keV, i.e. the energy obtained by the electrons passing through a voltage difference of 1,000-15,000 volts. The light is produced when they collide with atoms of the upper atmosphere, typically at altitudes of 80-150 km. It tends to be dominated by emissions of atomic oxygen – the greenish line at 557.7 nm and (especially with electrons of lower energy and higher altitude) the dark-red line at 630.0 nm. Both these represent "forbidden" transitions of atomic oxygen from energy levels which (in absence of collisions) persist for a long time, accounting for the slow brightening and fading (0.5-1 sec) of auroral rays. Many other lines can also be observed, especially those of molecular nitrogen, and these vary much faster, revealing the true dynamic nature of the aurora.

Auroras can also be observed in the ultra-violet (UV) light, a very good way of observing it from space (but not from ground – the atmosphere absorbs UV). The Polar spacecraft even observed it in X-rays. The image is very rough, but precipitation of high-energy electrons can be identified. Typically the aurora appears either as a diffuse glow or as "curtains" that approximately extend in the east-west direction. At some times, they form "quiet arcs", at others ("active aurora") they evolve and change constantly. Each curtain consists of many parallel rays, each lined up with the local direction of the magnetic field lines, suggesting that aurora is shaped by the Earth's magnetic field, Indeed, satellites show auroral electrons to be guided by magnetic field lines, spiraling around them while moving earthwards.

The curtains often show folds called "striations". When the field line guiding a bright auroral patch leads to a point directly above the observer, the aurora may appear as a "corona" of diverging rays, an effect of perspective. In 1741 Hiorter and Celsius first noticed other evidence for magnetic control, namely, large magnetic fluctuations occurred whenever the aurora was observed overhead. This indicates (it was later realized) that large electric currents were associated with the aurora, flowing in the region where auroral light originated. Kristian Birkeland (1903. deduced that the currents flowed in the east-west directions along the auroral arc, and such currents, flowing from the dayside towards (approximately) midnight were later named "auroral electrojets".

Still more evidence for a magnetic connection are the statistics of auroral observations. Elias Loomis (1860) and later in more detail Hermann Fritz (1881) established that the aurora appeared mainly in the "auroral zone", a ring-shaped region of approx. radius 2500 km around the magnetic pole of the Earth, not its geographic one. It was hardly ever seen near that pole itself. The instantaneous distribution of auroras ("auroral oval", Yasha Feldstein 1963) is slightly different, centreed about 3-5 degrees nightward of the magnetic pole, so that auroral arcs reach furthest equatorward around midnight. The aurora is a common occurrence in the ring-shaped zone. It is occasionally seen in temperate latitudes, when a strong magnetic storm temporarily expands the auroral oval. Large magnetic storms are most common during the peak of the 11-year sunspot cycle, or during the 3 years after that peak. However, within the auroral zone the likelihood of an aurora occurring depends mostly on the slant of IMF lines (known as Bz), being greater with southward slants.

Geomagnetic storms that ignite auroras actually happen more often during the months around the equinoxes. It is not well understood why geomagnetic storms are tied to the Earth's seasons when polar activity is not. It is known, however, that during spring and autumn, the earth's and the interplanetary magnetic field link up. At the magnetopause, Earth's magnetic field points north. When Bz becomes large and negative (i.e., the IMF tilts south) it can partially cancel Earth's magnetic field at the point of contact. South-pointing Bz's open a door through which energy from the solar wind can reach Earth's inner magnetosphere.

The peaking of Bz during this time is a result of geometry. The interplanetary magnetic field comes from the Sun and is carried outward the solar wind. Because the Sun rotates the IMF has a spiral shape. Earth's magnetic dipole axis is most closely aligned with the Parker spiral in April and October. As a result, southward (and northward) excursions of Bz are greatest then. However, Bz is not the only influence on geomagnetic activity. The Sun's rotation axis is tilted 7 degrees with respect to the plane of Earth's orbit. Because the solar wind blows more rapidly from the Sun's poles than from its equator, the average speed of particles buffeting Earth's magnetosphere waxes and wanes every six months. The solar wind speed is greatest — by about 50 km/s, on average — around Sept. 5th and March 5th when Earth lies at its highest heliographic latitude. Still, neither Bz nor the solar wind can fully explain the seasonal behaviour of geomagnetic storms. Those factors together contribute only about one-third of the observed semi-annual variation.

Autogenic succession Succession where the plant community causes the environment to change and this modification drives the succession.

Automatic picture transmission (APT) System developed to make real-time reception of satellite images possible whenever an APT-equipped satellite passes within range of an environmental satellite ground station. Transmission (analogue video format) consists of an amplitude-modulated audible tone that can be displayed as an image on a computer monitor when received by an appropriate ground station.

APT images are transmitted by polar-orbiting satellites such as the TIROS-N/NOAA satellites and Russia's METEOR, which orbit 500-900 miles above the Earth, and offer both visible and infrared images. An APT image has thousands of squares called picture elements or pixels. Each pixel represents a four-km square.

Autotroph An organism that produces food molecules inorganically by using a light or chemical based sources of external energy. This organism does not require outside sources of organic food energy for survival.

Autumn The season of the year which occurs as the sun approaches the winter solstice, and characterized by decreasing temperatures in the mid-latitudes. Customarily, this refers to the months of September, October, and November in the North Hemisphere and the months of March, April, and May in the Southern Hemisphere. Astronomically, this is the period between the autumnal equinox and the winter solstice.

Autumnal equinox One of the two periods when the declination of the sun is at the equator. The autumnal equinox occurs on September 22 or 23.

Available water Portion of the capillary water that is available for plant root uptake.

Average global temperature Average annual temperature of the Earth's entire surface atmosphere.

AVHRR Acronym for advanced very high resolution radiometre.

AWIPS Acronym for advanced weather interactive processing system.

Azimuth The direction, in degrees referenced to true north, that an antenna must be pointed to receive a satellite signal (compass direction). The angular distance is measured in a clockwise direction.

Azonal soil A soil without developed horizons.

Azores high A semi-permanent, subtropical area of high pressure in the North Atlantic Ocean that migrates east and west with varying central pressure. Depending on the season, it has different names. In the Northern Hemispheric winter and early spring, when the Icelandic Low dominates the North Atlantic, it is primarily centred near the Azores Islands. When it is displaced westward, during the summer and fall, the centre is located in the western North Atlantic, near Bermuda and is known as the Bermuda High. A comparable system is the North Pacific High.

B

B horizon Soil horizon normally found below the A horizon and above the C horizon. This layer is characterized by the following features: 1. Enrichment of clay because of illuviation from the A horizon.

2. Enrichment of iron and aluminum oxides because of illuviation of these materials from the A horizon. In some cases the precipitation of iron can cause the development of a hardpan.

3. Accumulation of calcium carbonate, calcium sulphate, and other salts.

4. Higher bulk density because of the illuvial deposition of clay particles.

Background extinction Normal extinction of species that occurs as a result of changes in local environmental conditions.

Backing A change in wind direction that shifts counterclockwise in the Northern Hemisphere at a certain location. In the Southern Hemisphere, it is clockwise. This can either happen in the horizontal or the vertical (with height). For example, the wind shifts from the northeast to the north to the northwest. It is the opposite of veering.

Backscatter A radar echo that is reflected, or scattered, at 180 degrees to the direction of the incident wave. Also the scattering of radiant energy into space before it reaches the earth's surface.

Backshore Area behind the shore. This coastal feature is located between the beach berm and the backshore slope.

Backshore slope Sloping bank landward of the shore. This coastal feature is composed of relatively non-mobile sediments.

Backswamp Marshy low lying area in a stream's floodplain. Commonly found behind levees.

Backwash The return water flow of swash. This sheet of water flows back to ocean because of gravity.

Bacteria Simple single celled prokaryotic organisms. Many different species of bacteria exist. Some species of bacteria can be pathogenic causing disease in larger more complex organisms. Many species of bacteria play a major role in the cycling of nutrients in ecosystems through aerobic and anaerobic decomposition. Finally, some species form symbiotic relationships with more complex organisms and help these life forms survive in the environment by fixing atmospheric nitrogen.

Badlands Term used to describe a semi-arid landscape that has been influenced by heavy fluvial erosion. Characterized by deep ravines and gullies, shape ridges, and a generally barren surface.

Bajada Consecutive series of alluvial fans forming along the edge of a linear mountain range. Surface of this feature undulates in a rolling fashion as one moves from the centre of one alluvial fan to another. Normally occur in arid climates.

Ball lightning A relatively rare form of lightning consisting of a reddish, luminous ball which moves rapidly along solid objects or remains floating in mid-air. Also known as globe lightning.

Band 1 In radio a continuous sequence of broadcasting frequencies within given limits.

2. In radiometry, a relatively narrow region of the electromagnetic spectrum to which a remote sensor responds; a multispectral sensor makes measurements in a number of spectral bands.

3. In spectroscopy spectral regions where atmospheric gases absorb (and emit) radiation, e.g., the 15 μm carbon dioxide absorption band, the 6.3 μm water vapour absorption band, and the 9.6 μm ozone absorption band.

Bandwidth The total range of frequency required to pass a specific modulated signal without distortion or loss of data. The ideal bandwidth allows the signal to pass under conditions of maximum AM or FM adjustment. (Too narrow a bandwidth will result in loss of data during modulation peaks. Too wide a bandwidth will pass excessive noise along with the signal.) In FM, radio frequency signal bandwidth is determined by the frequency deviation of the signal.

Bank-caving Collapse of stream bank material into a stream channel.

Bar 1. Coarse grained deposit of sediment from a stream or ocean currents.

2. A unit measurements for quantifying force. Equivalent to 1,000,000 dynes per square centimetre.

Barchan dune Crescent shaped sand dune that has its long axis transverse to the wind and its crescent tips pointed downwind.

Baroclinity The state of stratification in a fluid in which surfaces of constant pressure intersect surfaces of constant density. An example is a baroclinic pattern.

Barograph An instrument that continuously records a barometre's reading of atmospheric pressure.

Barometre An instrument used to measure atmospheric pressure. A standard mercury barometre has a glass column about 30 inches long, closed at one end, with a mercury-filled reservoir. Mercury in the tube adjusts until the weight of the mercury column balances the atmospheric force exerted on the reservoir. High atmospheric pressure forces the mercury higher in the column. Low pressure allows the mercury to drop to a lower level in the column. An aneroid barometre uses a small, flexible metal box called an aneroid cell. The box is tightly sealed after some of the air is removed, so that small changes in external air pressure cause the cell to expand or contract.

Barometric pressure The pressure exerted by the atmosphere at a given point. Its measurement can be expressed in several ways. One is in millibars. Another is in inches or millimetres of mercury (Hg). Also known as atmospheric pressure.

Barotropy The state of a fluid in which surfaces of constant density or temperature are coincident with surfaces of constant pressure. It is considered zero baroclinity. An example is a barotropic pattern.

Barrier beach A long and narrow beach of sand and/or gravel that runs parallel to the coastline and is not submerged by the tide.

Barrier island Long, narrow islands of sand and/or gravel that are usually aligned parallel to the shore of some coasts.

Barrier winds Refers to the westerly flow of air along the northern slope of the Brooks Range in northern Alaska that precedes the arrival of colder air from the north.

Basal sliding The sliding of a glacier over the surface it rests on. Caused by the gradient of the slope and the weight of the glacier's mass.

Basalt A dark coloured fine grained igneous rock formed from mafic magma.

Basalt plateau Extensive continental deposits of basaltic volcanic rock.

Basaltic magma Mafic magma that forms basaltic igneous rocks.

Base A substance that forms a salt when it reacts with acid. A base is a substance that removes hydrogen ions (protons) from an acid and combines with them in a chemical reaction.

Base flow Rate of discharge in a stream where only the throughflow and groundwater flow from subsurface aquifers contribute to the overall flow.

Base level The lowest level to which a stream can erode its bed. The ultimate base level of all streams is, of course, the sea.

Base maps Provide the background upon which thematic data is overlayed and analyzed. As inputs into a GIS, the term base map is usually applied to those sources of information about relatively permanent, sometimes timeless, features including topography, pedology, geology, cadastral divisions, and political divisions. Within a GIS database, such information may become part of a land base to which other information is indexed and referenced.

Basement rock Very old granite and metamorphic rocks found in continental crust. These rocks make up the continental shield.

Basic Substance having a pH greater than 7.

Basic solution Any water solution that is basic (pH greater than 7) or has less hydrogen ions (H^+) than hydroxide ions (OH^-).

Basin A topographic rock structure whose shape is concave downwards.

Batholith A batholith is a large emplacement of igneous intrusive (also called plutonic) rock that forms from cooled magma deep in the Earth's crust. Batholiths are almost always made mostly of felsic or intermediate rock-types, such as granite, quartz monzonite, or diorite. Although they may appear uniform, batholiths are in fact structures with complex histories and compositions. They are composed of multiple masses, or plutons, of magma that traveled toward the surface from a zone of partial melting at the base of the Earth's crust. While moving, these plutons of relatively buoyant magma are called plutonic diapirs. Because the diapirs are liquefied and very hot, they tend to rise

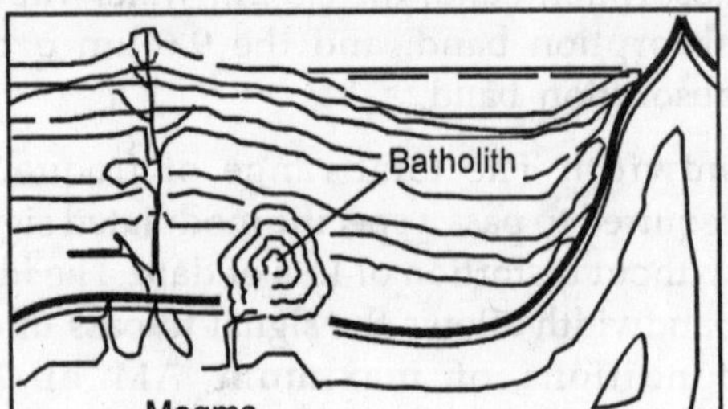

Fig. Batholith

through the surrounding country rock, pushing it aside and partially melting it. Most diapirs do not reach the surface to form volcanoes, but instead slow down, cool and usually solidify 5 to 30 kilometres underground as plutons. A batholith is formed when many plutons converge together to form a huge expanse of granitic rock. Some batholiths are mammoth, paralleling past and present subduction zones and other heat sources for hundreds of kilometres in continental crust. One such batholith is the Sierra Nevada Batholith, which is a continuous granitic formation that forms much of the Sierra Nevada in California. An even larger batholith, found predominantly in the mountains of western Canada, extends for 1,800 kilometres and reaches into southeastern Alaska.

There is also an important geographic usage of the term batholith. For a geographer, a batholith is an exposed area of mostly continuous plutonic rock that covers an area larger than 100 square kilometres. Areas smaller than 100 kilometres are called stocks. However, the majority of batholiths visible at the surface (via outcroppings) have areas far greater than 100 square kilometres. These areas are exposed to the surface through the process of erosion accelerated by continental uplift acting over many tens of millions to hundreds of millions of years. This process has removed several tens of kilometres of overlying rock in many areas, exposing the once deeply buried batholiths. Batholiths exposed at the surface are also subjected to huge pressure differences between their former homes deep in the earth and their new homes at or near the surface. As a result, their crystal structure expands slightly and over time. This manifests itself by a form of mass wasting called exfoliation. This form of erosion causes convex and relatively thin sheets of rock to slough off the exposed surfaces of batholiths (a process accelerated by frost wedging). The result is fairly clean and rounded rock faces. A famous example of the result of this process is Half Dome, which located in the world-famous Yosemite Valley.

Bathymetric map A bathymetric map is a map of the bottom of a body of water, noting the depth contours (these are called isobaths).

Bathymetry Science of measuring water depths (usually in the ocean) to determine bottom topography. Bathymetry is the underwater equivalent to altimetry. A bathymetric map gives the depth contours of the soil, rock, sand, etc. at the bottom of a body of water such as an ocean or a lake. In addition to this, a bathymetric chart provides navigational information. The data used to make bathymetric maps typically comes from an echosounder (sonar) mounted beneath or over the side of a boat, 'pinging' a "beam," or sound, downward at the seafloor. The amount of time it takes for the sound to make it through the water, bounce off the seafloor and come back up to the sounder, tells the equipment how far down the seafloor is.

Years ago, the occasional pings of a single-beam sounder might be averaged to make a map. Today, a multibeam echosounder may be used, featuring dozens of very narrow adjacent beams arranged in a fan-like swath of perhaps 90 to 180 degrees across. The tightly packed array of narrow individual beams provides very high resolution detail and accuracy. In general, the wide swath, which is depth dependent, allows the boat to map more seafloor in less time by making fewer passes. The beams update many times per second (typically 1-40 Hz depending on water depth), allowing faster boat speed while maintaining 100% coverage of the seafloor. Motion sensors correct for the boat's roll, pitch and yaw on the ocean surface, and a gyrocompass provides accurate heading information to correct for vessel yaw. Global positioning systems specify exactly where the boat is. Accurate sound velocity profiles (speed of sound in water) of the water column correct for refraction or "ray-bending" of the sound waves owing to non-uniform water column characteristics such as temperature, conductivity, and pressure. A computer system processes all the data, correcting for all of the above factors as well as for the angle of each individual beam. In the end, a map is semi-automatically generated from this massive trove of data.

Bathythermograph A device used to obtain a record of temperature against depth (pressure) in the ocean. May be referred to as a B.T.

Baud Unit of signaling speed. The speed in bauds is the number of discrete conditions or signal events per second. If each signal event represents only one bit condition, baud is the same as bits per second.

Bay A wide area of water extending into land from a sea or lake.

The measure of the "width" of an antenna pattern, measured in degrees of arc. Generally an antenna with low gain has a wide pattern, receiving signals well from a number of different directions.

Bayhead beach An extensive deposit of sand and/or gravel in the form of a beach at the back of a bay.

Bay-mouth bar A narrow deposit of sand and/or gravel found across the mouth of a bay.

Beach The terrestrial interface area in between land and a water body where there are accumulations of unconsolidated sediments like sand and gravel. These deposits are laid down by the action of breaking waves.

Beach drift The lateral movement of sediments on a beach when the angles of swash and backwash differ.

Bearing The combination of antenna azimuth and elevation required to point (aim) an antenna at a spacecraft. The bearing for geostationary (i.e., GOES) satellites is constant. The bearing for polar-orbiting satellites varies continuously.

Beaufort wind scale A system of estimating and reporting wind speeds. It is based on the Beaufort Force or Number, which is composed of the wind speed, a descriptive term, and the visible effects upon land objects and/or sea surfaces. The scale was devised by Sir Francis Beaufort (1777-1857), hydrographer to the British Royal Navy.

Bed Sedimentary structure that usually represents a layer of deposited sediment.

Bed load Portion of the stream load that is carried along the stream bed without being permanently suspend in the flowing water.

Bedding plane A layer in a series of sedimentary beds that marks a change in the type of deposits.

Bedrock Rock at or near (beneath soil and regolith) the Earth's surface that is solid and relatively unweathered.

Bellot winds Refers to the winds in the Canadian Arctic that blow through the narrow Bellot Strait between Somerset Island and the Boothia Peninsula, connecting the Gulf of Boothia and Franklin Strait.

Bench mark Relatively permanent material object, natural or artificial, bearing a marked point whose elevation above or below an adopted datum is known.

Benchmark tests Are standardized routines for comparing the performance of competing hardware and software under conditions intended to reflect common patterns of use. Widely used benchmark tests have been developed to compare the speed of computers. But more specific benchmark tests also may be developed by a single organization to test new equipment under conditions close to those which will be encountered in day-to-day use.

Benthos The plant and animal organisms that live on the sea floor. Often divided into two categories: deep-sea benthos, below 200 metres and the littoral benthos, from 200 metres to the high-water spring tide level.

Bergschrund A Bergschrund is a split or crevasse in the ice of a glacier, where the glacier detaches itself from the mountain's rock. It is often a serious obstacle to a mountaineer wishing to gain access to the mountain. A bergschrund is formed when the glacier moves downslope while its borders are firmly frozen to the rock, comparable to the forming of geological faults. In a corrie or cirque, the bergschrund is positioned at the rear, parallel to the back wall of the corrie. It is caused by the rotational movement of the glacier. In a longitudinal glacier, the bergschrund is at the top end of the glacier at a right angle to the flow of the glacier. It is caused by the downwards flow of the glacier.

Bergschrunds extend to the bedrock and can have a depth of well over 100 metres. In winter, Bergschrunds are often filled by snow from avalanches from the mountain above it. In later summer, due to melting, it lies open and can present a very difficult obstacle to alpinists. The bergschrund is distinct from the randkluft (also called rimaye) which is the crevasse of which one face is the rock, back wall of the corrie. The randkluft arises in part from the melting of the ice arising from the presence of the warmer rock face. However, the randkluft is sometimes called a bergschrund.

Berm Low hill of sand that forms along coastal beaches.

Bermuda high A semi-permanent, subtropical area of high pressure in the North Atlantic Ocean that migrates east and west with varying central pressure. Depending on the season, it has different names. When it is displaced westward, during the Northern Hemispheric summer and fall, the centre is located in the western North Atlantic, near Bermuda. In the winter and early spring, it is primarily centreed near the Azores in the eastern part of the North Atlantic. Then it may be referred to as the Azores High.

Bermuda Triangle The Bermuda Triangle (also known as Devil's Triangle) is a nearly half-million square-mile (1.2 million km^2) area of ocean roughly defined by Bermuda, Puerto Rico, and Fort Lauderdale, Florida. The Bermuda Triangle has become popular through representation by the mass media, in which it is a paranormal site in which the known laws of physics are violated and altered.

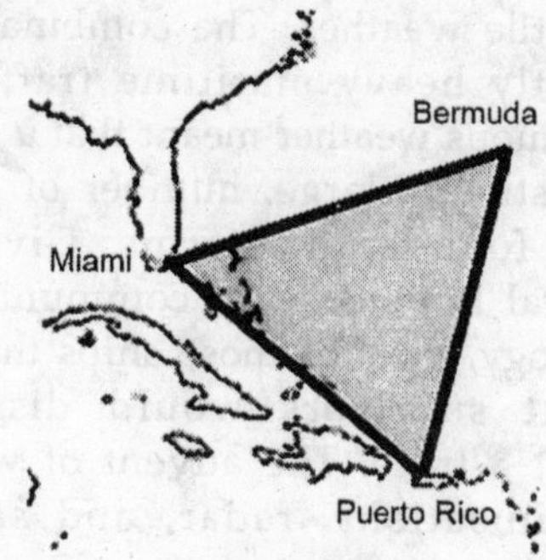

Fig. Bermuda Triangle

While there is a common belief that a number of ships and airplanes have disappeared under highly unusual circumstances in this region, the United States Coast Guard and others disagree with that assessment, citing statistics demonstrating that the number of incidents involving lost ships and aircraft is no larger than that of any other heavily travelled region of the world. Many of the alleged mysteries have proven not so mysterious or unusual upon close examination, with inaccuracies and misinformation about the cases often circulating and recirculating over the decades.

The triangle is an arbitrary shape, crudely marking out a corridor of the Atlantic Ocean, stretching northward from the West Indies, along the North American seaboard, as far as the Carolinas. In the Age of Sail, ships returning to Europe from parts south would sail north to the Carolinas, then turn east for Europe, taking advantage of the prevailing wind direction across the North

Atlantic. Even with the development of steam and internal-combustion engines, a great deal more shipping traffic was (and still is) found nearer the US coastline than towards the empty centre of the Atlantic. The Triangle also loosely conforms with the course of the Gulf Stream as it leaves the West Indies, and has always been an area of volatile weather. The combination of distinctly heavy maritime traffic and tempestuous weather meant that a certain, also distinctly large, number of vessels would founder in storms. Given the historical limitations of communications technology, most of those ships that sank without survivors would disappear without a trace. The advent of wireless communications, radar, and satellite navigation meant that the unexplained disappearances largely ceased at some point in the 20th Century. The occasional vessel still sinks, but rarely without a trace.

Bernoulli's theorem A statement of the conservation of energy for a steady, non-viscous, incompressible level flow. It is an inverse relationship in which pressures are least where velocities are greatest. An example is the Santa Ana winds. Theorized by Daniel Bernoulli (1700-1782), a Swiss mathematician and physicist.

Beta particle Electron emitted from the nucleus of a radioactive isotope.

Biennial plant Plant species that completes its life in two growing seasons.

Bifurcation ratio Quantitative ratio determined between the parts of systems that display branching. For example, trees have a main stem that bifurcates into smaller and smaller branches. The ratio between the branches that are derived from a larger branch or main stem is the bifurcation ratio.

Big bang Theory that suggests that about 15 billion years ago all of the matter and energy in the Universe was concentrated into an area smaller than a atom. At this instant, matter, energy, space and time did not exist. Then suddenly, the Universe began to expand at an incredible rate and matter, energy, space and time came into being. As the Universe expanded, matter began to coalesce into gas clouds, and then stars and planets. Some scientists believe that this expansion is finite and will one day cease. After this point in time, the Universe will begin to collapse until a Big Crunch occurs.

Big crunch Collapse of the Universe into its original form before the Big Bang. At the end of this process matter, energy, space, and time will not exist.

Bilingual The ability to use either one of two languages, especially when speaking.

Binary Code and binary files are information and commands stored and used by hardware and software in their most elemental form—strings of on-off signals to an electronic processor. Many systems of binary encoding of data and commands are proprietary and unique to particular hardware and software systems. They are usually the most compact means of storing data, and commands stored in binary form can execute very rapidly, but binary files often are difficult to transfer between differing computer systems. Often binary files are translated into ASCII form for transfer between computers.

Bioassay A measurement of the effects of a substance on living organisms.

Biodegradation Decomposition of material by micro organisms.

Biodiversity The diversity of different species (species diversity), genetic variability among individuals within each species (genetic diversity), and variety of ecosystems (ecosystem diversity). Abbrevi-ation of biological diversity.

Biogeochemical cycles Movements through the Earth system of key chemical constituents essential to life, such as carbon, nitrogen, oxygen, and phosphorus.

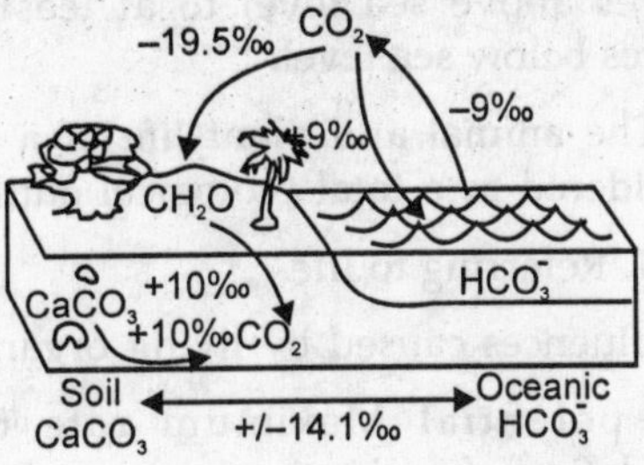

Fig. Biogeochemical cycles

Biogeography Field of physical geography that studies the spatial pattern of living organisms.

Biological amplification Increase in concentration of toxic fat-soluble chemicals in organisms at successively higher trophic levels of a grazing food chain or food web because of the consumption of organisms at lower trophic levels.

Biological diversity A concept recognizing the variety of life forms in an area of the Earth and the ecological interdependence of these life forms.

Biological weathering The disintegration of rock and mineral due to the chemical and/or physical agents of an organism.

Biomass The weight of living tissues usually measured per unit area over a particular time interval. Can include the dead parts of organisms like bark, hair, and nails.

Biome Largest recognizable assemblage of animals and plants on the Earth. The distribution of the biomes is controlled mainly by climate.

Bioregion A unique region on the Earth that has distinct soils, landforms, watersheds, climates, native plants, and animals, and/or other particular natural characteristics.

Biosphere Zone of the Earth where life is found. The biosphere is the part of a planet's outer shell — including air, land, surface rocks and water — within which life occurs, and which biotic processes in turn alter or transform. From the broadest geophysiological point of view, the biosphere is the global ecological system integrating all living beings and their relationships, including their interaction with the elements of the lithosphere (rocks), hydrosphere (water), and atmosphere (air). Our planet Earth is the only place where life is known to exist. This biosphere is generally thought to have evolved, beginning through a process of biogenesis or biopoesis, at least some 3.5 billion years ago. The term "biosphere" was coined by geologist Eduard Suess in 1875. While this concept has a geological origin, it is an indication of the impact of both Darwin and Maury on the earth sciences. The biosphere's ecological context comes from the 1920s, preceding the 1935 introduction of the term "ecosystem" by Sir Arthur Tansley. Vernadsky defined ecology as the science of the biosphere. It is an interdisciplinary concept for integrating astronomy, geophysics, meteorology, biogeography, evolution, geology, geochemistry, hydrology and, generally speaking, all life and earth sciences.

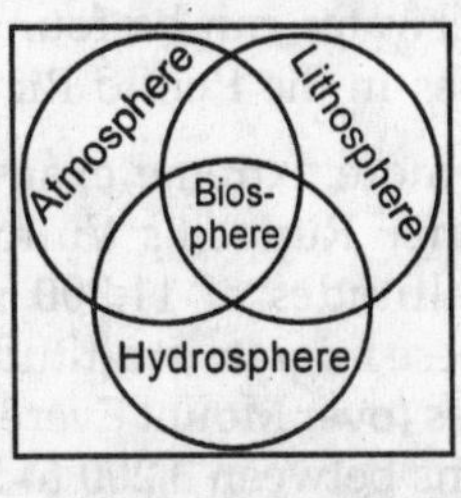

Fig. Biosphere

Some life scientists and earth scientists use biosphere in a more limited sense. For example, geochemists define the biosphere as being the total sum of living organisms (the "biomass" or "biota" as referred to by biologists and ecologists). In this sense, the biosphere is but one of four separate components of the geochemical model, the other three being lithosphere, hydrosphere, and atmosphere. The narrow meaning used by geochemists is one of the consequences of specialization in modern science. Some might prefer the word ecosphere, coined in the 1960s, as all encompassing of both biological and physical components of the planet. The Second International Conference on Closed Life Systems defined biospherics as the science and technology of analogs and models of Earth's biosphere; i.e., artificial Earth-like biospheres. Others may include the creation of artificial non-Earth biospheres – for example, human-centred biospheres or a native Martian biosphere – in the field of biospherics. Some theorists have postulated that the Earth is poorly suited to life, although nearly every part of the planet, from the polar ice caps to the Equator, supports life of some kind. Indeed, recent advances in microbiology have demonstrated that microbes live deep beneath the Earth's terrestrial surface, and that the total mass of microbial life in so-called "uninhabitable zones" may, in biomass, exceed all animal and plant life on the surface. The actual thickness of the biosphere on earth is hard to measure. Birds typically fly at altitudes of 650 to 2000 metres, and fish that live deep underwater can be found down to -8,372 metres in the Puerto Rico Trench.

There are more extreme examples for life on the planet: Rüppell's Vulture has been found at altitudes of 11,300 metres; Bar-headed Geese migrate at altitudes of at least 8,300 metres (over Mount Everest); Yaks live at elevations between 3,200 to 5,400 metres above sea level; mountain goats live up to 3,050 metres. Herbivorous animals at these elevations depend on lichens, grasses, and herbs but the biggest tree is the Tine palm or mountain coconut found 3,400 metres above sea level. Microscopic organisms (e.g., bacteria) live at such extremes that, taking them into consideration puts the thickness of the biosphere much greater, but at minimum it extends from 5,400 metres above sea level to at least 9,000 metres below sea level.

Biota The animal and plant life of a region considered as a total ecological entity.

Biotic 1. Referring to life.

2. Influences caused by living organisms.

Biotic potential Maximum rate that a population of a given species can increase when there are no limits on its rate of growth.

Biotite Rock forming mineral of the mica group.

Bird Group of warm blooded vertebrate animals whose body is covered with feathers.

Bit A contraction of "binary digit." The basic element of a two-element (binary) computer language.

Bit rate The speed at which bits are transmitted, usually expressed in bits per second.

Bituminous A soft coal that, when heated, yields considerable volatile matter.

Black blizzard A local term for a violent dust-storm on the south-central Great Plains that darkens the sky and casts a pall over the land. Also called a black roller.

Black body Is a body that emits electro-magnetic radiation, at any temperature, at the maximum possible rate per unit surface area. This body also absorbs all electromagnetic radiation that is intercepted by it.

Black ice Thin, new ice on fresh or salt water that appears dark in colour because of its transparency.

Blizzard A blizzard is a severe winter storm condition characterized by low temperatures, strong winds, and heavy snow. They are caused when a high pressure area meets a low pressure area. Because the factors involving classification of winter storms are complex, there are many different definitions of blizzard. A major consensus is that in order to be classified as a blizzard, as opposed to merely a winter storm, the weather must meet several conditions. The storm must decrease visibility to a quarter of a mile or 400 metres for three consecutive hours, including snow or ice as precipitation, and have wind speeds of at least 35 miles per hour or 56 kilometres per hour. Another standard, according to Environment Canada, is that the winter storm must have winds of 40 km/h (25 mph) or more, have snow or blowing snow, visibility less than 1 km (about 5D 8 mile), a wind chill of less than "25 °C ("13 °F), and all of these conditions must last for 4 hours or more, before the storm can be properly called a blizzard. When all of these conditions persist after snow has stopped falling, meteorologists refer to the storm as a ground blizzard. An extreme form of blizzard is a whiteout, when downdrafts coupled with snowfall become so severe that it is impossible to distinguish the ground from the air. People caught in a whiteout can quickly become disoriented, losing their sense of direction. This poses difficulty for aviation flying in the altitude of the storm.

Blocking high The development of a warm ridge or cutoff high aloft at high latitudes which becomes associated with a cold high at the surface, causing a split in the westerly winds. Such a high will move very slowly, tending to move westward during intensification and eastward during dissipation. It prevents the movement of migratory cyclones across its latitudes. Two examples are a cut-off high and an Omega block.

Blowing dust Dust that is raised by the wind to heights of six feet or greater. It is reported as "bldu" in an observation and on the METAR.

Blowing sand Sand that is raised by the wind to heights of six feet or greater. It is reported as "blsa" in an observation and on the METAR.

Blowing snow Snow that is raised by the wind to heights of six feet or greater. It is reported as "blsn" in an observation and on the METAR.

Blowing spray Salt spray that is raised by the wind to heights of six feet or greater. It is reported as "blpy" in an observation and on the METAR.

Blowout depression Saucer shaped depressions created by wind erosion. At the leeward end of the feature there usually is a deposit of sand. Blowouts are found in coastal beach areas and in arid and semiarid regions of the world. These features are smaller than a deflation hollow.

Blue norther Refers to a swift-moving cold frontal passage in the southern Great Plains, marked by a dark, blue-black sky with strong winter winds from the northwest or north and temperatures that may drop 20 to 30 degrees Fahrenheit in a few minutes. Also known as a Texas Norther.

Body wave Type of seismic wave that travels through the interior of Earth.

Bog A habitat that consists of waterlogged spongy ground. Common vegetation are sedges and sphagnum moss. Bogs are common in Canada, Russia, and Scandinavia.

Boiling point The temperature at which a liquid changes to a vapourous state. The temperature at which the equilibrium vapour pressure between a liquid and its vapour is equal to the external pressure on the liquid. The boiling point of pure water at standard pressure is 100 degrees Celsius or 212 degrees Fahrenheit.

Boll weevil A small, greyish beetle of the southeastern United States with destructive larvae that hatch in and damage cotton bolls.

Bolson Is a closed desert basin with no drainage outlet, surrounded by mountains.

Bora Term used to describe a katabatic wind in Yugoslavia.

Boreal forest High to mid-latitude biome dominated by coniferous forest. Predominant vegetation of this biome is various species of spruce, fir, pine, and cedars. Also called Taiga.

Bottomset bed Horizontal deltaic deposit of alluvial sediment composed of fine silt and clay.

Boulder Large fragment of rock that has a diameter greater than 256 millimetres (200 millimetres in the United Kingdom).

Boulder wind A local name referring to an extremely strong downslope wind in the Front Range of the Rocky Mountains near Boulder, Colourado.

Boundaries Lines indicating the limits of countries, states, or other political jurisdictions, or different air masses.

Boundary layer The lowest layer of the earth's atmosphere, usually up to 3,300 feet, or one kilometre, from the earth's surface, where the wind is influenced by the friction of the earth's surface and the objects on it. May also be called the surface boundary layer or the friction layer.

Boundary monument Material object placed on or near a boundary line to preserve and identify the location of the boundary line on the ground.

Boundary survey Survey made to establish or to re-establish a boundary line on the ground, or to obtain data for constructing a map or plat showing a boundary line.

Bow echo A radar echo signature often associated with severe thunderstorms, especially those that produce wind damage. It is bent outward in a "bow" shape.

Bowen reaction series Model that explains the origin of the various types of igneous rocks. It suggests that the presence or absence of particular minerals in igneous rocks depends on the temperature of crystallization and on the magma's original chemical composition.

Boyle's law States that when the temperature is held constant, the volume of a gas is inversely proportional to its pressure. Therefore, if the pressure increases, the volume decreases and visa versa. For example, if the volume if halved, then the pressure is doubled. If the temperature is held constant, it becomes an isothermal process. Discovered by Robert Boyle (1627-1691), an Irish physicist and chemist and co-founder of the Royal Society.

Brackish Environment that is influenced by seawater with a salinity less than 35 parts per thousand (usually caused by the presence of an inflow of fresh water).

Braided stream Shallow stream channel that is subdivided into a number of continually shifting smaller channels that are separated by bar deposits.

Break in bulk point Commonly, a transfer point on a transport route where the mode of transport (or type of carrier) changes and where large-volume shipments are reduced in size. For example, goods may be unloaded from a ship and transferred to trucks at an ocean port.

Breaker The quick collapse of an overextended water wave as it approaches the shoreline. The collapse occurs when the ratio of wave height to wavelength exceeds 1:7. This phenomenon also produces swash.

Breccia Coarse grained sedimentary rock composed of cemented angular rock fragments.

Bright band A narrow, intense radar echo due to water-covered ice particles at the melting level where reflectivity is at its greatest.

Brine Seawater with a salinity greater than 35 parts per thousand. Usually occurs in isolated bodies of seawater that have high amounts of water loss due to evaporation.

British thermal unit (**BTU**) Measurement unit for heat. It is the amount of energy required to raise the temperature of one pound of water one degree from 62 to 63° Fahrenheit. One Btu is equal to 252 calories and to 1055 joules.

Broken The amount of sky cover for a cloud layer between 5/8ths and 7/8ths, based on the summation layer amount for that layer.

Bromeliad Plants of the bromeliad family (Bromeliaceae). These plants grow from the dry deserts of the subtropics to equatorial tropical rain forests. Many bromeliads grow high up on the branches and trunks of trees in the tropical rainforest. Based on growth habits and other characteristics, Bromeliaceae is divided into the subfamilies Pitcairnioideae, Tillan-dsioideae, and Bromelioideae.

Brunisol soil Soil order (type) of the Canadian System of Soil Classification. This soil type is associated with forest vegetation. It is usually poorly developed and immature. The most identifying trait of this soil is the presence of a brown B horizon.

Bubble high A small high that may be created by precipitation and vertical instability associated with thunderstorm activity. A product of downdrafts, it is relatively cold and often has the characteristics of a different air mass. Unstable, overrunning air may form squall lines on its leading edge. Also known as a meso high.

Bundled and unbundled Refers to the way software and hardware is sold. Often software is sold together--bundled—with hardware as a complete ready-to-use turnkey system. It is more common recently to purchase hardware and software separately, in unbundled form.

Bus The basic frame of a satellite system that includes the propulsion and stabilization systems, but not the instruments or data systems.

Butte An isolated hill or mountain with steep or precipitous sides, usually having a smaller summit area than a mesa.

Buys ballot's law Describes the relationship of the horizontal wind direction to the pressure distribution. In the Northern Hemisphere, if one stands with one's back to the wind, the pressure on one's left is lower than the pressure on one's right. It is reversed in the Southern Hemisphere. This law was named after the Dutch meteorologist, Buys Ballot, who developed the formula in 1857.

BWER Acronym for Bounded Weak Echo Region. Refers to radar echo signatures with low reflectivity in the centre, surrounded by higher reflectivity. It is associated with strong updrafts and is found in the inflow region of a thunderstorm. Also known as a vault.

C

C horizon Soil horizon normally found below the B horizon and above the R horizon. This layer is composed of weathered bedrock that has not been yet significantly affected by the pedogenic processes.

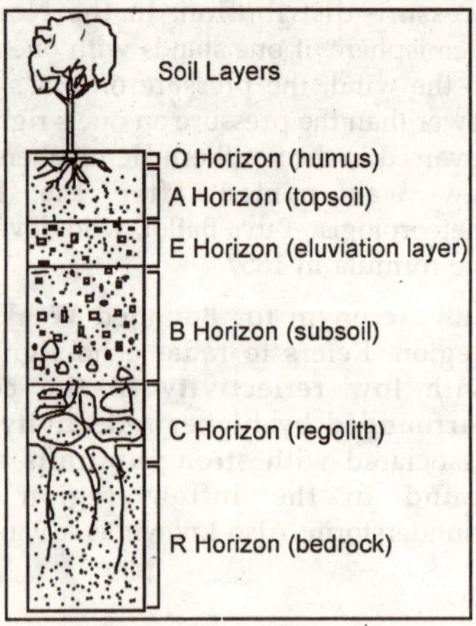

Fig. C horizon

Cadastral survey Survey relating to land boundaries, made to create units suitable for title transfer or to define the limitations of title. Derived from "cadastre" meaning a register of land quantities, values, and ownership used levying taxes, the term may properly be applied to surveys of a similar nature outside the public lands, such surveys are more commonly called "land surveys" or "property surveys.".

Calcification A dry environment soil-forming process that results in the accumulation of calcium carbonate in surface soil layers.

Calcite The carbonate mineral calcite is a calcium carbonate corresponding to the formula $CaCO_3$ and is one of the most widely distributed minerals on the Earth's surface. It is a common constituent of sedimentary rocks, limestone in particular. It is also the primary mineral in metamorphic marble. It also occurs as a vein mineral in deposits from hot springs, and also occurs in caverns as stalactites and stalagmites. Calcite is often the primary constituent of the shells of marine organisms, e.g., plankton (such as coccoliths and planktic foraminifera), the hard parts of red algae, some sponges, Brachiopoda, echinoderms, most Bryozoa, and parts of the shells of some bivalves, such as oysters and rudists). Calcite represents the stable form of calcium carbonate; aragonite will change to calcite at 470°C. Calcite crystals are hexagonal-rhombohedral, though actual calcite rhombohedrons are rare as natural crystals. However, they show a remarkable variety of habit including acute to obtuse rhombohedrons, tabular forms, prisms, or various scalenohedrons. Calcite exhibits several twinning types adding to the variety of observed forms. It may occur as fibrous, granular, lamellar, or compact. Cleavage is usually in three directions parallel to the rhombohedron form. Its fracture is conchoidal, but difficult to obtain.

It has a Mohs hardness of 3, a specific gravity of 2.71, and its lustre is vitreous in crystallized varieties. Colour is white or colourless, though shades of gray, red, yellow, green, blue, violet, brown, or even black can occur when the mineral is charged with impurities. Calcite is transparent to opaque and may occasionally show phosphorescence or fluorescence. It is perhaps best known because of its power to produce strong double refraction of light, such that objects viewed through a clear piece of calcite appear doubled in all of their parts - a phenomenon first described by Rasmus Bartholin. A beautifully transparent variety used for optical purposes comes from Iceland, called Iceland spar. Acute scalenohedral crystals are sometimes referred to as "dogtooth spar".

Calcium carbonate Compound consisting of calcium and carbonate. Calcium carbonate has the following chemical structure $CaCO_3$.

Caldera A large circular depression in a volcano.

Caldera volcano A caldera is a volcanic feature formed by the collapse of a volcano into itself. Calderas may be filled with water, creating crater lakes. The word 'caldera' comes from a Spanish word meaning "cauldron". A caldera collapse is usually triggered by the emptying of the magma chamber beneath the volcano, often as the result of a large volcanic eruption. If enough magma is erupted, the emptied chamber will not be able to support the weight of the volcanic edifice (the mountain) above. Fractures will form around the edge of the chamber, usually in a roughly circular shape. These ring fractures may in fact serve as volcanic vents. As the magma chamber empties, the centre of the volcano within the ring fractures and begins to collapse. The collapse may occur as the result of a single massive eruption, or it may occur in stages as the result of a series of eruptions. The total amount of collapse may be hundreds or thousands of metres.

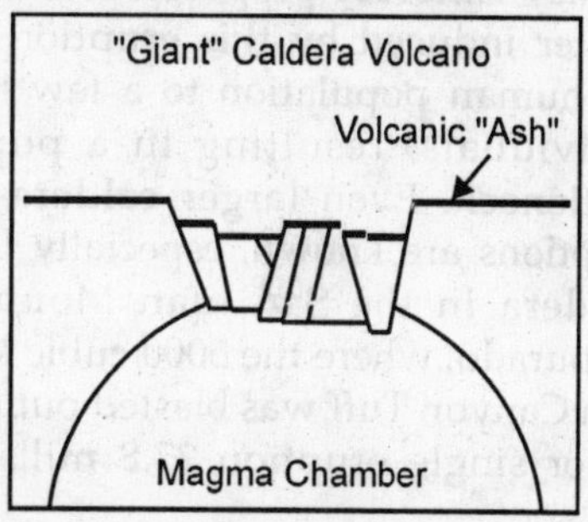

Fig. Caldera volcano

If the magma is rich in silica, the caldera is often filled in with ignimbrite, tuff, rhyolite, and other igneous rocks. Silica-rich magma is very viscous. As a result, gases tend to become trapped at high pressure within the magma. When the magma gets near the surface of the Earth, the gas expands quickly, causing explosions and spreading volcanic ash over wide areas. Further lava flows may be erupted, and the centre of the caldera is often uplifted in the form of a resurgent dome by subsequent intrusion of magma. A silicic or rhyolitic caldera may erupt hundreds or even thousands of cubic kilometres of material in a single event. Even small caldera-forming eruptions, such as Krakatoa in 1883 or Mount Pinatubo in 1991, may result in significant local destruction and a noticeable drop in temperature around the world. Large calderas may have even greater effects.

When Yellowstone Caldera erupted 640,000 years ago it released 1000 cubic kilometres of material, covering half of North America in up to two metres of debris. By comparison, when Mount St. Helens erupted in 1980, it released 1.2 cubic kilometres of ejecta. The ecological effects of the eruption of a large caldera can be seen in the record of the Lake Toba eruption in Indonesia. About 75,000 years ago, this volcano released 2800 cubic kilometres of ejecta, the largest known eruption within the Quaternary Period (last 1.8 million years). In the late 1990s, archeologist Stanley Ambrose proposed that a volcanic winter induced by this eruption reduced the human population to a few thousand individuals, resulting in a population bottleneck. Even larger caldera-forming eruptions are known, especially La Garita Caldera in the San Juan Mountains of Colourado, where the 5000 cubic kilometre Fish Canyon Tuff was blasted out in a truly major single eruption 27.8 million years ago.

At some points in geologic time, rhyolitic calderas have appeared in distinct clusters. The remnants of such clusters may be found in places such as the San Juan Mountains of Colourado (erupted during the Tertiary Period) or the Saint Francois Mountain Range of Missouri (erupted during the Proterozoic). Some volcanoes, such as Kilauea on the island of Hawaii, form calderas in a different fashion. In the case of Kiluaea, the magma feeding the volcano is relatively silica poor. As a result, the magma is much less viscous than the magma of a rhyolitic volcano. Such calderas are also known as subsidence calderas. The magma chamber is drained by large lava flows rather than by explosive events. Kilauea Caldera has an inner crater known as Halema'uma'u, which has often been filled by a lava lake. The largest volcano on Earth, Mauna Loa is also capped by a subsidence caldera called Moku'âweoweo Caldera. It is possible, although rare, for a caldera-like formation to be created by erosion rather than volcanism. It is believed that the Caldera de Taburiente on La Palma in the Canary Islands is an example of this.

Calibration Act of comparing an instrument's measuring accuracy to a known standard.

Caliche An accumulation of calcium carbonate at or near the soil surface.

Calm Atmospheric conditions devoid of wind or any other air motion. In oceanic terms, it is the apparent absence of motion of the water surface when there is no wind or swell.

Calorie In meteorology, it is the amount of heat required to raise the temperature of one 1. gram of water one 1. degree Celsius. It is a unit of heat energy.

Calving The loss of glacier mass when ice breaks off into a large water body like an ocean or a lake.

Cambrian Geologic period that occurred from 570 to 505 million years ago. During this period, invertebrates become common in the oceans and the Burgess Shale was formed.

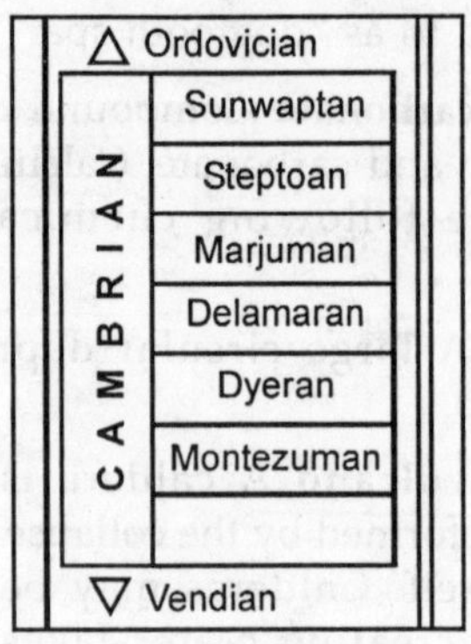

Fig. Cambrian

Cambrian explosion Great diversification of multicellular life forms in the Earth's oceans that started during the Cambrian about 570 million years ago.

Canadian high High pressure system that develops in winter over central North America.

Canadian shield Very old igneous and metamorphic shield rock that covers much of northern Canada. Created more than two to three billion years ago.

Canadian system of soil classification A hierarchical system that is used in Canada to classify soils. This system has five levels: order, great group, subgroup, family, and series. At the order level, nine types of soils are recognized: brunisol, chernozem, cryosol, gleysol, luvisol, organic, podzol, regosol, and solonetzic.

Canal A canal is a man-made waterway used for transportation or irrigation.

Canopy The layer formed naturally by the leaves and branches of trees and plants.

Canopy drip Redirection of a proportion of the rain or snow falling on a plant to the edge of its canopy.

Canyon A canyon is a deep valley with very steep sides - often carved from the Earth by a river.

Canyon A canyon or gorge is a deep valley often carved from the Earth by a river. Most canyons originate by a process of long-time erosion from a plateau level, with a stream gradually carving out its valley. The cliffs form because harder rock strata that are resistant to erosion and weathering remain exposed on the valley walls. Canyons are much more common in arid areas than in wetter areas because weathering has a lesser effect in arid zones. Canyons' walls are often formed of resistant sandstones or granite. Submarine canyons are those which form underwater, generally at the mouths of rivers. The word canyon is Spanish in origin. The word canyon is generally used in the United States, while the word gorge is more common in Europe and Australasia, though it is also used in some parts of the United States and Canada. A famous example is the Grand Canyon in Arizona. In the south-western United States, canyons are important archeologically because of the many cliff-dwellings built there, largely by the Ancient Pueblo Peoples. Sometimes large rivers run through canyons as the result of gradual geologic uplift. These are called entrenched rivers, because they are unable to easily alter course. The Colorado River and the Snake River in the north-western United States are two examples of this. Canyons often form in areas of limestone rock. Limestone is to a certain extent soluble, so cave systems form in the rock. When these collapse a canyon is left, for example in the Mendip Hills in Somerset and Yorkshire Dales in Yorkshire, England.

Two canyons usually attributed as the world's largest are the:

1. Grand Canyon in Arizona, USA, and the

2. Copper Canyon in Chihuahua, Mexico.

However, this definition used are rather imprecise, as a canyon can be "large" by its depth, length, or the total area of the canyon system. Also the inaccessibility of the major canyons in the Himalaya contributes to the fact that they are not often known or regarded as candidates for the biggest canyon. The deepest canyon in the world, Yarlung Tsangpo Canyon, along the Brahmaputra in Tibet, China, even slightly longer than Grand Canyon, is also regarded by many as be the world's largest canyon, followed by the Kali Gandaki Canyon in Nepal, and Polung Tsangpo Canyon in Tibet.

Cap Composed of a layer of warmer, dryer air aloft which may suppress or delay the development of thunderstorms. As an air parcel rises into this area, it becomes too cool to rise further. Also referred to as a lid.

Cape A cape is a pointed piece of land that sticks out into a sea, ocean, lake, or river.

CAPE Acroymn for convective available potential energy. The amount of energy available to create convection, with higher values increasing the possibility for severe weather.

Cape Verde islands A group of volcanic islands in the eastern Atlantic Ocean off the coast of West Africa. A Cape Verde hurricane originates near here.

Capillary action Movement of water along microscopic channels. This movement is the result of two forces: the adhesion and absorption of water to the walls of the channels; and cohesion of water molecules to each other.

Capillary water Water that moves horizontally and vertically in soils by the process of capillary action. This water is available for plant use.

Caprock A strata of erosion-resistant sedimentary rock (usually limestone) found in arid areas. Caprock forms the top layer of most mesas and buttes.

Carbohydrate Is an organic compound composed of carbon, oxygen, and hydrogen atoms. Some examples are sugars, starch, and cellulose.

Carbon cycle All parts (reservoirs) and fluxes of carbon. The cycle is usually thought of as four main reservoirs of carbon interconnected by pathways of exchange. The reservoirs are the atmosphere, terrestrial biosphere (usually includes freshwater systems), oceans, and sediments (includes fossil fuels). The annual movements of carbon, the carbon exchanges between reservoirs, occur because of various chemical, physical, geological, and biological processes. The ocean contains the largest pool of carbon near the surface of the Earth, but most of that pool is not involved with rapid exchange with the atmosphere.

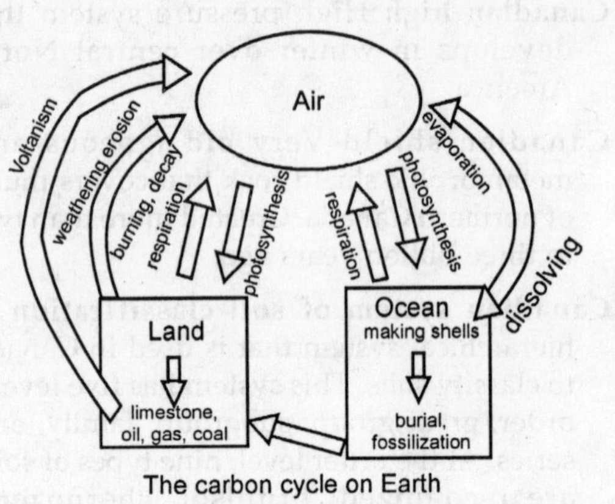

Fig. Carbon cycle

Carbon dioxide (CO_2) A minor but very important component of the atmosphere, carbon dioxide traps infrared radiation. Atmospheric CO_2 has increased about 25 percent since the early 1800s, with an estimated increase of 10 percent since 1958 (burning fossil fuels is the leading cause of increased CO_2, deforestation the second major cause). The increased amounts of CO_2 in the atmosphere enhance the greenhouse effect, blocking heat from escaping into space and contributing to the warming of Earth's lower atmosphere.

Carbon monoxide A colourless, odourless, and tasteless gas that is produced by the incomplete burning of fossil fuels. The chemical formula for carbon dioxide is CO.

Carbonate Compound consisting of a single atom of carbon and three atoms of oxygen. Carbonate has the following chemical structure CO_3.

Carbonation Is a form of chemical weathering where carbonate and bicarbonate ions react with minerals that contain calcium, magnesium, potassium, and sodium.

Cardinal points The four main navigational directions (North, East, South, and West) found on a compass or a map.

Carnivore Heterotrophic organism that consumes living animals or the parts of living animals.

Carrier Radio frequency capable of being modulated with some type of information.

Carrying capacity The number of people that an area can support given the quality of the natural environment and the level of technology of the population.

Cartesian coordinates Are a system of positional reference in which location is measured along two or three orthogonal (perpendicular) axes. Every location can be defined uniquely by its X, Y, and Z coordinates. Locations in the coordinate system can be established using any unit of measurement such as metres, feet, and miles. Cartesian coordinates differ from latitude- longitude coordinates in that the latter comprise a spherical (rather than planar) reference system.

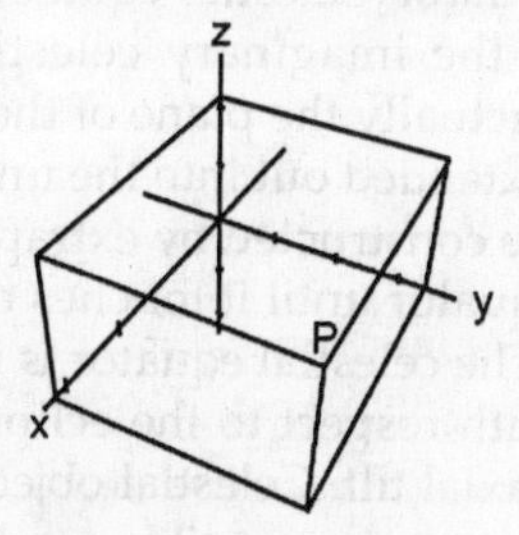

Fig. Cartesian coordinates

Cartographer A person who draws or makes maps or charts.

Cartography Cartography or mapmaking is the study and practice of making maps or globes. Maps have traditionally been made using pen and paper, but the advent and spread of computers has revolutionized cartography. Most commercial quality maps are now made with map making software that falls into one of three main types; CAD, GIS, and specialized map illustration software. Maps function as visualization tools for spatial data. Spatial data is acquired from measurement and can be stored in a database, from which it can be extracted for a variety of purposes. Current trends in this field are moving away from analogue methods of mapmaking and toward the creation of increasingly dynamic, interactive maps that can be manipulated digitally.

The cartographic process rests on the premise that the world is measurable and that we can make reliable representations or models of that reality. Mapmaking involves advanced skills and attitudes, particularly the use of symbols to represent certain geographic phenomena, as well as the ability to visualize the world in an abstract and scaled down form. In cartography, technology has continually changed in order to meet the demands of new generations of mapmakers and map users. The first maps were manually constructed with brushes and parchment and therefore varied in quality and were limited in distribution. The advent of magnetic devices, such as the compass and much later magnetic storage devices, allowed for the creation of far more accurate maps and the ability to store and manipulate them digitally.

Advances in mechanical devices such as the printing press, quadrant and vernier allowed for the mass production of maps and the ability to make accurate reproductions from more accurate data. Optical technology, such as the telescope, sextant and other devices that use telescopes, allowed for accurate surveying of land and the ability of mapmakers and navigators to find their latitude by measuring angles to the North Star at night or the sun at noon. Advances in photochemical technology, such as the lithographic and photochemical processes, have allowed for the creation of maps that have fine details, do not distort in shape and resist moisture and wear. This also eliminated the need for engraving which further shortened the time it takes to make and reproduce maps. In the mid to late 20th century advances in electronic technology have led to a new revolution in cartography. Specifically computer hardware devices such as computer screens, plotters, printers, scanners (remote and document) and analytic stereo plotters along with visualization, image processing, spatial analysis and database software, have democratized and greatly expanded the making of maps.

Cascading system This is a system where we are primarily interested in the flow of energy and/or matter from one element to another and understand the processes that cause this movement. In a cascading system, we do not fully understand quantitative relationships that exist between elements related to the transfer of energy and/or matter.

Catalina eddy A weak low pressure circulation that may form off the Southern California coast.

Catalogue number A five-digit number assigned to a catalogued orbiting object. This number may be found in the NASA Satellite Situation Report and on the NASA Prediction Bulletins.

Catastrophism General theory that suggests that certain phenomena on the Earth are the result of catastrophic events. For example, the Biblical Flood is responsible for sedimentary rock formations and the extinction of the dinosaurs.

Cation An ion carrying a positive atomic charge.

Cation exchange Chemical trading of cations between the soil minerals and organic matter with the soil solution and plant roots.

Cation exchange capacity The capacity of a soil to exchange cations with the soil solution. Often used as a measure of potential soil fertility.

Cave A cave is a large hole in the ground or in the side of a hill or mountain.

Cavitation Process of intense erosion due to the surface collapse of air bubbles found in constricted rapid flows of water. Causes the detachment of material from a surface.

Ceiling The lowest cloud layer that is reported as broken or overcast. If the sky is totally obscured, then it is the height of the vertical visibility.

Ceiling light An instrument consisting of a drum and an optical system that projects a narrow vertical beam of light onto a cloud base.

Ceilometre An instrument that is used to measure the angular elevation of a projected light on the base of a cloud. It measures the angle of the cloud base included by the observer (or machine), the ceiling light and the illuminated spot on the cloud.

Celestial equator Celestial equator is a great circle on the imaginary celestial sphere, which is actually the plane of the terrestrial equator extended out into the universe (i.e., it could be constructed by extrapolating the Earth's equator until it touches the celestial sphere). The celestial equator is inclined by ~23.5°, with respect to the ecliptic plane; a result of axial tilt. Celestial objects near the celestial equator are visible worldwide, but they culminate the highest in the sky in the tropics.

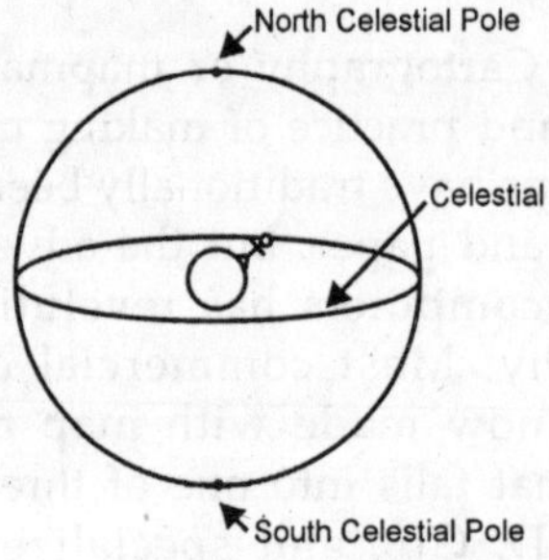

Fig. Celestial equator

Celestial sphere The apparent sphere of infinite radius having the earth as it centre. All heavenly bodies (planets, stars, etc.) appear on the "inner surface" of this sphere and the sun moves along the ecliptic.

Cell A cell is the smallest self-functioning unit found in living organisms. Each cell is enclosed by an outer membrane or wall and contains genetic material (DNA) and other parts to carry out its life functions. Some organisms such as bacteria consist of only one cell, but most of the organisms found on the Earth are made up of many cells.

Cellular Composed of cells.

Cellulose A type of carbohydrate. Primary component used in the construction of plant cell walls.

Celsius scale Scale for measuring temperature. In this scale, water boils at 100° and freezes at 0°.

Celsius temperature scale A temperature scale where water at sea level has a freezing point of 0 degrees C (Celsius) and a boiling point of +100 degrees C. More commonly used in areas that observe the metric system of measurement. Created by Anders Celsius in 1742. Same as Centigrade. In 1948, the Ninth General Conference on Weights and Measures replaced "degree centigrade" with "degree Celsius.".

Cenozoic Geologic era that occurred from 65 million years ago to today.

Centigrade Temperature scale proposed by Swedish astronomer Anders Celsius in 1742. A mixture of ice and water is zero on the scale; boiling water is designated as 100 degrees. A degree is defined as one hundredth of the difference between the two reference points, resulting in the term centigrade (100th part).

To convert centigrade to Fahrenheit; multiply the centigrade temperature by 1.8 and add 32º. F = 9/5C + 32.

To convert Fahrenheit to centigrade: subtract 32o from the Fahrenheit temperature and divide the quantity by 1.8. C = (F -32. / 1.8)

Central meridian A central meridian is a meridian that passes through the centre of a projection. The central meridian is often a straight line that is an axis of symmetry of the projection.

Central pressure The atmospheric pressure at the centre of a high or low. It is the highest pressure in a high and the lowest pressure in a low, referring to the sea level pressure of the system on a surface chart.

Central vent The main passage way by which volcanic magma travels to the Earth's surface.

Centrifugal force The apparent force in a rotating system that deflects masses radially outward from the axis of rotation. This force increases towards the equator and decreases towards the poles. This force on the earth and in the atmosphere due to the rotation about the earth's axis is incorporated with the field of gravitation to form gravity.

Centripetal force Force required to keep an object moving in a circular pattern around a centre of rotation. This force is directed towards the centre of rotation. Common in meteorological phenomena like tornadoes and hurricanes.

Centroid Is the term given to the centre of an area, region, or polygon. In the case of irregularly shaped polygons, the centroid is derived mathematically and is weighted to approximate a sort of "centre of gravity." Centroids are important in GIS because these discrete X-Y locations are often used to index or reference the polygon within which they are located. Sometimes attribute information is "attached," "hung," or "hooked" to the centroid location.

Chain reaction (nuclear) A large number of nuclear fissions, taking place within a certain mass of a fissionable isotope, that release a great quantity of energy in a short time.

Chain Unit of length equal to 66 feet, used especially in the U.S. public land surveys. The original measuring instrument

(Gunter's chain) was literally a chain consisting of 100 iron links, each 7.92 inches long. Steel-ribbon tapes began to supersede chains around 1900, but surveying tapes are often still called "chains" and measuring with a tape is often called "chaining." The chain is a convenient unit in cadastral surveys because 10 square chains equal 1 acre.

Chalk Form of limestone. This sedimentary rock is composed of the shells and skeletons of marine microorganisms.

Channel A channel is a narrow body of water that connects two larger bodies of water (like the English Channel). A channel is also a part of a river or harbor that is deep enough to let ships sail through.

Chaparral A type of plant community common to areas of the world that have a Mediterranean climate (for example, California and Italy). It is characterized by shrubs, shrubby thickets and small trees that are adapted to seasonal dry conditions. Also called Mediterranean Scrubland.

Charles' law States that when the pressure is held constant, the volume of a gas varies directly with the temperature. Therefore, if the pressure remains constant, the volume of a gas will increase with the increase of temperature. It was developed by Jacques Charles and is also known as the Charles-Guy-Lussac Law.

Chart Special-purpose map designed for navigation or to present specific data or information. The term "chart" is applied chiefly to maps made primarily for nautical and aeronautical navigation, and to maps of the heavens, although the term is sometimes used to describe other special-purpose maps.

Chart, aeronautical Charts designed to meet requirements of aerial navigating, produced in several series, each on a specified map projection and differing in scale, format, and content, for use as dictated by type of aircraft and whether flight is to be conducted under visual or instrument flight rules.

Chart, nautical Representation of a portion of the navigable waters of the Earth and adjacent coastal areas on a specified map projection and designed specifically to meet requirements for marine navigation. Included on most nautical charts are depths of water, characteristics of the bottom, elevations of selected topographic features, general configurations and characteristics of the coast, the shoreline (usually the mean high water line), dangers, obstructions and aids to navigation limited tidal data, and information about magnetic variation in the charted area.

Chelate Organic substances that cause the chemical process of chelation.

Chelation Chemical weathering process that involves the extraction or metallic cations from rocks and minerals by chelates.

Chemical One of the millions of different elements and compounds found naturally and synthesized by humans.

Chemical autotroph Organism that uses the external energy found in chemical compounds to produce food molecules. The process used to produce food by these organisms is known as chemosynthesis.

Chemical energy Energy used or released in chemical reactions.

Chemical reaction Reaction between chemicals where there is a change in the chemical composition of the elements or compounds concerned.

Chemical weathering Breakdown of rock and minerals into small sized particles through chemical decomposition.

Chemosphere A vaguely defined region of the upper atmosphere in which photochemical reactions take place. It includes the top of the stratosphere, all of the mesosphere, and sometimes the lower part of the thermosphere.

Chemosynthesis Process in which specific autotrophic organisms extract inorganic compounds from their environment and convert them into organic nutrient compounds without the use of sunlight.

Chernozem soil 1. Soil order (type) of the Canadian System of Soil Classification. This soil is common on the Canadian Prairies.

2. Type of soil commonly found in grassland environments. These soils are often black in colour and have a well developed A horizon rich in humus.

Cheyenne fog An upslope fog formed by the westward flow of air from the Missouri River Valley, producing fog on the eastern slopes of the Rockies.

Chinook Chinook winds, often just called chinooks, are a variety of Föhn winds pattern observed in the interior West of North America, where the Canadian Prairies and Great Plains end and the mountains begin. These wind patterns are named for the Chinook Indians. In popular myth, Chinook is supposed to mean "snow eater", as a strong Chinook can make a foot of snow all but vanish inside of one day. The snow partially melts, and partially evaporates in the dry wind. The true origin of the name is that "Chinook Wind" in the local argot of the fur trade era meant that the wind came from the direction of the country of the Chinooks (the lower Columbia River), i.e. from across the Rocky Mountains.

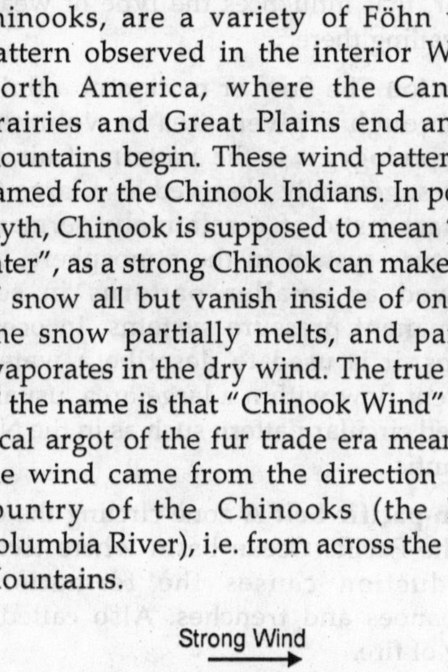

Fig. Chinook

Chinook winds have been observed to elevate winter temperatures, often from below "20°C ("4°F) to as high as 10°C to 20°C (50°F to 68°F), for a few hours or days, at the end of which, the temperatures plummet to their base levels. One of the most dramatic examples of the Chinook winds occurred on January 15, 1972, in Loma, Montana. The temperature rose from "47°C ("54°F) to 9°C (49°F); the greatest temperature change ever recorded during a 24-hour period. A "Chinook Wind" or simply "a Chinook", has a different meaning on the British Columbia and Pacific Northwest coasts where the term originates. At these locations west of the Rockies, a Chinook is a warm, moist wind from the southwest, likely to bring rain (or snow at higher elevations). This was the origin of the name that spread via the fur trade to the Prairies - the wind from the country of the Chinooks, i.e. from the southwest. It is also pronounced differently on the Pacific Coast, with Chinook pronounced as in English church, while on the Prairies it is pronounced a la française, Shinook - because it was the French-speaking voyageurs of the fur companies who brought the term back across the mountains.

Chlorofluorocarbon (CFC) A family of compounds of chlorine, fluorine, and carbon, entirely of industrial origin. CFCs include refrigerants, propellants for spray cans (this usage is banned in the U.S., although some other countries permit it) and for blowing plastic-foam insulation, Styrofoam packaging, and solvents for cleaning electronic circuit boards. The compounds' lifetimes vary over a wide range, exceeding 100 years in some cases.

CFCs' ability to destroy stratospheric ozone through catalytic cycles is contributing to the depletion of ozone worldwide. Because CFCs are such stable molecules, they do not react easily with other chemicals in the lower atmosphere. One of the few forces that can break up CFC molecules is ultraviolet radiation, however, the ozone layer protects the CFCs from ultraviolet radiation in the lower atmosphere. CFC molecules are then able to migrate intact into the stratosphere, where the molecules are bombarded by ultraviolet rays, causing the CFCs to break up and release their chlorine atoms. The released chlorine atoms participate in ozone destruction, with a single atom of chlorine able to destroy ozone molecules over and over again.

International attention to CFCs resulted in a meeting of diplomats from around the world in Montreal in 1987. They forged a treaty that called for drastic reductions in the production of CFCs. In 1990, diplomats met in London and voted to significantly strengthen the Montreal Protocol by calling for a complete elimination of CFCs by the year 2000.

Chlorophyll Green pigment found in plants and some bacteria used to capture the energy in light through photosynthesis.

Chloroplast Organelle in a cell that contains chlorophyll and produces organic energy through photosynthesis.

Chocolatta north A West Indian gale that blows from the northwest.

Chromosome Organic structure that carries an organism's genetic code (DNA).

Chromosphere A thin layer of relatively transparent gases above the photosphere of the sun. It is observed best during a total eclipse of the sun.

Cinder cone volcano A small volcano, between 100 and 400 metres tall, made up of exploded rock blasted out of a central vent at a high velocity. These volcanoes develop from magma of basaltic to intermediate composition.

Circadian rhythm The cyclical changes in physiological processes and functions that are related to the 24-hour diurnal cycle.

Circle of illumination A line that bisects areas on the Earth receiving sunlight and those areas in darkness. Cuts the spherical Earth into lighted and dark halves.

Circuit The complete path of an electric current; an assemblage of electronic elements; a means of two-way communication between two points—comprised of associated "go" and "return" channels.

Circularly polarized RF Radio frequency transmissions where the wave energy is divided equally between a vertically- and a horizontally-polarized components.

Circulation cells Large areas of air movement created by the rotation of the earth and the transfer of heat from the equator polar ward. Circulation is confined to a specific region, such as the tropics, temperate, or polar, that influences the type of weather prevailing there.

Circulation The flow or motion of a fluid in or through a given area or volume. In meteorology, it is the pattern of air as it moves, generally observed as a large flow characteristic of a relatively permanent pressure system in the atmosphere. Also viewed as smaller patterns in semi-permanent pressure systems. In oceanic terms, it is used to describe a water in current flow within a large area, usually a closed circular pattern such as in the North Atlantic.

Circum-pacific belt A zone circling the edge of the Pacific Ocean basin where tectonic subduction causes the formation of volcanoes and trenches. Also called the ring of fire.

Cirque Glacially eroded rock basin found on mountains. Most alpine glaciers originate from a cirque.

Cirque glacier Small glacier that just occupies a cirque.

Cirriform Clouds composed of small particles, mostly ice crystals. Because the particles are fairly widely dispersed, this usually results in relative transparency and whiteness, often producing a halo phenomena not observed in other clouds forms. These clouds generally have bases above 20,000 feet in the mid-latitudes, and are classified as high clouds. They include all varieties of cirrus, cirrocumulus, and cirrostratus clouds.

Cirrocumulus A cirriform cloud with vertical development, appearing as a thin sheet of small white puffs which give it a rippled effect. It often creates a "mackerel sky", since the ripples may look like fish scales. Sometimes it is confused with altocumulus, however, it has smaller individual masses and does not cast a shadow on other elements. It is also the least common cloud type, often forming from cirrus or cirrostratus, with which it is associated in the sky.

Cirrostratus A cirriform cloud that develops from cirrus spreading out into a thin layer, creating a flat sheet like appearance. It can give the sky a slightly milky or veiled look. When viewed from the surface of the earth, these ice crystals can create a halo effect around the sun or moon. This cloud is a good precursor of precipitation, indicating it may occur within 12 to 24 hours.

Cirrus One of the three basic cloud forms (the others are cumulus and stratus). It is also one of the three high cloud types. Cirrus are thin, wispy clouds composed of ice crystals and often appear as veil patches or strands. In the mid-latitudes, cloud bases are usually found between 20,000 to 30,000 feet, and it is the highest cloud that forms in the sky, except for the tops, or anvils, of cumulonimbus, which occasionally build to excessive heights.

Civil twilight The time between the moment of sunset, when the sun's apparent upper edge is just at the horizon, until the centre of the sun is 6 degrees directly below the horizon.

Clarke belt A belt 22,245 miles (35,800 kilometres) directly above the equator where a satellite orbits the Earth at the same speed the Earth is rotating. Science fiction writer and scientist Arthur C. Clarke wrote about this belt in 1945, hence the name.

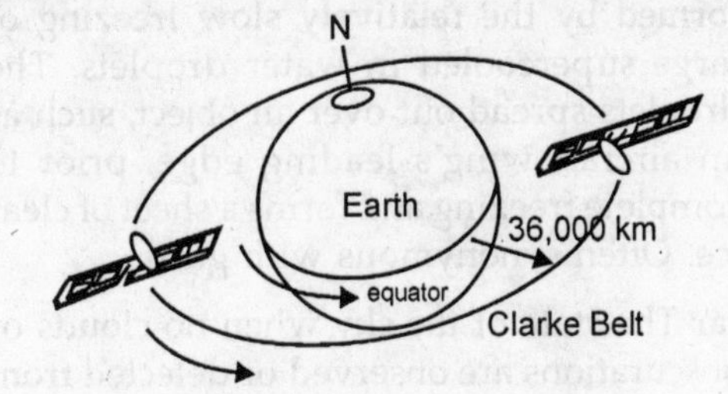

Fig. Clarke belt

Classification Process of grouping things into categories.

Clastic sedimentary rock Sedimentary rocks that are formed by the lithification of weathered rock debris that has been physically transported and deposited.

Clay Mineral particle with a size less than 0.004 millimetres in diameter.

Clear air turbulence Clear-Air Turbulence (often abbreviated CAT and sometimes colloquially referred to as "air pockets") is the erratic movement of air masses in the absence of any visual cues (such as clouds). Clear-Air Turbulence is caused when bodies of air moving at widely different speeds meet; at high altitudes (7,000-12,000 metres/23,000-39,000 feet) this is frequently encountered around jet streams or sometimes near mountain ranges. Clear-Air Turbulence is impossible to detect either with the naked eye or with radar, meaning that it is difficult to avoid. However, it can be remotely detected with instruments that can measure turbulence with optical techniques, such as scintillometres. This kind of turbulence creates a hazard for air navigation. The rapid changes in the speed and direction of the air mass cause the lift created by an aircraft's wings to vary quickly and unpredictably, making for a rough flight. Cabin crew and passengers on airliners have been injured (and in a small number of cases, killed) when thrown around inside an aircraft experiencing turbulence.

Clear ice A glossy, clear, or translucent ice formed by the relatively slow freezing of large supercooled in water droplets. The droplets spread out over an object, such as an aircraft wing's leading edge, prior to complete freezing and forms a sheet of clear ice. Often synonymous with glaze.

Clear The state of the sky when no clouds or obscurations are observed or detected from the point of observation.

Cleavage The tendency of some minerals or rocks to break along planes of weakness. This weakness occurs because of the nature of the bonds between mineral grains.

Cliff A cliff is a steep face of rock and soil.

Climate The climate is commonly considered to be the weather averaged over a long period of time, typically 30 years. Somewhat more precisely, the concept of "climate" also includes the statistics of the weather - such as the degree of day-to-day or year-to-year variation expected. The Intergovernmental Panel on Climate Change (IPCC) glossary definition is: .

Climate in a narrow sense is usually defined as the "average weather", or more rigorously, as the statistical description in terms of the mean and variability of relevant quantities over a period of time ranging from months to thousands or millions of years. The classical period is 30 years, as defined by the World Meteorological Organization (WMO). These quantities are most often surface variables such as temperature, precipitation, and wind. Climate in a wider sense is the state, including a statistical description, of the climate system.

In the most succinct words, weather is the combination of events in the atmosphere and climate is the overall accumulated weather in a certain location. The exact boundaries of what is climate and what is weather are not well defined and depend on the application. For example, in some senses an individual El Niño event could be considered climate; in others, as weather. When the original conception of climate as a long-term average came to be considered, perhaps towards the end of the 19th century, the idea of climate change was not current, and a 30 year average seemed reasonable. Given the current availability of long-term trends in the temperature record, it is harder to give a definition of climate to suit all purposes: over a 30 year period, averages may shift; over a shorter period, the statistics are less stable. In a given geographical region, the climate generally does not vary over time on the scale of a human life span. However, over geological time, climate can vary considerably for a given place on the Earth. For example, Scandinavia has been through a number of ice ages over hundreds of thousands of years (the last one ending about 10,000 years ago). Paleoclimatology is the study of these past climates, their origin, and by extension, the origin of today's climate.

In the original ancient Greeks sense, 'Clime,' now our modern day climate was a concept used to divide the world into regions sharing similar climatic attributes, such as temperature and length of the solar day (sunlight), and in particular especially latitudinal distance from the equator into a so called Climatic zone. The more common modern equivalent and technically accepted term is Climate regions, as latitudinal concepts of antiquity don't hold up well against modern science data and analysis, which show climate is a much more localized phenomenon affected by the presence of significant geographically affective features such as local hills or mountain ranges, large lakes, broad plains, forests or their lack; and seas or oceans, etcetera.

Consequently, today's Climate regions are classified on the basis of temperature and precipitation alone. Examples of such climate schemes are the Köppen climate classification or the Thornthwaite climate classification schemes. In broad terms, climate zones range in latitude from tropical (hot, no real seasons based on temperature) to mid-latitude (moderate; conventional winter-spring-summer-autumn seasons) to polar (cold; strong seasonality; long winters). They also divide into Continental (high contrast between summer and winter) and Maritime (ocean influence moderates temperature extremes, but generally contribute extra moisture to local weather patterns). The mid-latitude 'conventional four seasons' only work well in the mid-latitudes. In other regions, the more useful divide is into "wet" and "dry" seasons based on the monsoon.

Climatic optimum Warmest period during the Holocene epoch. This period is dated from about 5,000 to 3,000 BC. During this time average global temperatures were 1 to 2° Celsius warmer than they are today.

Climatology Climatology is the study of climate, or past weather, and is a branch of the atmospheric sciences. In contrast to meteorology, which studies short term weather systems lasting up to a few weeks, climatology studies the frequency with which these weather systems occurred in the past. It studies the periodicity of weather events over years to millennia, as well as changes in long-term average weather patterns, in relation to atmospheric conditions. Climatologists, those who practice climatology, study both the nature of climates - local, regional or global - and the natural or human-induced factors that cause climates to change. Climatology considers the past and can help predict future climate change. Phenomena of climatological interest include the atmospheric boundary layer, circulation patterns, heat transfer (radiative, convective and latent), interactions between the atmosphere and the oceans and land surface (particularly vegetation, land use and topography), and the chemical and physical composition of the atmosphere. Related disciplines include chemistry, ecology, geology, geophysics, glaciology, hydrology, oceanography, and volcanology. Climatology is approached in a variety of ways. Paleoclimatology seeks to reconstruct past climates by examining records such as ice cores and tree rings (dendroclimatology). The study of contemporary climates incorporates meteorological data accumulated over many years, such as records of rainfall, temperature and atmospheric composition. Knowledge of the atmosphere and its dynamics is also embodied in models, either statistical or mathematical, which help by integrating different observations and testing how they fit together. Modelling is used for understanding past, present and potential future climates. Climate research is made difficult by the large scale, long time periods, and complex processes which govern climate. It is generally accepted that climate is governed by differential equations based on physical laws, but what, exactly, are these equations, and what can be concluded from them, is still subject to debate. Climate is sometimes modelled as a stochastic process but this is generally accepted as an approximation to processes that are otherwise too complicated to analyze.

Climax community Plant community that no longer undergoes changes in species composition due to succession.

Climax vegetation The vegetation that would exist in an area if growth had proceeded undisturbed for an extended period. This would be the "final" collection of plant types that presumably would remain forever, or until the stable conditions were somehow disturbed.

Climograph Two dimensional graph that plots a location's air temperature and precipitation on times scales that range from a 24 hour period to a year.

Clinometre An instrument used to measure angles of inclination. Used in conjunction with a ceiling light, it determines cloud height at night, based on the angle of a projected light on the clouds, the observer, and the ceiling light.

Clone 1. A group of genetically similar plants that have originated by vegetative asexual reproduction from a single parent.

2. Replication of an individual who is genetically identical to its parent.

Closed low A low pressure system that is completely encircled by an isobar. They often move slowly, as they displaced south of the main westerlies. Although, strictly speaking, all lows are closed, this term distinguishes a low from a trough on a surface chart, and on upper level charts, it accentuates that the circulation is "closed." A form of cut-off low.

Closed system Is is a system that transfers energy, but not matter, across its boundary to the surrounding environment. Our planet is often viewed as a closed system.

Closed talik Is a form of localized unfrozen ground (talik) in an area of permafrost. It is completely enclosed by permafrost in all directions.

Cloud A cloud is a visible mass of condensed droplets or ice crystals suspended in the atmosphere above the surface of the Earth or another planetary body. The branch of meteorology in which clouds are studied is nephology. On Earth, the condensing substance is water vapour, which forms small droplets of water (typically 0.01 mm of ice crystals) that, when surrounded with billions of other droplets or crystals, are visible as clouds. Dense deep clouds exhibit a high reflectance (70 to 95%) throughout the visible range of wavelengths: they thus appear white, at least from the top. Cloud droplets tend to scatter light very efficiently, so that the intensity of the solar radiation decreases with depth into the cloud, hence the grey or even sometimes dark appearance of the clouds at their base. Thin clouds may appear to have acquired the colour of their environment or background, and clouds illuminated by non-white light, such as during sunrise or sunset, may be coloured accordingly. In the near-infrared range, however, clouds would appear very dark because the water that constitutes the cloud droplets strongly absorb solar radiation at these wavelengths.

Clouds form when the invisible water vapour in the air condenses into visible water droplets or ice crystals. The water in a typical cloud can have a mass of up to several million tonnes. However, the volume of a cloud is correspondingly high, and the net density of the relatively warm air holding the droplets is low enough that air currents below and within the cloud are capable of keeping it suspended. As well, conditions inside a cloud are not static: water droplets are constantly forming and re-evaporating. A typical cloud droplet has a radius on the order of 1×10^{-5} m and a terminal velocity of about 1-2 cm/s. This give these droplets plenty of time to re-evaporate as they fall into the warmer air beneath the cloud.

Most water droplets are formed when water vapour condenses around a *condensation nucleus*, a tiny particle of smoke, dust, ash, or salt. In supersaturated conditions, water droplets may act as condensation nuclei. Water droplets large enough to fall to the ground are produced in two ways. The most important means is through the Bergeron Process, theorized by Tor Bergeron, in which supercooled water droplets and ice crystals in a cloud interact to produce the rapid growth of ice crystals;

these crystals precipitate from the cloud and melt as they fall. This process typically takes place in clouds with tops cooler than -15°C. The second most important process is the collision and wake capture process, occurring in clouds with warmer tops, in which the collision of rising and falling water droplets produces larger and larger droplets, which are eventually heavy enough to overcome air currents in the cloud and the updraft beneath it and fall as rain. As a droplet falls through the smaller droplets which surround it, it produces a "wake" which draws some of the smaller droplets into collisions, perpetuating the process. This method of raindrop production is the primary mechanism in low stratiform clouds and small cumulus clouds in trade winds and tropical regions and produces raindrops of several millimetres diametre. The actual form of cloud created depends on the strength of the uplift and on air stability. In unstable conditions convection dominates, creating vertically developed clouds. Stable air produces horizontally homogeneous clouds. Frontal uplift creates various cloud forms depending on the composition of the front (ana-type or kata-type warm or cold front). Orographic uplift also creates variable cloud forms depending on air stability, although cap cloud and wave clouds are specific to orographic clouds.

Clouds form when water vapour rises, cools and condenses out of the air as micro-droplets. These tiny particles of water are relatively dense, and sunlight cannot penetrate far into the cloud before it is reflected out, giving a cloud its characteristic white colour. As a cloud matures, the droplets may combine to produce larger droplets, which may themselves combine to form droplets large enough to fall as rain. In this process of accumulation, the space between droplets becomes larger and larger, permitting light to penetrate much farther into the cloud. If the cloud is sufficiently large, and the droplets within are spaced far enough apart, it may be that very little light which enters the cloud is able to be reflected back out before it is absorbed (Think of how much farther one can see in a heavy rain as opposed to how far one can see in a heavy fog). This process of reflection/ absorption is what leads to the range of cloud colour from white through grey through black. For the same reason, the undersides of large clouds and heavy overcasts appear various degrees of grey; little light is being reflected or transmitted back to the observer.

Other colours occur naturally in clouds. Bluish-grey is the result of light scattering within the cloud. In the visible spectrum, blue and green are at the short end of light's visible wavelengths, while red and yellow are at the long end. The short rays are more easily scattered by water droplets, and the long rays are more likely to be absorbed. The bluish colour is evidence that such scattering is being produced by rain-sized droplets in the cloud. A more ominous colour is the one seen frequently by severe weather observers. A greenish tinge to a cloud is produced when sunlight is scattered by ice. A cumulonimbus cloud which shows green is a pretty sure sign of imminent heavy rain, hail, strong winds and possible tornados. Yellowish clouds are rare, but may occur in the late spring through early fall months during forest fire season. The yellow colour is due to the presence of smoke. Red, orange and pink clouds occur almost entirely at sunrise/ sunset and are the result of the scattering of sunlight by the atmosphere itself. The clouds themselves are not that colour; they are merely reflecting the long (and unscattered) rays of sunlight which are predominant at those hours. The effect is much the same as if one were to shine a red spotlight on a white sheet. In combination with large, mature thunderheads, this can produce blood-red clouds. The evening before the Edmonton, Alberta tornado in 1987, Edmontonians observed such clouds - deep black on their dark side and intense red on their sunward side. In this case, the adage "red sky at night, sailor's delight" was clearly incorrect.

In addition to being the colloquial term sometimes used to describe dry ice, hot ice is the name given to a surprising phenomenon in which water can be turned into ice *at room temperature* by supplying an electric field of the order of 1 million volts per metre. The effect of such electric fields has been suggested as an explanation of cloud formation. This theory, however, is highly controversial and is not, by any means, widely accepted as being the actual mechanism of cloud formation. The first time cloud ice forms around a clay particle, it requires a temperature of -10°C, but subsequent freezing around the same clay particle requires a temperature of just -5°C, suggesting some kind of "ice memory".

Clouds are divided into two general categories: layered and convective. These are named stratus clouds (or stratiform, the Latin *stratus* means layer) and cumulus clouds (or cumuliform; *cumulus* means piled up). These two cloud types are divided into four more groups that distinguish the cloud's altitude. Clouds are classified by the cloud base height not the cloud top.

High clouds These generally form above 16,500 feet (5,000 m), in the cold region of the troposphere. However, in Polar regions, they may form as low as 10,000 ft (3,048 m). They are denoted by the prefix *cirro-* or cirrus. At this altitude, water almost always freezes so clouds are composed of ice crystals. The clouds tend to be wispy, and are often transparent.

1. Cirrus
2. Cirrus uncinus
3. Cirrus Kelvin-Helmholtz colombia
4. Cirrostratus
5. Cirròcumulus
6. Pileus
7. Contrail, a long thin cloud which develops as the result of the passage of an aircraft at high altitudes.

Middle clouds These develop between 6,500 and 16,500 feet (between 2,000 and 5,000 m) and are denoted by the prefix *alto-*. They are made of water droplets and are frequently supercooled.

1. Altostratus
2. Altostratus undulatus
3. Altocumulus
4. Altocumulus undulatus
5. Altocumulus mackerel sky
6. Altocumulus castellanus
7. Altocumulus lenticularis.

Low clouds These are found up to 6,500 feet (2,000 m) and include the stratus (dense and grey). When stratus clouds contact the ground, they are called fog.

1. Stratus
2. Nimbostratus
3. Cumulus humilis
4. Cumulus mediocris
5. Stratocumulus

Vertical clouds These clouds can have strong up-currents, rise far above their bases and form at many heights.

1. Cumulonimbus (associated with heavy precipitation and thunderstorms)
2. Cumulonimbus incus
3. Cumulonimbus calvus
4. Cumulonimbus with mammatus
5. Cumulus congestus
6. Pyrocumulus

Cloud bank A well-defined cloud mass that can be observed at a distance. It covers the horizon, but is not directly overhead.

Cloud streets Lines or rows of cumuliform clouds.

Cloudburst A sudden, heavy rainfall of a showery nature.

Coal Sedimentary rock composed of the compacted, lithified and altered remains of plants. Coal is a solid, combustible mixture of organic compounds, hydrocarbons, with 30 % to 98 % carbon by weight, mixed with various amounts of water and small amounts of sulphur and nitrogen compounds. It is formed in several stages as the remains of plants are subjected to heat and pressure over millions of years.

Coalescence The merging of two water drops into a single larger drop.

Coastal dune Sand dune that forms in coastal areas. The sand for its formation is supplied from a beach.

Coastal wetland Wetland habitat found along a coastline and is covered with ocean salt water for all or part of the year. Examples of this type of habitat include tidal marshes, bays, lagoons, tidal flats, and mangrove swamps.

Coastal zone Relatively nutrient-rich, shallow part of the ocean that extends from the high-tide mark on land to the edge of the continental shelf.

Coastal zone colour scanner (CZCS) The first spacecraft instrument devoted to measurement of ocean colour. Although instruments on other satellites have sensed ocean colour; their spectral bands, spatial resolution, and dynamic range were optimized for geographical or meteorological use. In the CZCS, every parameter is optimized for use over water to the exclusion of any other type of sensing. The CZCS flew on the Nimbus-7 spacecraft.

Coastline The line that separates a land surface from an ocean or sea.

Coefficient of determination Statistic that measures the proportion of the variation in the dependent variable that is associated with the statistical regression of an independent variable. Can be calculated by taking the square if the correlation coefficient.

Coevolution The coordinated evolution of two or more species that interact and exert selective pressures on each other that can cause each species to undergo associated adaptations.

Cogo Is an abbreviation for coordinate geometry, a method of data input and analysis similar in principle to metes-and-bounds surveying. Input begins at a point, moves at a given angle or in a certain direction for a set distance, and continues in the same fashion until a geographic feature is completely outlined.

Col Saddle like depression found between two mountain peaks. Formed when two opposing cirque glaciers back erode an arête.

Cold A condition marked by low or decidedly subnormal temperature. The lack of heat.

Cold advection The horizontal movement of colder air into a location. Contrast with warm advection.

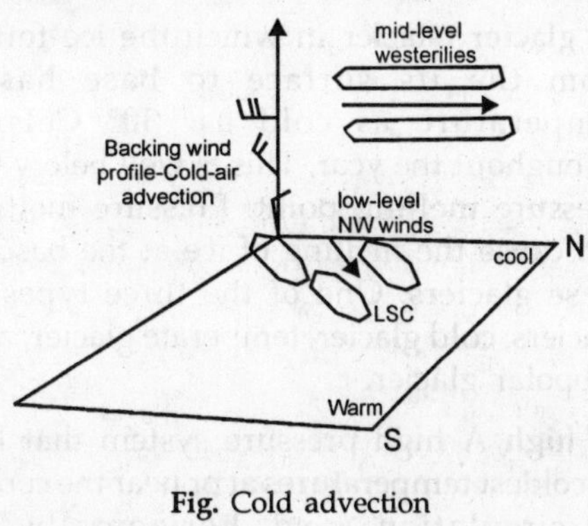

Fig. Cold advection

Cold air funnel Funnel clouds, usually short-lived, that develop from relatively small showers or thunderstorms when the air aloft is very in cold. Cold air funnels may touch down briefly, but in general are less violent than most other types of tornadoes.

Cold core thunderstorms Thunderstorms formed primarily due to steep lapse rates, especially when very cold air aloft overlies warmer surface air.

Cold desert Desert found in the high latitudes and at high altitudes where precipitation is low. Surface air temperatures are generally cold in these dry environments.

Cold front A cold front is defined as the leading edge of a mass of air which is colder than the air in front of it. The colder air, being denser, wedges under the less dense warmer air, lifting it, causing the formation of mostly cumuliform (puffy, cotton-ball-like) clouds. The passage of a cold front usually results in velocity changes in winds and creates vertical movement of air (turbulence) and can set off atmospheric disturbances such as rainshowers, thunderstorms, squall lines, tornadoes, and snowstorms ahead of and immediately behind the moving cold front. The air behind the cold front is generally drier and cooler than that which it is replacing. On weather maps, the surface position of the cold front is marked with the symbol of a blue line of triangles/spikes (pips) pointing in the direction of travel.

Cold glacier Glacier in which the ice found from the its surface to base has a temperature as cold as -30° Celsius throughout the year. This is well below the pressure melting point. Pressure melting can cause the melting of ice at the base of these glaciers. One of the three types of glaciers: cold glacier; temperate glacier; and subpolar glacier.

Cold high A high pressure system that has its coldest temperatures at or near the centre of circulation, and horizontally, is thermally barotropic. It is shallow in nature, as circulation decreases with height. Associated with cold Arctic air, it is usually stationary. Also known as a cold core high. Contrast with a warm high.

Cold low A low pressure system that has its coldest temperatures at or near the centre of circulation, and horizontally, is thermally barotropic. Both warmth and in circulation increase with height and usually there is a warm low aloft to support it. Also known as a cold core low. A cut off low is an example, where an isolated pool of colder air is located south of the main westerlies.

Cold wave A rapid fall in temperature within twenty-four hours to temperatures requiring substantially increased protection to agriculture, industry, commerce, and social activities. National Weather Service criteria includes the rate of temperature fall and the minimum to which it falls, depending on the region of the country and time of the in year. The Weather Channel uses the following criteria for a cold wave: a cold spell of two days or more with below normal temperatures in at least fifteen states, with at least five of them more than fifteen degrees below normal.

Collada A strong, steady wind blowing from the north or northwest in the upper part of the Gulf of California and from the northeast in the lower part.

Colonization Movement of individuals or propagules of a species to a new territory.

Colour separation Process of preparing a separate drawing, engraving, or negative for each colour required in the printing production of a map or chart.

Colorado low A low pressure disturbance that forms in the lee of the Rocky Mountains, usually in south-eastern Colorado.

Comet A large mass of ice and dust that has an orbit around a star.

Comma cloud A feature seen on satellite photographs with a distinctive comma-shape. This is indicative of a synoptic cloud pattern associated with large, well-developed low pressure systems.

Commensialism Biological interaction between two species where one species benefits in terms of fitness while they other experiences no effect on its fitness.

Community Refers to all the populations of interacting species found in a specific area or region at a certain time.

Community boundary Spatial edge of a unique community.

Compass Navigation instrument that uses the Earth's magnetic field to determine direction.

Fig. Compass

Compass roseA compass rose is a design on a map that shows direction. It points which way is north, south, east, west, and some intermediate directions on the map.

Competition Interaction where two or more organisms in the same space require the same resource (e.g., food, water, nesting space, and ground space) which is in limiting supply to the individuals seeking it. Competition can occur at the interspecific or intraspecific biotic levels. Competition may also be the result of two different processes: exploitation or interference.

Competitive exclusion Situation where no two competitively interacting species can occupy exactly the same fundamental niche indefinitely because of resource limitations. The outcome of this process is the local extinction the species that is a poorer competitor.

Compilation Preparation of a new or revised map or chart, or portion thereof, from existing maps, aerial photographs, field surveys, and other sources.

Composite volcano Volcano created from alternate layers of flows and exploded rock. Their height ranges from 100 to 3,500 metres tall. The chemistry of the magma of these volcanoes is quite variable ranging from basalt to granite.

Composites Plants of the compositae family (Asteraceae). Common examples of these flowering plants are thistles, dandelion, and sunflowers.

Compound A compound is the atoms of different elements joined together.

Concrete space Actual geographic space in the real world. Geographers approximate this space when they try to represent it in a model, or map. This approximation is referred to as abstract space.

Condensation Condensation is the change in matter of a substance to a denser phase,

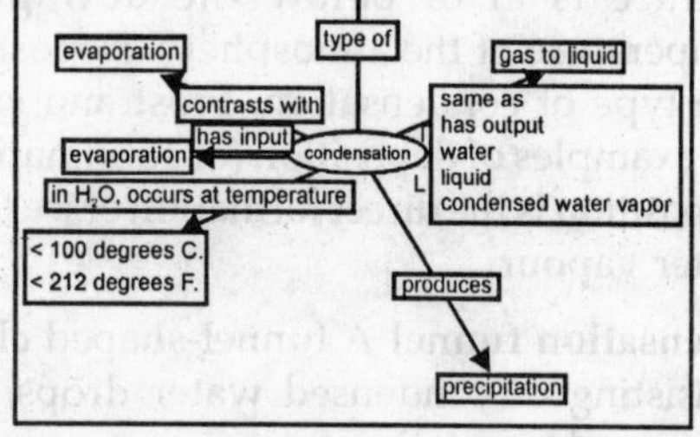

Fig. Condensation

such as a gas (or vapour) to a liquid. Condensation commonly occurs when a vapour is cooled to a liquid, but can also occur if a vapour is compressed (i.e., pressure on it increased) into a liquid, or undergoes a combination of cooling and compression. Liquid which has been condensed from a vapour is called condensate. A device or unit used to condense vapours into liquid is called a condenser. Condensers are typically coolers or heat exchangers which are used for various purposes, have various designs, and come in many sizes ranging from rather small (hand-held) to very large.

Condensation of vapour into liquid is the opposite of evaporation or boiling and is an exothermic process, meaning it releases heat. The water seen on the outside of a cold glass on a hot day is condensation. Water vapour will only condense onto another surface when that surface is cooler than the temperature of the water vapour, or when the water vapour equilibrium in air, i. e. saturation humidity, has been exceeded. When water vapour condenses onto a surface, a net warming occurs on that surface. The water molecule brings a parcel of heat with it. In turn, the temperature of the atmosphere drops very slightly. In the atmosphere, condensation of water vapour is what produces clouds. The dew point of an air parcel is the temperature to which it must cool before condensation in the air begins to form. Also, a net condensation of water vapour occurs on surfaces when the temperature of the surface is at or below the dew point temperature of the atmosphere. Deposition is a type of condensation. Frost and snow are examples of deposition (or sublimation). Deposition is the direct formation of ice from water vapour.

Condensation funnel A funnel-shaped cloud consisting of condensed water drops that has possible rotation.

Condensation nuclei Microscopic particle of dust, smoke or salt that allows for condensation of water vapour to water droplets in the atmosphere. Nucleus for the formation of a rain drop. Condensation normally occurs on these particles when relative humidity becomes 100 %. Some condensation nuclei, like salt, are hygroscopic and water can condense on them at relative humidities lower than 100 %.

Conditional instability Stable unsaturated air that will result in instability in the event or on the condition that the air becomes saturated. If the air is saturated, it is considered unstable; if air is unsaturated, it is considered stable.

Conduction The transfer of heat from one substance to another by direct contact. Denser substances are better conductors; the transfer is always from warmer to colder substances.

Cone of depression Cone shaped depression occurring horizontally across a water table. Causes by excessive removal of groundwater by a surface well.

Confined aquifer Aquifer between two layers of relatively impermeable earth materials, such as clay or shale.

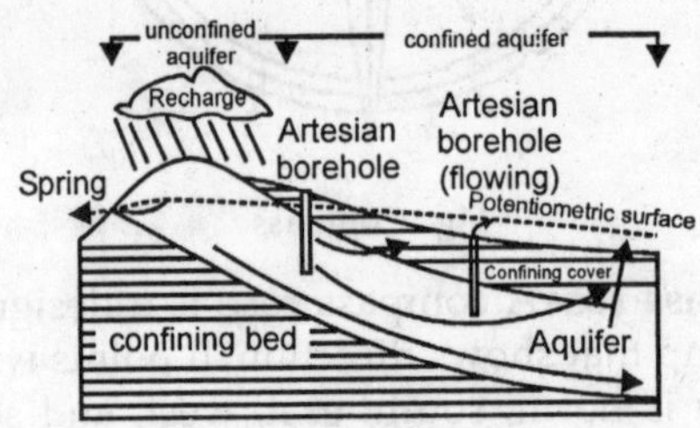

Fig. Confined aquifer

Confined groundwater Groundwater trapped between two impervious layers of rock. Groundwater is water located beneath the ground surface in soil pore spaces and in the fractures of geologic formations. A formation of rock/soil is called an aquifer when it can yield a useable quantity of water. The depth at which soil pore spaces become saturated with water is called the water table. Groundwater is recharged from, and eventually flows to, the surface naturally; natural discharge often occurs at springs and seeps and can form oases or wetlands. Groundwater is also often withdrawn for agricultural, municipal and industrial use by constructing and operating extraction wells. The study of the distribution and movement of groundwater is hydrogeology. Groundwater can be a long-term 'reservoir' of the natural water cycle (with residence times from days to millennia), as opposed to short-term water reservoirs like the atmosphere and fresh surface water (which have residence times from minutes to years). The figure shows how deep groundwater (which is quite distant from the surface recharge) can take a very long time to complete its natural cycle. Groundwater is naturally replenished by surface water from precipitation, streams, and rivers when this recharge reaches the water table. It is estimated that the volume of groundwater is fifty times that of surface freshwater; the icecaps and glaciers are the only larger reservoir of fresh water on earth.

Usable groundwater is contained in aquifers, which are subterranean areas (or layers) of permeable material (like sand and gravel) that channel the groundwater's flow. Aquifers can be confined or unconfined. If a confined aquifer follows a downward grade from a recharge zone, groundwater can become pressurized as it flows. This can create artesian wells that flow freely without the need of a pump. The top of the upper unconfined aquifer is called the water table or phreatic surface, where water pressure is equal to atmospheric pressure. Typically groundwater is thought of as liquid water flowing through shallow aquifers, but technically it can also include soil moisture, permafrost (frozen soil), immobile water in very low permeability bedrock, and deep geothermal or oil formation water. Groundwater is believed to provide lubrication and buoyancy which allow thrust faults to move. Nearly any point in the Earth's subsurface has water in it, to some degree (it may be very dry or mixed with other fluids). Groundwater is not confined only to the Earth, either; subsurface water on Mars is believed to have given rise to some of the landforms observed there. Groundwater makes up about twenty percent of the world's fresh water supply, which is about .61 percent of the entire world's water supply. The temperature of groundwater averages out climactic fluctuations to maintain a relatively steady temperature. Increasingly this factor is used by various means to heat and cool structures. In many cases, during hot weather, groundwater is cool enough to be used as is, say to be simply pumped through radiators in a home, then returned to the ground in another well. During cold seasons, the water, because it has a high Specific heat capacity can be used as a source of heat for heat pumps that is much more efficient than using air.

Confluence A pattern of air flow in which wind direction converges along an axis oriented parallel to the flow along the axis. Considered a form of convergence.

Conglomerate Coarse grained sedimentary rock composed of rounded rock fragments cemented in a mixture of clay and silt.

Conic projection A conic projection is a type of map in which a cone is wrapped around a sphere (the globe), and the details of the globe are projected onto the conic surface. Then, the cone is unwrapped into a flat surface.

Coniferous vegetation Cone-bearing vegetation of middle and high latitudes that are mostly evergreen and that have needle-shaped or scale like leaves. Compare with deciduous vegetation.

Connectivity Is a topological property relating to how geographical features are attached to one another functionally, spatially, or logically. In an water distribution system, connectivity would refer to the way pipes, valves, and reservoirs are attached, implying that water could be "traced" from its source in the network, from connection to connection, to any given final point. Functional, spatial, and logical connectivity are examples of relationships that can be represented and analyzed in a GIS database.

Conservation biology Multidisciplinary science that deals with the conservation of genes, species, communities, and ecosystems that make up Earth's biodiversity. It generally investigates human effects on biodiversity and tries to develop practical approaches to preserving biodiversity and ecological integrity.

Constant pressure chart A chart of a constant pressure surface in which atmospheric pressure is uniform everywhere at any given moment. Elements may include analyses of height above sea level, wind, temperature, and humidity.

Constant pressure surface A surface along which the atmospheric pressure is equal everywhere.

Consumer Organism who gets its organic nutrient from the consumption of tissues of producers and other consumers because it cannot synthesize the organic nutrients in requires. Several different kinds of consumers have been recognized including: carnivores, herbivores, and detritivores.

Contact metamorphism Is the small scale metamorphic alteration of rock due to localized heating. It is usually cause by an igneous intrusion like a sill or a dyke.

Continent One of the large, continuous areas of the Earth into which the land surface is divided. The six geographically defined continents are politically defined as seven; Africa, Asia, Australia, Europe, North America, South America, and Antarctica.

Continental air mass An air mass with continental characteristics. It is a secondary characteristic of an air mass classification, signified by the small "c" before the primary characteristic, which is based on source region. For example, cP is an air mass that is continental polar in nature.

Continental arctic air mass (A) Air mass that forms over extensive landmass areas of the high latitudes. In the Northern Hemisphere, these system form only in winter over Greenland, northern Canada, northern Siberia, and the Arctic Basin. Continental Arctic air masses are very cold and extremely dry. These air masses are also very stable.

Continental climate A continental climate is the climate typical of the middle-latitude interiors of the large continents of the Northern Hemisphere in the zone of westerly winds; similar climates exist along the east coasts and southwest coasts of the same continents, and also at higher elevations in certain other parts of the world. This climate is characterized by winter temperatures cold enough to support a fixed period of stable snow cover each year, and relatively low precipitation occurring mostly in summer, although east coast areas (chiefly in North America) may show an even distribution of precipitation. Only a few areas in Iran, adjacent Turkey and Central Asia show a winter maximum in precipitation, which typically melts in early spring to give short-lived floods.

These regions generally have either forest or tall-grass prairie as natural groundcover and include some of the most productive farmlands in the world. All such climates have at least three months of temperatures in excess of 10°C (50°F) and winters with at least one month below 0°C (32°F) (although some classifications have a lower threshold for winter based on snow cover, in the Köppen climate classification -3°C (27.4°F) is used). Most such areas fit Köppen classifications of Dfa, Dwa (cold winters, long hot summers; "w" indicating very dry winters characteristic especially of China) or Dfb or Dwb (cold winters and long, mild summers, same distinction for winter dryness).

Continental climates exist where cold air masses infiltrate during the winter and warm air masses form in summer under conditions of high sun and long days. Places with continental climates are as a rule either far from any moderating effects of oceans (example: Omaha, Nebraska, USA) or are so situated (example: Boston, Massachusetts, USA) that prevailing winds tend to head offshore. Such regions get quite warm in the summer, achieving temperatures characteristic of tropical climates but are much colder than any other climates of similar latitude in the winter. These climates grade off toward subtropical climates equatorward where winters are less severe and semiarid climates where precipitation becomes inadequate for tall-grass prairies. In Europe these climates may grade off into oceanic climates in which the influence of moderating air masses is more marked toward the west. The sub polar climate, with very cold, long and dry winters, but with at least one month above 10°C, might be considered a sub-type of the continental climate.

The Midwestern United States, parts of China, and most of Russia are examples of areas of the world with continental climates, which do not exist at all in the Southern Hemisphere due to the lack of broad land masses at middle latitudes, the southernmost parts of Africa and Australia being under marine influences and southern South America being too narrow in breadth to allow air masses as cold as those in corresponding latitudes in North America and Asia to form in the winter. Antarctica, of course, lies completely outside the middle latitudes.

Continental crust Granitic portion of the Earth's crust that makes up the continents. Thickness of the continental crust varies between 20 to 75 kilometres.

Continental divide A continental divide is a line of elevated terrain which forms a border between two watersheds such that water falling on one side of the line eventually travels to one ocean or body of water, and water on the other side travels to another, generally on the opposite side of the continent. Because the exact border between bodies of water is usually not clearly defined, the continental divide is not always definite for any continent. Moreover, some rivers empty into deserts or inland seas, and thus do not end up in the oceans.

Continental drift The concept of continental drift was first proposed by Alfred Wegener. In 1912 he noticed that the shapes of continents on either side of the Atlantic Ocean seem to fit together (for example, Africa and South America). Francis Bacon, Antonio Snider-Pellegrini, Benjamin Franklin, and others had noted much the same thing earlier. The similarity of southern continent fossil faunae and some geological formations had led a relatively small number of Southern hemisphere geologists to conjecture as early as 1900 that all the continents had once been joined into a supercontinent known as Pangaea. The concept was initially ridiculed by most geologists, who felt that an explanation of how a continent drifted was a prerequisite and that the lack of one made the idea of drifting continents wholly unreasonable. The theory received support through the controversial years from South African geologist Alexander Du Toit as well as from Arthur Holmes. The idea of continental drift did not become widely accepted as theory until the 1950s in Europe. By the 1960s, geological research conducted by Robert S. Dietz, Bruce Heezen, and Harry Hess along with a rekindling of the theory including a mechanism by J. Tuzo Wilson led to acceptance among North American geologists.

Continental effect The effect that continental surfaces have on the climate of locations or regions. This effect results in a greater range in surface air temperature at both daily and annual scales.

Continental glacier Largest type of glacier with a surface coverage in the order of 5 million square kilometres.

Continental ice sheet See continental glacier.

Continental margin The area between a continent's shoreline and the beginning of the ocean floor. It includes the continental shelf, continental rise, and continental slope.

Continental plate A rigid, independent segment of the lithosphere composed of mainly granite that floats on the viscous plastic asthenosphere and moves over the surface of the Earth. The Earth's continental plates are an average 125 kilometres thick and were formed more than 3 billion years ago.

Continental polar air mass (CP) Air mass that forms over extensive landmass areas of middle to high latitudes. In North America, these system form over northern Canada. Continental Polar air masses are cold and very dry in the winter and cool and dry in the summer. These air masses are also atmospherically stable in both seasons.

Continental rise Thick layers of sediment found between the continental slope the ocean floor.

Continental shelf The continental shelf is the extended perimeter of each continent, which is covered during interglacial periods such as the current epoch by relatively shallow seas (known as shelf seas) and gulfs. The shelf usually ends at a point of increasing slope (called the shelf break). The sea floor below the break is the continental slope. Below the slope is the continental rise, which finally merges into the deep ocean floor, the abyssal plain. The width of the continental shelf varies significantly. It is common for an area to have virtually no shelf at all, especially where the forward edge of an advancing oceanic plate dives beneath continental crust in an offshore subduction zone such as off the coasts of Chile or the west coast of Sumatra. The largest shelf—the Siberian shelf in the Arctic Ocean—stretches to 1500 kilometres in width. The South China Sea lies over another extensive area of continental shelf, the Sunda shelf, which joins Borneo, Sumatra, and Java to the Asian mainland. Other familiar bodies of water that overlie continental shelves are the North Sea and the Persian Gulf. The average width of continental shelves is about 80 kilometres. The depth of the shelf also varies, but is generally limited to water shallower than 150 m. The slope of the shelf is usually quite low, on the order of 0.5°; vertical relief is also minimal, at less than 20 m. Though the continental shelf is treated as a physiographic province of the ocean, it is not part of the deep ocean basin proper, but the flooded margins of the continent. Passive continental margins such as most of the Atlantic coasts have wide and shallow shelves, comprised of thick sedimentary wedges derived from long erosion of a neighbouring continent. Active continental margins have narrow, relatively steep shelves, due to frequent earthquakes that move sediment to the deep sea .

Continental shelf break Boundary zone between the continental shelf and slope.

Continental slope Steeply sloping portion of continental crust found between the continental shelf and continental rise.

Continental tropical air mass (CT) Air mass that forms over extensive landmasses areas of the low latitudes. In North America, these system form over southwestern United States and northern Mexico. Continental Tropical air masses are warm and dry in the winter and hot and dry in the summer. These air masses are also generally unstable in the winter and stable in the summer.

Continentality The quality or state of being a continent.

Continuous map is a means of storing a digital map as if it represented a continuous, unbroken surface. Many large GIS databases are broken into sub files which represent smaller sections of a larger area. These separate sub files are referred to as tiles or facets. A continuous map allows the user to move in any and all directions without moving from file to file. Maintaining the illusion of a continuous map in a computer is a challenging problem because, of course, computers typically store data in strings, in linear order, on magnetic or optical storage media.

Continuous permafrost Form of permafrost that exists across a landscape as an unbroken layer.

Continuous tone Image not broken into dots by photographic screen; contains unbroken gradient tones from black to white, and may be either in negative or positive form. Aerial photographs are examples of continuous-tone prints. Contrasted with halftone (screened) and line copy.

Contour Imaginary line on ground, all points of which are at the same elevation above or below a specific datum.

Contour interval Difference in elevation between two successive contour lines. The interval at which contours are drawn on a map depends on the amount of the relief depicted and the scale of the map.

Contrails Condensation trails. Artificial clouds made by the exhaust of jet aircraft.

Control mapping Points of established position or elevation, or both, which are used to fix references in positioning and correlating map features. Fundamental control is provided by stations in the national networks of triangulation and traverse (horizontal control) and leveling (vertical control). Usually it is necessary to extend geodetic surveys, based on fundamental stations, over the area to be mapped, to provide a suitable density and distribution of control points. Supplemental control points are those needed to relate the aerial photographs used for mapping with the system of ground control. These points must be positively photoidentified; that is, the points must be positively correlated with their images on the photographs.

Control station Point on the ground whose position (horizontal or vertical) is known and can be used as a base for additional survey work.

Control system A system that is intelligently controlled by the activities of humans. For example, a dam on a river.

Conurbation An extensive urban area formed when two or more cities, originally separate, coalesce to form a continuous metropolitan region.

Convection The rising of warm air and the sinking of cool air. Heat mixes and moves air. When a layer of air receives enough heat from the Earth's surface, it expands and moves upward. Colder, heavier air flows under it which is then warmed, expands, and rises. The warm rising air cools as it reaches higher, cooler regions of the atmosphere and begins to sink. Convection causes local breezes, winds, and thunderstorms.

Convection current The movement of a gas or a fluid in chaotic vertical mass motions because of heating.

Convectional lifting The vertical lifting of parcels of air through convective heating of the atmosphere. This process can initiate adiabatic processes inside the air parcel.

Convectional precipitation The formation of precipitation due to surface heating of the air at the ground surface. If enough heating occurs, the mass of air becomes warmer and lighter than the air in the surrounding environment, and just like a hot air balloon it begins to rise, expand and cool. When sufficient cooling has taken place saturation occurs forming precipitation. This process is active in the interior of continents and near the equator forming cumulus clouds and possible later thunderstorms. Rain is usually the precipitation type that is formed, and in most cases this moisture is delivered in large amounts over short periods of time in extremely localized areas.

Convective Condensation Level (CCL) The height at which a parcel of air, if heated sufficiently from below, will rise adiabatically until it is just saturated.

Convergence Wind movement that results in a horizontal net inflow of air into a particular region. Convergent winds at lower levels are associated with upward motion. Contrast with divergence.

Convergence precipitation The formation of precipitation due to the convergence of two air masses. In most cases, the two air masses have different climatological characteristics. One is usually warm and moist, while the other is cold and dry. The leading edge of the latter air mass acts as an inclined wall or front causing the moist warm air to be lifted. Of course the lifting causes the warm moist air mass to cool due to expansion resulting in saturation. This precipitation type is common at the mid-latitudes where cyclones form along the polar front. Also called frontal precipitation.

Convergent lifting The vertical lifting of parcels of air through the convergence of opposing air masses in the atmosphere. This process can initiate adiabatic processes inside the air parcel.

Conversion Is the process of transferring data derived from existing records and maps to a digital database. Conversion is a major input problem and can consume the greatest share of time in a GIS project.

Cooling degree day A cooling degree day is given for each degree that the daily mean temperature departs above the baseline of 75 degrees Fahrenheit. It is used to estimate the energy requirements, and is an indication of fuel consumption for air conditioning or refrigeration. Refer to degree day or heating degree day.

Coordinated Universal Time (UTC) Current official world time reference for civil and scientific purposes. Coordinated Universal Time is measured from six standard atomic clocks at the international bureau of weights and measures (BIPM) in Paris, France. Implemented in 1964.

Coordinates Linear and (or) angular quantities that designate the position of a point in relation to a given reference frame.

Coordinates, origin of Points in a system of coordinates which serves as a zero point in computing the system's elements or in prescribing its use.

Coral Simple marine animals that live symbiotically with algae. In the symbiotic relationship, the algae provides the coral with nutrients, while the coral provide the algae with a structure to live in. Coral animals secrete calcium carbonate to produce a hard external skeleton.

Coral bleaching Situation where coral lose their colourful symbiotic algae. Thought to be caused by unusually warm water, changes in salinity of ocean seawater, or excessive exposure to ultraviolet radiation.

Coral reef Ridge of limestone found generally below the ocean surface. This marine feature is produced by numerous colonies of tiny coral animals, called polyps, that create calcium carbonate structures around themselves for protection. When the corals die, their vacant exterior skeletons form layers that cause the reef to grow. Coral reefs are found in the coastal zones of warm tropical and subtropical oceans.

Core The core is a layer rich in iron and nickel found in the interior of the Earth. It is composed of two sub-layers: the inner core and outer core. The core is about 7,000 kilometres in diameter.

Core area The portion of a country that contains its economic, political, intellectual, and cultural focus. It is often the centre of creativity and change.

Coriolis force The apparent tendency of a freely moving particle to swing to one side when its motion is referred to a set of axes that is itself rotating in space, such as Earth. The acceleration is perpendicular to the direction of the speed of the article relative to the Earth's surface and is directed to the right in the northern hemisphere. Winds are affected by rotation of the Earth so that instead of a wind blowing in the direction it starts, it turns to the right of that direction in the northern hemisphere; left in the southern hemisphere.

Coromell The prevailing evening land breeze which takes place from November to May in the vicinity of La Paz, at the southern tip of Baja California, Mexico.

Corona A pastel halo around the moon or sun created by the diffraction of water droplets. The droplets in the cloud, such as cirrostratus, and the cloud layer itself must be almost perfectly uniform in order for this phenomena to occur. The colour display sometimes appears to be iridescent.

Corposant A luminous, and often audible, electric discharge that is intermediate in nature. It occurs from objects, especially pointed ones, when the electrical field strength near their surfaces attains a value near 1000 volts per centimetre. It often occurs during stormy weather and might be seen on a ship's mast or yardarm, aircraft, lightning rods, and steeples. Also known as corona discharge or St. Elmo's Fire.

Correlation coefficient Statistic that measures the degree of linear association between two variables. Its values vary from between -1 and 1. Perfect positive (the dependent variable increases with an increase in the independent variable) linear association has a correlation coefficient of 1. Perfect negative (the dependent variable decreases with an increase in the independent variable) linear association has a correlation coefficient of -1. Absolutely no association between variables has a value of zero.

Coulee 1. Steep-sided flow of volcanic lava that has solidified.

2. Abandoned glacial meltwater channel.

3. Term used in the United States to describe a steep-sided stream valley.

Counter-radiation Redirection of the Earth's longwave radiation back to the surface because of the greenhouse effect.

Coupled system Two or more processes that affect one another;

Cove A cove is small, horseshoe-shaped body of water along the coast; the water is surrounded by land formed of soft rock.

Coverage Is a term applied in the ARC system to the various layers of geographic features.

Crater Circular depression in the ground surface created by volcanic activity or asteroid impact.

Craton Stable foundation core of the Earth's various plates of continental crust. Composed of the shield and platform.

Creep 1. Slow mass movement of soil downslope. Occurs where the stresses on the slope material are too small to create a rapid failure.

2. Another term used to describe traction.

Crepuscular rays Contrasting, alternating bright and dark rays in the sky. Sunlight is scattered by molecules and particles rendering these bright rays visible. Contrast is enhanced by haze, dust, or mist. These rays are more likely to be seen in the late afternoon, as cumulus clouds come between the sun and the observer. A similar effect occurs when the sun shines though a break in a layer of clouds.

Cretaceous Geologic period that occurred roughly 65 to 144 million years ago. During this period, the first flowering plant species appear and dinosaurs are at their greatest diversity. Dinosaurs die out at the end of this period.

Crevasse 1. Opening on a levee that allows for the drainage of water from the floodplain to the stream channel.

2. Fracture on the brittle surface of a glacier.

Critical entrainment velocity Velocity required to entrain a particular sized particle into the moving medium of air or water.

Crop calendar The schedule of the maturing and harvesting of seasonal crops.

Crop-lien system A farm financing scheme whereby money is loaned at the beginning of a growing season to pay for farming operations, with the subsequent harvest used as collateral for the loan.

Crust Earth's outer most layer of solid rock. Between 7 to 70 kilometres thick. Two types of crust exist: oceanic crust and continental crust.

Cryosol soil Soil order (type) of the Canadian System of Soil Classification. This soil is common to high latitude tundra environments. The main identifying feature of this soil is a layer of permafrost within one metre of the soil surface.

Cryosphere One of the interrelated components of the Earth's system, the cryosphere is frozen water in the form of snow, permanently frozen ground (permafrost), floating ice, and glaciers. Fluctuations in the volume of the cryosphere cause changes in ocean sea-level, which directly impact the atmosphere and biosphere.

Cryostatic pressure Pressure exerted on a substance by ice at rest.

Cryotic Something that is frozen.

Crystalization The process of a substance going directly from a vapour form (water vapour) to a solid (ice) at the same temperature, without going through the liquid phase (water). The opposite of sublimation.

Culmination The point at which a satellite reaches its highest position or elevation in the sky relative to an observer (aka the closest point of approach).

Culture Features constructed by man that are under, on, or above the ground which are delineated on a map. These include roads, trails, buildings, canals, sewer systems, and boundary lines. In a broad sense, the term also applies to all names, other identification, and legends on a map.

Culture hearth The area from which the culture of a group diffused.

Culture The accumulated habits, attitudes, and beliefs of a group of people that define for them their general behaviour and way of life; the total set of learned activities of a people.

Cumuliform Clouds composed of water droplets that exhibit vertical development. The density of the droplets often blocks sunlight, casting shadows on the earth's surface. With increasing vertical height, they are often associated with convection. Bases of these clouds are generally no more than 3,000 feet above the ground, but they can develop past the troposphere in both temperate and tropical latitudes. They are classified as low clouds, and include all varieties of cumulus and cumulonimbus. The opposite in type are the horizontal development of stratiform clouds.

Cumulonimbus cloud A well developed vertical cloud that often has top shaped like an anvil. These clouds are very dense with condensed and deposited water. Weather associated with this cloud includes: strong winds; hail; lightning; tornadoes; thunder; and heavy rain. When this weather occurs these clouds are then thunderstorms. Can extend in altitude from a few hundred metres above the surface to more than 12,000 metres.

Cumulus One of the three basic cloud forms (the others are cirrus and stratus). It is also one of the two low cloud types. A cloud that develops in a vertical direction from the base (bottom) up. They have flat bases and dome- or cauliflower-shaped upper surfaces. The base of the cloud is often no more than 3,000 feet above the ground, but the top often varies in height. Small, separate cumulus are associated with fair weather (cumulus humilis). With additional heating from the earth's surface, they can grow vertically throughout the day. The top of such a cloud can easily reach 20,000 or more into the troposphere. Under certain atmospheric conditions, these clouds can develop into larger clouds, known as towering cumulus (cumulus congestus), and may produce a rain shower. Further development may create a cumulonimbus.

Cumulus congestus A strongly sprouting cumulus cloud with generally sharp outlines and often with great vertical development. It may occur as tower-like clouds with cauliflower tops. These clouds may produce abundant showers and may develop further into cumulonimbus. Also known as towering cumulus.

Cumulus fractus Cumulus clouds that appear in irregular fragments, as if they had been shred or torn. Also appears in stratus clouds (called stratus fractus), but not in cirrus clouds.

Cumulus humilis Cumulus clouds with little or no vertical development characterized by a generally flat appearance. Their growth is usually limited by a temperature inversion, which is marked by the unusually uniform height of the clouds. Also called fair-weather cumulus.

Cumulus mediocris Cumulus clouds characterized by moderate vertical development with upper protuberances not very marked in appearance. This cloud does not produce precipitation, but could develop into towering cumulus or cumulonimbus which do.

Current A horizontal movement of water, such as the Gulf Stream off the east coast of North America, or air, such as the jet stream.

Cuspate foreland Is a triangular accumulation of sand and/or gravel located along the coastline. This feature is formed by the joining of two spits.

Cut and sew industry The manufacture of basic ready-to-wear clothing. Such facilities usually have a small fixed investment in the manufacturing facility.

Cut off high A warm high which has become displaced and is on the polar ward side of the jet stream. It occurs mostly during the spring and is most frequent over north-eastern Siberia, Alaska, and Greenland. It is an example of a blocking high.

Cut off low A cold low which has become displaced and is on the equator ward side of the jet stream. It frequently occurs during the spring and is often located over the south-western United States and along the north-western coast of Africa.

Cyanobacteria Bacteria that have the ability to photosynthesize.

Cyclogenesis The process that creates a new low pressure system or cyclone, or intensifies a pre-existing one. It is also the first appearance of a trough.

Cyclone In meteorology, a cyclone is the rotation of a volume of air around an area of low atmospheric pressure. Cyclones are responsible for a wide variety of different meteorological phenomena such as tropical cyclones and tornadoes. Because of this, most weather forecasters avoid using the term cyclone without a qualifying term. The terms hurricane and typhoon are regionally specific names for a strong tropical cyclone —a non-frontal, synoptic-scale, warm-core low-pressure system with cyclonic surface wind circulation. The centre of a cyclone is a low-pressure region. Near the centre of the cyclone the pressure gradient force, from high- to low-pressure regions, and the Coriolis force must be in an approximate balance (or else the cyclone would collapse in on itself under the pressure gradient). The wind flow around a large cyclone is counter clockwise in the northern hemisphere, and clockwise in the southern hemisphere, due to the Coriolis effect.

1. Tropical cyclones-Tropical cyclones (also known as tropical storms, hurricanes and typhoons) are storm systems that form over warm (generally tropical) ocean waters and draw energy from condensation of moist air. A tropical cyclone's low-pressure centre is at sea level; a strong tropical cyclone develops a high-pressure area over it high in the atmosphere. Tropical cyclones are associated with strong thunderstorms, high winds, and flooding from storm surge.

2. Extratropical cyclones-Extratropical cyclones (or low-pressure cells) have characteristics of tropical cyclones and mid-latitude cyclones, drawing some energy from evaporation and condensation over warm ocean waters, and some from atmospheric temperature gradients. They move away from tropical regions toward polar regions, bringing precipitation and high winds. They often form quickly along cold fronts that stall in regions of warm, moist air, which overruns the denser cold air along and behind a front and develops a shallow cyclonic circulation. Extratropical cyclones also evolve from tropical cyclones that move into non-tropical regions and lose strictly tropical characteristics .

3. Subtropical cyclones-A subtropical cyclone is a weather system that has some characteristics of a tropical cyclone and some characteristics of an extratropical cyclone. They can form in a wide band of latitude, from the equator to 50°.

4. Mid-latitude cyclones-Mid-latitude cyclones are generated by the temperature difference between warm and cold air masses, with warm water at high latitudes generally providing that differential. These storms have a cold core (low-pressure centre), unlike tropical and extratropical cyclones. Similar storms may appear at very high latitudes; these very cold storms are called subarctic (or subantarctic) cyclones.

5. Polar low-Polar lows are similar in behaviour and size to tropical cyclones, although generally much shorter lived. Polar lows are typically several hundred kilometres in diameter, generally have strong winds (usually less than hurricane intensity) and last one to two days on average. Unlike most typical cyclones, they develop extremely rapidly, reaching peak intensity within 24 hours. They generally form under cold upper-level lows when cold arctic air flows over a warm body of water. On satellite imagery, a polar low appears very similar to a hurricane, with an eye and convective bands wrapping around the centre. Research aircraft data suggests that these "arctic hurricanes" may be warm-core systems. Lack of weather data makes forecasting of polar lows difficult. Most predictions in this area are more subjective than the prediction of tropical cyclones.

6. Arctic cyclone-Arctic cyclones are vast areas of low pressure in polar regions that have a weak cyclonic rotation.

7. Mesocyclones-A mesocyclone is an area of vertical atmospheric rotation, typically 2-6 miles across. They are often found in the right-rear flank of supercell thunderstorms, and are visible as a hook echo on Doppler weather radar. The presence of a mesocyclone can only be truly verified by radar, although visual clues such as curved inflow bands may be

present. Mesocyclones form when strong changes of wind speed and/or direction with height (wind shear) sets the lower part of the atmosphere spinning horizontally. The updraft of a thunderstorm can then draw this area of spinning air from horizontal to vertical.

8. Tornadoes-In North America, tornadoes are sometimes described as cyclones because they involve powerful winds around a low-pressure vortex. However, they differ from other cyclones by their very local nature; most cyclones are massive storms, while tornadoes are comparatively small but extremely powerful. Tornadoes occur on too local a scale for the Coriolis effect to determine the direction of rotation; for this reason tornado winds sometimes flow anticyclonically, or opposite the direction dictated by the Coriolis effect.

Cyclonic flow Winds that blow in and around a cyclone, that is counterclockwise in the Northern Hemisphere, clockwise in the Southern Hemisphere.

Cylindrical projection A cylindrical projection is a type of map in which a cylinder is wrapped around a sphere (the globe), and the details of the globe are projected onto the cylindrical surface. Then, the cylinder is unwrapped into a flat surface, yielding a rectangular-shaped map. Cylindrical maps have a lot of distortion in the polar regions (that is, the size of the polar regions is greatly exaggerated on these maps).

D

D value The deviation of actual altitude along a constant pressure surface from the standard atmosphere altitude of that surface.

Daily mean The average temperature for a day computed by averaging either the hourly readings or, more commonly, the maximum and minimum temperatures.

Dalton's law States that the total pressure exerted by a mixture of gases is equal to the sum of the partial pressures of the gases. Formulated by John Dalton, an English physicist.

Data buoys Buoys placed throughout the Gulf of Mexico and along the Atlantic and Pacific coasts of the United States that relay information on air and water temperature, wind speed, air pressure, and wave conditions via radio signals.

Data Collection System (DCS) DCS units are flown on both GOES and NOAA polar-orbiting spacecraft. They gather and relay data from both mobile and stationary platforms at various locations. DCS units on NOAA satellites can also determine the precise location of moving platforms at the time the data were acquired.

Data rate The amount of information transmitted per unit time.

Datum In surveying, a reference system for computing or correlating the results of surveys. There are tow principal types of datums: vertical and horizontal. A vertical datum is a level surface to which heights are referred. In the United States, the generally adopted vertical datum for levelling operations is the National Geodetic Vertical Datum of 1929. The horizontal datum is used as a reference for position. The North American Datum of 1927 is defined by the latitude and longitude of an initial point (Meade's Ranch in Kansas), the direction of a line between this point and a specified second point, and two dimensions that define the spheroid. The new North American Datum of 1983 is based on a newly defined spheroid (GRS80); it is an Earth-centred datum having no initial point or initial direction.

Dawn The first appearance of light in the eastern sky before sunrise. It marks the beginning of morning twilight. The visual display is created by the scattering of light reaching the upper atmosphere prior to the sun's rise to the observer's horizon. Also known as daybreak.

Day Considered a basic unit of time as defined by the earth's motion. It represents the time needed for one complete revolution of the earth about its own axis. Also know as a sidereal day, it is approximately equal to 23 hours, 56 minutes, and 4.09 seconds. Different definitions of the day are based on the apparent motion of the Sun across the sky. The reason for this apparent motion is the rotation of the Earth around its axis, as well as the revolution of the Earth in its orbit around the Sun. A day, as opposed to night, is commonly defined as the period during which sunlight directly reaches the ground, assuming that there are no local obstacles. Two effects make days on average longer than nights. The Sun is not a point, but has an apparent size of about 32 minutes of arc. Additionally, the atmosphere refracts sunlight in such a way that some of it reaches the ground even when the Sun is below the horizon by about 34 minutes of arc. So the first light reaches the ground when the centre of the Sun is still below the horizon by about 50 minutes of arc. The difference in time depends on the angle at which the Sun rises and sets (itself a function of latitude), but amounts to almost seven minutes at least.

Ancient custom has a new day start at either the rising or setting of the Sun on the local horizon (Italian reckoning, for example). The exact moment of, and the interval between, two sunrises or two sunsets depends on the geographical position (longitude as well as latitude), and the time of year. This is the time as indicated by ancient hemispherical sundials. A more constant day can be defined by the Sun passing through the local meridian, which happens at local noon (upper culmination) or midnight (lower culmination). The exact moment is dependent on the geographical longitude, and to a lesser extent on the time of the year. The length of such a day is nearly constant (24 hours ± 30 seconds). This is the time as indicated by modern sundials.

A further improvement defines a fictitious mean Sun that moves with constant speed along the celestial equator; the speed is the same as the average speed of the real Sun, but this removes the variation over a year as the Earth moves along its orbit around the Sun (due to both its velocity and its axial tilt). The Earth's day has increased in length over time. The original length of one day, when the Earth was new about 4.5 billion years ago, was about six hours as determined by computer simulation. It was 21.9 hours 620 million years ago as recorded by rhythmites (alternating layers in sandstone). This phenomenon is due to tides raised by the Moon which slow Earth's rotation. Because of the way the second is defined, the mean length of a day is now about 86,400.002 seconds, and is increasing by about 1.7 milliseconds per century (an average over the last 2700 years).

Day length Period of time for a location on the Earth when insolation from the sun is being received.

Daylight savings time The setting of time so it is one hour ahead starting in the spring and one hour back beginning in the fall in the Northern Hemisphere. In Canada and the United States the dates for these events is the first Sunday in April (spring ahead) and the last Sunday in October (fall back).

De facto segregation The spatial and social separation of populations that occurs without legal sanction.

De jure segregation The spatial and social separation of populations that occurs as a consequence of legal measures.

Debris cloud Considered a rotating cloud of debris or dust that is on the ground or near the ground. The debris cloud appearing beneath a thunderstorm will most likely confirm the presence of a tornado.

Debris flow A type of mass movement where there is a downslope flow of a saturated mass of soil, sediment, and rock debris.

Decibel (DB) A tenth of a bel. A unit used to measure the volume of a sound, equal to ten times the common logarithm of the ratio of the intensity of the sound to the intensity of an arbitrarily chosen standard sound. The decibel also is used to measure relative strengths of antenna and amplified signals and always refers to a ratio or difference between two values.

Deciduous forest Forests in which the trees lose their leaves each year.

Deciduous vegetation Type of vegetation that sheds its leaves during winter or dry seasons. Compare with coniferous vegetation.

Declination In astronomy, the angular distance of a celestial body above (north, plus) or below (south, minus) the celestial Equator. Magnetic declination is the angular difference between magnetic north and true (geographic) north at the point of observation; it is not constant but varies with time because of the "wandering" of the magnetic north pole.

Decomposition 1. To chemically or physically breakdown a mass of matter into smaller parts or chemical elements.

2. Breakdown of organic matter into smaller parts or inorganic constituents by decomposing organisms.

Deduction Inference in which the conclusion about particulars follows necessarily from general theory. In a science like Physical Geography, deductive reasoning would involve stating a theory first and then trying to find facts that reject this idea.

Deepening Used in describing the history of a low pressure system or an area of cyclonic circulation, it means a decrease in the central pressure of the system. Although it usually describes the action of a pressure system on a constant pressure chart, it also means a surface low is increasing in cyclonic circulation and acquiring more energy. The opposite of filling.

Deflation Process where wind erosion creates blowout depressions or deflation hollows by removing and transporting sediment and soil.

Deflation hollow A surface depression or hollow commonly found in arid and semiarid regions caused by wind erosion.

Deforestation The clearing of forested areas as part of a commercial forestry enterprise or for some other economic purpose such as the expansion of settlement or the development of agriculture. Deforestation has traditionally been associated with economic development, but it is no longer characteristic of developed nations. In such areas, all the trees that are likely to be removed have already been removed. There are exceptions, however. In Canada, for example, large areas are cleared every year in the commercial logging of the boreal forest. There and in other areas of the developed world, where forestry continues to be an important element in the economy, good forest management requires that deforested areas be regenerated, but the process is not always effective. Currently, deforestation is most extensive in the world's less-developed nations, particularly those in the tropics. Clear cutting and burning of the tropical rainforest in South and Central America, Africa and South East Asia is proceeding rapidly, although the exact rate is difficult to assess accurately. Estimates for the Amazon region in Brazil, for example, have ranged from 13,800 to 21,000 sq km per annum. Whatever the true value, given the demographic and economic pressures on these areas, the regeneration of such a large area of rainforest is unlikely.

The impact of deforestation extends beyond the destruction of the trees. It destroys the habitat for other plants and for forest-dwelling animals. By opening up the environment, and altering such elements as radiation, temperature, wind and moisture, deforestation also produces significant microclimatological changes. The higher levels of solar radiation reaching the surface cause temperatures to rise, accelerating the decomposition of organic matter. Temperature ranges and wind speeds increase following deforestation and the intensity of the rainfall reaching the surface is higher. The latter causes more rapid leaching of the nutrients released from the organic matter. Runoff also increases, creating a greater potential for soil erosion, particularly in hilly areas. On a larger scale, deforestation may contribute to the enhancement of the greenhouse effect. Much of the deforestation in tropical regions is achieved through burning which releases carbon dioxide (CO_2) directly into the atmosphere. In addition, growing forests are significant carbon (C) sinks, and once they are removed, the recycling of carbon dioxide is reduced, allowing it to remain in the atmosphere to enhance global warming.

Degradation Readjustment of the stream profile where the stream channel is lowered by the erosion of the stream bed. Usually associated with high discharges.

Degree A unit of angular measure represented by the symbol o. The circumference of a circle contains 360 degrees. When applied to the roughly spherical shape of the Earth for geographic and cartographic purposes, degrees are each divided into 60 minutes.

Degree day A measure of the departure of the mean daily temperature from a given standard. That is one degree day for each degree (Fahrenheit or Celsius) of departure above or below the standard during one day. Refer to cooling degree day and heating degree day.

Delta A depositional landform created when a river or stream flows into a body of standing water, such as a lake or sea. The sudden reduction in the velocity of the stream which results causes it to deposit the sediment it is carrying. Flocculation of clay particles in salt water also increases the rate of deposition when a river flows directly into the sea. Although deposition is generally rapid, it is not disorganized and deltaic sediments exhibit well-defined bedding patterns. Coarser sediments tend to be deposited first, forming the foreset beds, while finer clays are carried further to form the bottomset beds. The rapid deposition of sediments tends to block the main stream channel, causing it to divide and subdivide into secondary channels called distributaries. The settling out of sediments as the distributary channels change helps to produce the topset beds that form the delta surface. The classic fan-shaped plan from which the landform derives its name (the Greek letter Ä) is well illustrated by the delta of the River Nile, where it flows into the Mediterranean Sea. However, during and following deposition, such factors as the volume and size distribution of the sediments, the form of the lake or seabed offshore and the impact of wind, waves and tides will influence the shape of the delta. With their large areas of flat, easily worked land, kept fertile by a regular supply of nutrients, the deltas of the world's major rivers, such as the Nile in Egypt and the Ganges/Brahmaputra in Bangladesh, have supported large agriculturally based populations for thousands of years. Despite regular floods, which can destroy crops and homes and cause major loss of life, the advantages apparently outweigh the disadvantages, and habitation has persisted. Such long-term occupation of the deltas, however, has ensured that they are no longer completely natural features. The construction of flood protection structures and the creation of permanent shipping channels have altered both the form of deltas and the processes involved in their formation. In addition, the construction of dams or the alteration of the stream channel upstream from the delta can have an impact on the entire delta environment by changing the amount and nature of the sediments reaching it. For example, reduced sediment yields on many deltas have also meant a reduction in the availability of natural nutrients. Replacement of the latter with chemical fertilizers has created previously unknown environmental problems.

Demodulation The process of retrieving information (data) from a modulated carrier wave, the reverse of modulation.

Demography The systematic analysis of population.

Dendritic Term used to describe the stream channel pattern that is completely random. Resembles the branching pattern of blood vessels or tree branches.

Dendrochronology The dating of past events and variations in the environment and the climate by studying the annual growth rings of trees. The approximate age of a temperate forest tree can be determined by counting the annual growth rings in the lower part of the trunk. The width of these annual rings is indicative of the climatic conditions during the period of growth; wide annual rings signify favourable growing conditions, absence of diseases and pests, and favourable climatic conditions, while narrow rings indicate unfavourable growing conditions or climate.

Dendroclimatology The study of past climates as revealed by variations in the widths of the annual growth rings of trees. Such variations reflect changes in growing conditions from one year to the next, and although the rate of growth represents the combination of a variety of factors, the role of climate is usually sufficiently strong to establish a correlation between tree growth and such climatic elements as temperature and precipitation. The most obvious relationships have been identified in areas where the trees are growing under some form of environmental stress or close to their climatological limits. For example, the pioneering work in dendroclimatology at the University of Arizona was carried out in the south-western part of the United States where temperatures are normally sufficiently high for growth, but moisture is a significant limiting factor. Elsewhere, along the northern tree-line in Europe and North America, for example, moisture levels are adequate, and summer temperatures correlate well with ring widths. In some areas, however, it may not be possible to identify a relationship between growth and a specific climate element, while in the tropics the technique is limited by the absence of distinct annual rings in many species.

Climate reconstruction using tree rings is now technically very advanced. The process of establishing a record involves a series of stages. Initially, the samples obtained in an area have to be treated statistically to remove the influence of non-climatic elements from the measured ring widths. As a tree ages, for example, the rings that it produces tend to become thinner. By comparing the actual width of a ring with the expected width for a tree of a particular age on the site being examined, standardized values which reflect only the influence of climate can be obtained. These standardized values can then be used to calibrate the sequence of rings. Using a period for which meteorological records are available, a statistical relationship is sought between tree ring width over the same period and a specific meteorological parameter such as temperature or precipitation. Once the relationship has been established, the other rings can be used to create a record of meteorological conditions outside the period for which numerical data are available. The results are then verified by comparing them with additional records from outside the calibration period or results from other sources. Using samples from bristlecone pines, which are among the oldest living trees in the world, it has been possible to reconstruct changing climates over the past

4000 years in the mountains of southern California. By overlapping samples from these living trees with ring sequences from dead trees, the record has been extended back more than 8000 years. Most records are much shorter, but climatic reconstructions based on tree rings are available for large areas of North America and Europe, providing information on such elements as temperature, precipitation and atmospheric pressure, both annually and seasonally.

Denitrification Conversion of nitrates into gaseous nitrogen and nitrous oxide.

Dense fog advisory Advisory issued when fog reduces visibility to 1/8 mile or less, creating possible hazardous conditions.

Density The ratio of the mass of a substance to the volume it occupies. In oceanography, it is equivalent to specific gravity and represents the ratio of the weight of a given volume of sea water to that of an equal volume of distilled water at 4.0 degrees Celsius or 39.2 degrees Fahrenheit.

Density altitude The altitude at which a given density is found in the standard atmosphere. Used in aviation, it is computed from the station pressure at takeoff and the virtual temperature at the particular altitude under consideration.

Denudation The wearing away, or literally the 'laying bare' of the land surface by the combined effects of weathering, mass movement and erosion. The result is a net reduction in the relief of the landscape. The main agent of denudation is water. Glacial ice also contributes to the process in mountainous areas and in high latitudes, and during the ice ages, glaciers had a major role in wearing down the landscape. Denudation rates are usually calculated by measuring sediment yield from drainage basins, expressed as millimetres per 1000 years. Average values may be only a few millimetres over that time period, but rates vary considerably in time and place. They are greatest in areas of strong relief with steep slopes, little vegetation and abundant precipitation. Warm, semi-arid regions, with a sparse vegetation cover, also exhibit high denudation rates, and rates are generally lowest on moist, level plains, where gradients are low and there is sufficient precipitation to keep the landscape well vegetated. Changing climatic conditions over time also cause variations in denudation rates. Human intervention in the environment through mining and quarrying, forest clear cutting and inappropriate agricultural practices has accelerated the natural rates of denudation in many areas.

Deoxyribonucleic Acid (DNA) Form of nucleic acid that is organized into a double-helix molecule. DNA is used by most organisms to chemically code their genetics and to direct the development and functioning of cells. This direction requires RNA which represents a copy of a portion of DNA. Found in the nucleus of cells.

Dependent variable Variable in a statistical test whose observation's values are thought to be controlled through cause and effect by another independent variable modeled in the test.

Deposition 1. The change in state of matter from gas to solid that occurs with cooling. Usually used in meteorology when discussing the formation of ice from water vapour. This process releases latent heat energy to the environment.

2. Laying down of sediment transported by wind, water, or ice.

Deposition nuclei Six-sided microscopic particle that allows for deposition of water as ice crystals in the atmosphere. Nucleus for the formation of snowflakes. Deposition normally occurs on these particles when relative humidity becomes 100 %.

Depositional landform Is a landform formed from the deposition of weathered and eroded surface materials. On occasion, these deposits can be compressed, altered by pressure, heat and chemical processes to become sedimentary rocks. This includes landforms with some of the following geomorphic features: beaches, deltas, floodplains, and glacial moraines.

Depression In meteorology, it is another name for an area of low pressure, a low, or trough. It also applies to a stage of tropical cyclone development and is known as a tropical depression to distinguish it from other synoptic features.

Depth curve Line on a map or chart connecting points of equal depth below the datum.

Deranged drainage Drainage pattern that is highly irregular. Areas that have experienced continental glaciations may have this type of drainage pattern.

Derecho A line of intense, widespread, and fast-moving thunderstorms that moves across a great distance. They are characterized by damaging straight-line winds over hundreds of miles. Spanish for straight.

Desalination The removal of dissolved salts from sea water or saline groundwater to provide fresh water for domestic, industrial or agricultural use. Distillation and reverse osmosis are the most common processes employed. In distillation, water vapour, driven off as steam when salt water is boiled, condenses as fresh water, leaving the solid salts behind. Reverse osmosis involves the pumping of salt water at high pressure through a semi-permeable membrane. Dissolved salts are retained on the membrane while fresh water passes through. Other less frequently used methods which accomplish the separation of the salts from the fresh water include electrodialysis - the passage of an electric current through the saline solution - and freezing, which causes the formation of freshwater ice and leaves behind a concentrated salt solution. Distillation is most widely used, accounting for between 85 and 90 per cent of the freshwater produced by desalination. The main concerns associated with desalination centre on the high energy use involved, and the potential environmental problems created by the disposal of large volumes of salt or brine. Currently, desalination plants provide less than 0.1 per cent of the world's fresh water. They are most commonly found where energy costs are relatively low and the costs of supplying water from any source are high - as in Kuwait and Saudi Arabia - or where other sources are limited and the demand is high enough to meet the costs - as in Florida and California in the United States.

Descending node The point in a satellite's orbit at which it crosses the equatorial plane from north to south.

Desert An area of permanent aridity in which precipitation is infrequent or irregular in its occurrence, and the resulting low annual rainfall totals are exceeded by high evapotranspiration rates. Such conditions are characteristic of sub-tropical regions between 25° and 30° north and south of the equator. There, the descending arms of the tropical Hadley Cells inhibit precipitation, and the high temperatures encourage high evapotranspiration rates which contribute to major soil moisture deficits. The Sahara Desert in North Africa - the world's biggest - is a product of such conditions. Elsewhere, deserts are created by the rain shadow effects of mountain systems that lie across the prevailing wind directions - as in North America, where the Mojave and Sonoran Deserts owe their aridity to the presence of the Sierra Nevada blocking the moist air masses from the Pacific Ocean. The rain shadow effect also applies to the Gobi and other deserts of Central Asia. The Himalayas prevent the moisture of the Indian monsoon from reaching the area, but the continentality of the region is also an important element. Its distance from the ocean means that the air masses reaching the area have lost most of their moisture before they arrive. The world's deserts cover some 5 million km^2, and an additional 40 million km2 are arid or semi-arid. In total, some 33 per cent of the earth's land surface exhibits desert characteristics to some degree or other. The area covered by desert is not static. In the past it has been larger and also smaller. At various times during the Pleistocene, many areas now considered deserts were much wetter. In the Sahara, for example, the abandoned shorelines of Lake Chad indicate that it was once larger and deeper than it is now. At other times, the deserts have been even drier and more extensive. In North America in the nineteenth century, parts of the Great Plains were so dry that the area was referred to as 'The Great American Desert'. At present the main concern is that the deserts are expanding again, with the desertification of sub-Saharan Africa having the potential to affect the lives and livelihood of millions of people. Despite their overall aridity, water may be available in deserts from time to time and from place to place. Occasional storms provide a short-term water supply and sometimes cause flash-floods, while rivers supplied from wetter areas outside the deserts provide a permanent source of water in some areas. The Nile in Egypt, the Tigris-Euphrates system in Iraq and the Colorado in the United States are examples of such rivers. In addition, there is often abundant groundwater beneath the desert. It may appear at the surface in the form of springs or at oases, but access to the groundwater supply in desert areas usually involves the drilling of wells. Since such subsurface water has taken thousands of years to accumulate, and the replenishment of aquifers in deserts is extremely slow, careful management of the groundwater resource is important. However, in most cases, even steady, controlled use leads to a progressive depletion of the groundwater supply. The typical desert landscape is often represented as a sea of sand, overlying a relatively subdued plain, with some relief provided by the presence of sand dunes. This is the so-called erg desert, which is well developed in the Sahara, although even there only 20 per cent of the surface is sand. Elsewhere, the combination of climate, rock type and relief has produced rocky or stony surfaces, bare rock pavements, deep valleys and steep slopes.

Although water is not common in desert areas, when it is available - for example, during a major storm - it can achieve spectacular amounts of erosion. This is made possible in part by the absence of plant cover, which encourages rapid runoff, and the availability of large quantities of easily eroded and transported unconsolidated sediments. The most obvious erosional landforms are the major, usually dry, river channels or wadis of the Middle East, but in areas of strong relief, smaller scale steep-sided canyons are common. The materials eroded from such features are deposited rapidly, often in the form of fans, when water is no longer available. Wind is also a potent erosional and depositional force in desert areas. Lifting loose sand from a surface, it can create deflation hollows and the moving sand can erode adjacent rocks in a form of natural sand-blasting. The redeposited sand may take the form of sheets or sand dunes which move according to the prevailing wind speed and direction.

Limited precipitation and high evapotranspiration limit plant growth, and desert surfaces are characterized by sparse, xerophytic vegetation, or, in some areas, by the complete absence of plant life. In places, the high evapotranspiration rates create very salty soils which also inhibit plant growth. However, even those surfaces that are apparently completely devoid of vegetation may bloom immediately following precipitation, as seeds and dormant plants respond rapidly to the availability of moisture. Those plants which survive in the desert environment have adapted to the conditions, for example, by reducing their water needs, and the same applies to desert animals. Many have developed nocturnal habits to avoid the heat of the day, and all have adapted to living with the lack of water. Despite such adaptability, the biomass productivity of desert areas remains low.

In the past, the human occupation of deserts was restricted to those areas where water was readily available, such as at oases or along river systems, for example, in Egypt and Mesopotamia. More recently, technological advancements and changing economic factors have allowed the development of new areas. Improved irrigation techniques, for example, have allowed the Israelis to bring parts of the Negev Desert into agricultural production, while in the United States the demand for vacation and retirement facilities in California and Arizona has made it economically feasible to bring water from the neighbouring mountains to supply new residential areas. The water is also used to irrigate high value crops, such as strawberries or out-of-season vegetables, and even golf-courses. Such situations are the exception, however, and the potential for further development of desert areas remains limited.

Desert pavement A veneer of coarse particles left on the ground after the erosion of finer particles by wind.

Desertification The expansion of desert or desert-like conditions into adjacent areas, initiated by natural environmental change, by human degradation of marginal environments or a combination of both. Most modern approaches to the definition of desertification recognize the combined impact of adverse climatic conditions and the stress created by human activity. Both have been accepted by the United Nations as the elements that must be considered in any working definition of the process. The United Nations Environmental Program (UNEP) has tended to emphasize the importance of the human impact over drought, but the relative importance of each of these elements remains very controversial. Some see drought as the primary element, with human intervention aggravating the situation to such an extent that the overall expansion of the desert is increased, and any recovery – for example, following a change in climatic conditions – is lengthier than normal. Others see direct human activities as instigating the process. In reality, there must be many causes that together bring desert-like conditions to perhaps as much as 60,000 sq km of the earth's surface every year and threaten up to a further 30 million sq km. The areas directly threatened are those adjacent to the deserts and semi-deserts on all continents. Africa is currently receiving much of the attention, but large sections of the Middle East, the central Asian republics of the former Soviet Union, China adjacent to the Gobi Desert, north-west India and Pakistan, along with parts of Australia, South America and the United States are also susceptible to desertification. Even areas not normally considered as threatened, such as Southern Europe from Spain to Greece, are not immune. At least 50 million people are directly at risk of losing life or livelihood in these regions. In a more graphic illustration of desertification, the United States Agency for International Development (USAID), at the height of the Sahelian drought in 1972, claimed that the Sahara was advancing southwards at a rate of as much as 30 miles (48 km) per year along a 2000 mile (3200 km) front. Although these specific numbers are not universally accepted – for example, Nelson has suggested that all such data be treated with a healthy scepticism – they do give an indication of the magnitude of the problem, and remain one of the reasons why there has been increasing cause for concern in recent years.

Human nature being what it is, when drought strikes an area, there is a natural tendency to hope that it will be short and of limited intensity. The inhabitants of drought-prone areas, therefore, may not react immediately to the increased aridity. They may continue to cultivate the same crops, perhaps even increasing the area under cultivation to compensate for reduced yields, or they may try to retain flocks and herds which have expanded during the times of plenty. If the drought is prolonged in the arable areas, the crops die and the bare earth is exposed to the ravages of soil erosion. The Dustbowl in the Great Plains developed in this way. Once the available moisture had evaporated and the plants had died, the wind removed the topsoil - the most fertile part of the soil profile - leaving a barren landscape, which even the most drought-resistant desert plants found difficult to colonize.

Prolonged drought in pastoral areas is equally damaging. It reduces the forage supply, and, if no attempt is made to reduce the animal population, the land may fall victim to overgrazing. The retention of larger herds during the early years of the Sahelian drought, for example, allowed the vegetation to be overgrazed to such an extent that even the plant roots died. When the wind blew, it lifted the exposed, loose soil particles and carried them away, taking with them the ability of the land to support plant and animal life. In combination, these human and physical activities seemed to be pushing the boundaries of the Sahara Desert inexorably southwards. Out of this grew the image of the shifting sands, which came to represent desertification in the popular imagination. As an image, it was evocative, but the reality of such a representation has been increasingly questioned in the 1990s.

Climatic variability clearly made a major contribution to desertification in both the Sahel and the Great Plains, and in concert with human activities created serious environmental problems. An alternative view sees human activity in itself capable of initiating desertification in the absence of increased aridity. For example, human interference in areas where the environmental balance is delicate might be sufficient to set in motion a train of events leading eventually to desertification. The introduction of arable agriculture into areas more suited to grazing, or the removal of forest cover, to open up agricultural land or to provide fuelwood, may disturb the ecological balance to such an extent that the quality of the environment begins to decline, and if nothing is done, the soil becomes highly susceptible to erosion. In such cases, desertification is initiated by human activities with little or no contribution from nature.

Although human activities have been widely accepted as causing desertification, and the processes involved have been observed, there is increasing concern that the human contribution has been overestimated. Current academic and popular attitudes to desertification owe a lot to the findings of a United Nations Conference on Desertification (UNCOD) held in Nairobi, Kenya in 1977. At the Conference, the role of human activities in land degradation was considered to be firmly established, and the contribution of drought was seen as secondary at best. Since human action had caused the problem, it seemed to follow that human action could solve it. In keeping with this philosophy, the United Nations Environment Program (UNEP) was given the responsibility for taking global initiatives to introduce preventive measures which would alleviate the problem of desertification. Fifteen years and US$6 billion later, few effective countermeasures have been taken, and the plan of action is widely seen as a failure.

The data upon which the UNEP responses were based are now considered by many researchers to be unrepresentative of the real situation. Nelson, for example, has suggested that the extent of irreversible desertification has been over-estimated, although he does not deny that it remains a serious concern in many parts of the world. Problems arising from the timing and method of collection of the data were aggravated by the UNEP premise that human activity was the main cause of the land degradation that produced desertification. Natural causes such as short-term drought and longer term climatic change were ignored or given less attention than they deserved, yet both can produce desert like conditions without input from society. The inclusion of areas suffering from short-term drought may well have inflated the final results in the UNEP accounting of land degradation. Failure to appreciate the various potential causes of desertification would also limit the response to the problem. Different causes would normally elicit different responses, and UNEP's application of the societal response to all areas, without distinguishing the cause, may in part explain the lack of success in dealing with the problem. The debunking of some of the myths associated with desertification and the realization that even after more than 15 years of study its nature and extent are inadequately understood, does not mean that desertification should be ignored. There are undoubtedly major problems of land degradation in many of the earth's arid lands. Perhaps sensing an increased vulnerability as a result of the current controversy, and certainly fearful of being left behind in the rush to deal with the problems of the developed world, the nations occupying the land affected appeared at the Rio Earth Summit in 1992 and proposed a Desertification Convention to address their problems. Negotiations continued after the Summit, and in 1994, 110 governments signed the UN Convention to Combat Desertification. A much more comprehensive treaty than earlier efforts, the Convention will seek innovative solutions through national action programmes and partnership agreements. Particular attention will be paid to Africa, where desertification is most severe.

Although obscured by the current controversy and lost in the complexity of attempts to define the issue of desertification more accurately, two questions remain of supreme importance to the areas suffering land degradation. Can desertification be prevented? Can the desertification that has already happened be reversed? In the past, the answer to both has always been a qualified yes, and seems likely to remain so, although some researchers take a more pessimistic view. In theory, society could work with the environment by developing a good understanding of environmental relationships in the threatened areas or by assessing the capability of the land to support certain activities and by working within the constraints that these would provide. In practice, non-environmental elements – such as politics and economics – may prevent the most ecologically appropriate use of the land.

The fight against desertification has been marked by a distinct lack of success. Recent reassessments of the problem, beginning in the late 1980s, suggest that this may be the result of the misinterpretation of the evidence and a poor understanding of the mechanisms that cause and sustain the degradation of the land. The additional research required to resolve that situation will further slow direct action against desertification, but it may be the price which has to be paid to ensure future.

Detachment One of three distinct processes involved in erosion. This process involves the disengagement of a particle from its surroundings.

Detector A device in a radiometer that senses the presence and intensity of radiation. The incoming radiation is usually modified by filters or other optical components that restrict the radiation to a specific spectral band. The information can either be transmitted immediately or recorded for transmittal at a later time.

Detrital rock Sedimentary rock that is composed of particles transported to their place of deposition by erosional processes. Examples of such rock include sandstone and shale.

Detritivore Heterotrophic organism that feeds on detritus. Examples of such organisms include earthworms, termites, slugs, snails, bacteria, and fungi.

Detritus Shed tissues, dead bodies and waste products of organisms. In most ecosystems, detritus accumulates at the soil surface and other types of surface sediments.

Detritus food chain Model describing the conversion of organic energy in a community or ecosystem into inorganic elements and compounds through decomposition. The organisms involved in this conversion are called detritivores.

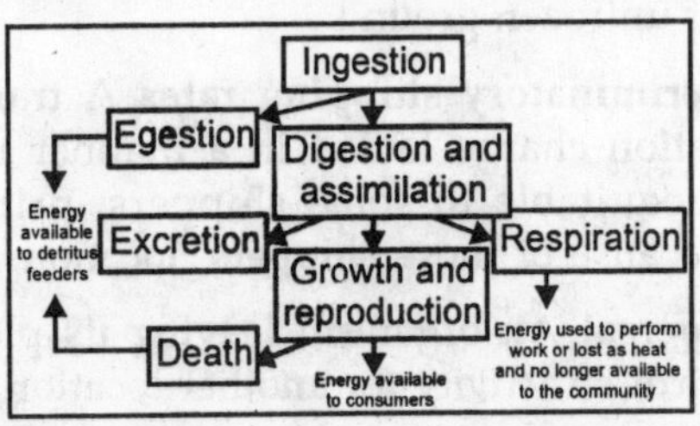

Fig. Detritus food chain

Deuterium Isotope of hydrogen, with a nucleus containing one proton and one neutron, and an atomic mass number of 2.

Devonian Geologic period that occurred roughly 360 to 408 million years ago. During this period, the first amphibians and trees appear.

Dew Atmospheric moisture that condenses after a warm day and appears during the night on cool surfaces as small drops. The cool surfaces cause the water vapour in the air to cool to the point where the water vapour condenses.

Dew point Dew point is the temperature at which water vapour saturates from an air mass into liquid or solid usually forming rain, snow, frost or dew. Dew point normally occurs when a mass of air has a relative humidity of 100 %. If the dew point is below freezing, it is referred to as the frost point.

Diablo winds Dry winds in the Diablo mountain range in central California that can exceed 60 miles per hour. Similar to the Santa Ana winds, they develop as the wind flows from high pressure over Nevada to lower pressure along the central California coast.

Diazo process Rapid method for copying documents in which the image is developed by exposure to ammonia.

Diffluence A pattern of air flow in which wind direction diverges along an axis oriented parallel to the flow along the axis. Considered a form of divergence. The opposite of confluence.

Diffraction The result of light waves interfering with other after passing through a narrow aperture, causing them to bend or spread.

Diffused solar radiation Solar radiation received by the Earth's atmosphere or surface that has been modified by atmospheric scattering.

Diffusion 1. Molecular mixing of one substance into another substance.

2. Redirection or refraction of solar insolation in many directions. Process cause

the beam of travelling radiation to become less intense.

Digital image An analogue image converted to numerical form so that it can be stored and used in a computer. The image is divided into a matrix of small regions called picture elements or pixels. At sub-satellite point each pixel represents a specific amount of area. For example, in APT each pixel represents 4.1 kilometres. Each pixel has a numerical value or data number value, quantifying the radiance of the image at that spot. The data number value of each pixel usually represents a value between black and white, i.e., shades of grey.

False colour can be applied to the image by assigning a graduated colour palette to represent the grey shades. The colour is "false" because it represents an assigned, not actual, colour.

Digital system A system in which information is transmitted in a series of pulses. The source is periodically sampled, analyzed, and converted or coded into numerical values and transmitted. Digital transmissions typically use the binary coding used by computers so most data is in appropriate form, but verbal and visual communication must be converted. Many satellite transmissions use digital formats because noise will not interfere with the quality of the end product, producing clear and higher-resolution imagery.

Digitize Is the process of converting source materials, prepared manually, into the digital codes stored and processed by computers. Digitizing involves tracing map features into a computer using a digitizing tablet, graphics tablet, mouse, or keyboard cursor.

Dike Bank of earth or stone used to form a barrier, frequently and confusingly interchanged with levee. A dike restrains water within an area that normally is flooded.

Diorite A coarse grained igneous rock of intrusive origin that is darker and chemically more mafic than granite.

Dip One of the directional properties of a geologic structure such as a fold or a fault. Dip is the inclination angle of the formation as measured at right angles to strike.

Diploid Cell that contains two sets of chromosomes.

Direct readout The capability to acquire data directly from environmental satellites via an Earth station. Data can be acquired from NOAA and other nations' environmental satellites, which offer weather information from geostationary and polar-orbiting satellites.

Direct solar radiation Solar radiation received by the Earth's atmosphere or surface which has not been modified by atmospheric scattering.

Directional shear The shear created by a rapid change in wind direction with height.

Director Parasitic element(s) of a VHF antenna located forward of the driven element.

Discontinuity Comparatively large contrast in meteorological elements over a relatively small distance or period of time. In oceanography, it is the abrupt change or jump of a variable at a line or surface.

Discontinuous permafrost Form of permafrost that contains numerous scattered pockets of unfrozen ground.

Discriminatory shipping rates A transportation charge levied in a manner that is inequitable to some shippers, primarily because of those shippers' location.

Dispersal An organism leaving its place or birth or activity for another location.

Dissociation Chemical process where a compound or molecule breaks up into simpler constituents.

Dissolution The process of a substance dissolving and dispersing into a liquid.

Dissolved load Portion of the stream load that is in solution in the flowing water.

Distance ratio Method for measuring the gradient of a slope. Simply involves dividing the vertical change in distance (rise) by horizontal change in distance (run) or rise/run. The measurement is usually presented as a percentage or relative to some unit distance travelled in the horizontal.

Distributary A smaller branching stream channel that flows away from a main stream channel. Common on deltas. Opposite of tributary.

Distributional limit Spatial boundary that defines the edge of a species geographical range.

Disturbance This has several applications. It can apply to a low or cyclone that is small in size and influence. It can also apply to an area that is exhibiting signs of cyclonic development. It may also apply to a stage of tropical cyclone development and is known as a tropical disturbance to distinguish it from other synoptic features.

Diurnal Pertaining to actions or events that occur during a twenty-four hour cycle or recurs every twenty-four hours. Meteorological elements that are measured diurnally include clouds, precipitation, pressure, relative humidity, temperature, and wind.

Diurnal arc The apparent arc described by heavenly bodies from their rising to their setting.

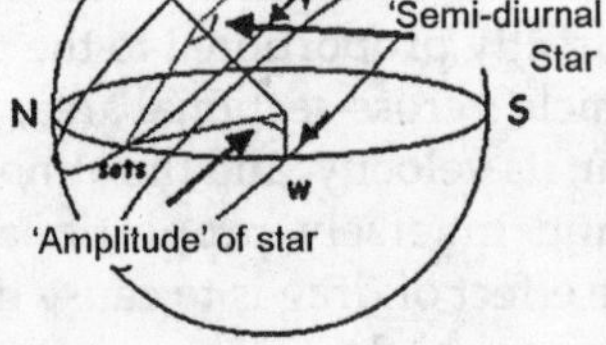

Fig. Diurnal arc

Diurnal tide Tides that have one high and one low water per tidal period.

Divergence Wind movement that results in a horizontal net outflow of air into a particular region. Divergence at lower levels is associated with a downward movement of air from aloft. Contrast with convergence.

Divergent evolution Creation of two or more unique species from one ancestral species through the differential evolution of isolated populations.

Diversity (species diversity) Number of different species in a given region.

Divide The topographic ridge that separates drainage basins.

DLG Stands for digital line graph, a form of digital map.

Dobson unit (DU) The standard way to express ozone amounts in the atmosphere. One DU is 2.7×10^{16} ozone molecules per square centimetre. One Dobson unit refers to a layer of ozone that would be 0.001 cm thick under conditions of standard temperature (0° C) and pressure (the average pressure at the surface of the Earth). For example, 300 Dobson units of ozone brought down to the surface of the Earth at 0° C would occupy a layer only 0.3 cm thick in a column. Dobson was a researcher at Oxford University who, in the 1920s, built the first instrument (called the Dobson metre) to measure total ozone from the ground.

Dog days The name given to the very hot summer weather that may persists for four to six weeks between mid-July through early September in the United States. In western Europe, this period may exist from the first week in July to mid-August and is often the period of the greatest frequency of thunder. Named for Sirius, the Dog Star, which lies in conjunction with the sun during this period, it was once believed to intensify the sun's heat during the summer months.

Doldrums Located between 30 degrees North and 30 degrees South latitudes in the vicinity of the equator, this area typically has calm or light and variable winds. Also a nautical term for the equatorial trough.

Dolomite 1. Sedimentary rock formed from $CaMg(CO_3)_2$.

2. Mineral with the chemical formula $CaMg(CO_3)_2$.

Dome An uplifted area of sedimentary rocks with a downward dip in all directions; often caused by molten rock material pushing upward from below. The sediments have often eroded away, exposing the rocks that resulted when the molten material cooled.

Doppler effect (Doppler shift) The apparent change in frequency of sound or light waves, varying with the relative velocity of the source and the observer. If the source and observer draw closer together; the frequency is increased. Named for Christian Doppler; Austrian mathematician and physicist (1803-1853).

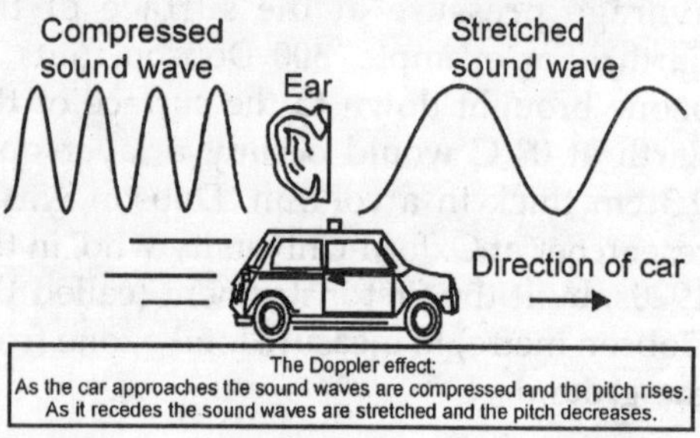

Fig. Doppler effect

Doppler radar Weather radar that measures direction and speed of a moving object, such as drops of precipitation, by determining whether atmospheric motion is horizontally toward or away from the radar. Using the Doppler effect, it measures the velocity of particles. Named for J. Christian Doppler, an Austrian physicist, who in 1842 explained why the whistle of an approaching train had a higher pitch than the same whistle when the train was going away.

Downburst A severe localized downdraft from a thunderstorm or shower. This outward burst of cool or colder air creates damaging winds at or near the surface. Sometimes the damage resembles tornadic damage.

Downconverter Any radio frequency circuit that converts a higher frequency to a lower frequency This enables signal processing by a receiver. A typical downconverter will feature one or more states of RE preamplification, a mixer where the frequency conversion occurs, a local oscillator chain, and often one or more intermediate frequency preamplifiers to minimize the effect of line losses between the converter and the receiver.

Downdraft A sudden descent of cool or cold air to the ground, usually with precipitation, and associated with a thunderstorm or shower. Contrast with an updraft.

Downpour A heavy rain.

Downslope effect The warming of an air flow as it descends a hill or mountain slope. Contrast with an upslope effect.

Downwelling current Ocean current that travels downward into the ocean because of the convergence of opposing horizontal currents or because of an accumulation of seawater.

Drag A retarding force caused by the Earth's atmosphere. Thus by definition, drag will act opposite to the vehicle's instantaneous velocity vector with respect to the atmosphere. The magnitude of the drag force is directly proportional to the product of the vehicle's cross-sectional area, its drag coefficient, its velocity, and the atmospheric density and inversely proportional to its mass. The effect of drag is to cause the orbit to decay or spiral downward. A satellite of very high mass and very low cross-sectional area, and in a very high orbit, may be very little affected by drag, whereas a large satellite of low mass, in a low altitude orbit may be affected very strongly by drag. Drag is the predominant force affecting satellite lifetime.

Drainage basin Land surface region drained by a length of stream channel.

Drainage divide Topographic border between adjacent drainage basins or watersheds.

Drainage network System of interconnected stream channels found in a drainage basin.

Drainage pattern Geometric pattern that a stream's channels take in the landscape. These patterns are controlled by factors such as slope, climate, vegetation, and bedrock resistance to erosion.

Drainage wind A wind common to mountainous regions that involves heavy cold air flowing from high to low elevations because of gravity.

Drift Normally used when referring to snow or sand, particles are deposited behind obstacles or irregularities of the surface or driven into piles by the wind.

Drifting snow Snow particles blown from the ground by the wind to a height of less than six feet.

Drizzle Slowly falling precipitation in the form of tiny water droplets with diameters less than 0.02 inches or 0.5 millimetres. It falls from stratus clouds and is often associated with low visibility and fog. It is reported as "DZ" in an observation and on the METAR.

Dropsonde A radiosonde dropped with a parachute from an aircraft rather than lifted by a balloon to measure the atmosphere below.

Drought Drought is a rather imprecise term with both popular and technical usage. To some, it indicates a long, dry spell, usually associated with lack of precipitation, when crops shrivel and reservoirs shrink. To others, it is a complex combination of meteorological elements, expressed in some form of moisture index. There is no widely accepted definition of drought. It is, however, very much a human concept, and many current approaches to the study of drought deal with moisture deficiency in terms of its impact on human activities, particularly those involving agriculture. Agricultural drought is defined in terms of the retardation of crop growth by reduced soil moisture levels. This, in turn, may lead to economic definitions of drought when, for example, dry conditions reduce yield or cause crop failure, leading to a reduction in income. It is also possible to define drought in purely meteorological terms, where moisture deficiency is measured against normal or average conditions which have been established through long-term observation.

The establishment of normal moisture levels also allows a distinction to be made between aridity and drought. Aridity is usually considered to be the result of low average rainfall, and is a permanent feature of the climatology of a region. In contrast, drought is a temporary feature, occurring when precipitation falls below normal or when near normal rainfall is made less effective by other weather conditions such as high temperature, low humidity and strong winds.

Aridity is not a prerequisite for drought. Even areas normally considered to be humid may suffer from time to time, but some of the worst droughts ever experienced have occurred in areas that include some degree of aridity in their climatological make-up. Along the desert margins in Africa, for example, annual precipitation is low, ranging between 100 and 400 mm, but, under normal conditions, this would allow sufficient vegetation growth to support pastoral agriculture. Some arable activity might also be possible if dry-farming techniques were employed, yet drought occurs with considerable regularity in these areas. The problem lies not in the small amount of precipitation, but rather in its variability. Mean values of 100 to 400 mm are based on long-term observations and effectively mask totals in individual years, which may range well above or below the values quoted. Weather records at Beijing, in drought-prone northern China, show that the city receives close to 600 mm of precipitation in an average year. However, the amount falling in the wetter years can be six to nine times that of the drier years. Only 148 mm were recorded in 1891, for example, and 256 mm in 1921, compared to a maximum of 1405 mm in 1956. Rainfall variability is now recognized as a major factor in the occurrence of drought, and a number of writers have questioned the use of 'normal' values in such circumstances. In areas of major rainfall variability, the nature of the environment reflects that variability rather than the so-called normal conditions, and any response to the problems which arise from drought conditions must take that into account. Drought is perhaps the most ubiquitous climatological problem society has to face. Its greatest impact is in areas that experience seasonal drought or contingent drought. The former is common in the subtropics, where the year includes distinct dry and wet seasons. Drought intensifies when the dry season is extended beyond its normal duration or the wet season provides less precipitation than usual. The extension of such conditions over a period of several years has produced some of the most catastrophic droughts on record, in areas such as the Sahel, East Africa, India, China and Australia, all of which are subject to seasonal drought. Contingent drought is characterized by irregular and variable precipitation in areas that normally have an adequate supply of moisture to meet their needs. The droughts of 1975–1976 and 1988–1992 in the normally humid UK were contingent droughts. The Great Plains of North America have suffered contingent drought for thousands of years, from prehistoric times up to the present. Arguably the most important of these, in terms of its extent, duration and impact on government policies and agricultural practices, was the drought of the 1930s. It created the Dustbowl and caused widespread hardship and misery for the inhabitants of the area, but it also brought about the realization that drought was an integral component of the environment of the plains and had to be treated as such.

Modern views of drought vary with time and place and with the nature of the event itself. It may be seen as a technological problem, an economic problem, a political problem, a cultural problem, or sometimes a multifaceted problem involving all of these. Whatever else it may be, however, it is always an environmental problem, and basic to any understanding of the situation is the relationship between society and environment in drought-prone areas .

Over thousands of years, certain plants and animals have adapted to life with the limited moisture. Their needs are met, therefore no drought exists. This is the theoretical situation in most arid areas. In reality it is much more complex, for although the flora and fauna may exist in a state of equilibrium with other elements in the environment, it is a dynamic equilibrium, and the balance can be disturbed. Fluctuations in weather patterns, for example, might reduce the amount of precipitation available, thereby changing the whole relationship. If plants and animals can no longer cope with the reduced water supply, they will suffer the effects of drought. Depending upon the extent of the change, plants may die from lack of moisture, they may be forced out of the area as a result of competition with species more suited to the new conditions, or they may survive, but at a reduced level of productivity. The situation is more complex for animals, but the response is often easier. In addition to requiring water, they also depend upon the plants for food, and their fate will therefore be influenced by that of the plants. They have one major advantage over plants, however. Being capable of movement, they can respond to changing conditions by migrating to areas where their needs can be met. Some degree of balance will eventually be attained again, although certain areas - such as the world's desert margins - can be in a continual state of flux for long periods of time.

The human animal, like other species, is forced to respond to such changing environmental conditions. In earlier times this often involved migration, which was relatively easy for small primitive communities, living by hunting and gathering, in areas where the overall population was small. As societies changed, however, this response was often no longer possible. In areas of permanent or even semi-permanent agricultural settlement, with their associated physical and socioeconomic structures, migration was certainly not an option - indeed, it was almost a last resort. The establishment of political boundaries, which took no account of environmental patterns, also restricted migration in certain areas. As a result, in those regions susceptible to drought, the tendency, perhaps even the necessity, to challenge the environment grew. If sufficient water was not available from precipitation, either it had to be supplied in other ways - for example, by well and aqueduct - or different farming techniques had to be adopted to reduce the moisture need in the first place. The success of these approaches depended very much on such elements as the nature, intensity and duration of the drought, as well as on various human factors, which included the numbers, stage of cultural development and technological level of the peoples involved. Technology has helped those living in drought-prone areas to cope with drought. In some cases drought prediction has allowed responses to be planned and the impacts mitigated, but there are concerns that technology will create greater problems in the future, either directly through the rapid depletion of groundwater supplies or indirectly through the activities that lead to global warming.

Drought prediction The attempt to forecast the occurrence of drought so that responses can be planned and consequences much reduced. The simplest approach to drought prediction is the actuarial forecast, which estimates the probability of drought based on past occurrences. Problems with the length and homogeneity of meteorological records often reduce the reliability of actuarial predictions. Links between meteorological variables and other physical or environmental variables have been assessed as possible predictors. There is, for example, a relationship between sunspot activity and precipitation. Drought on the Great Plains has been correlated with the minimum of the 22-year double sunspot cycle. The correlation has no physical theory to explain it, however, and the relationship does not appear to apply outside the western United States. Most modern attempts at drought prediction involve teleconnection. Global sea-surface temperatures (SSTs) have been correlated with drought in the Sahel, for example, and ENSO events have been recognized as precursors of drought in Brazil, Australia, Indonesia and India. The time lag between the occurrence of a specific SST or an ENSO event and the onset of drought allows the prediction to be made.

Drumlin A hill shaped deposit of till. The shape of these features resembles an elongated teaspoon laying bowl down. The tapered end of the drumlin points to the direction of glacial retreat. Drumlins come in assorted sizes. Lengths can range from

100 to 5,000 metres and heights can be as great as 200 metres.

Dry adiabat The lifting of dry air, or air that does not contain water vapour. As a parcel lifts, its pressure decreases, and since pressure decreases with height, the air temperature falls due to expansion. When air is dry and is lifted adiabatically, then the temperature decreases at a rate of 1 degree Celsius per 100 metres (5.5 degrees Fahrenheit per 1,000 feet). Contrast with a moist adiabat.

Dry adiabatic lapse rate (DALR) The rate of decline in the temperature of a rising parcel of air before it has reached saturation. This rate of temperature decline is 9.8° Celsius per 1000 metres because of adiabatic cooling.

Dry bulb thermometre A thermometer used to measure the ambient temperature. The temperature recorded is considered identical to air temperature. One of the two thermometers that make up a psychrometer.

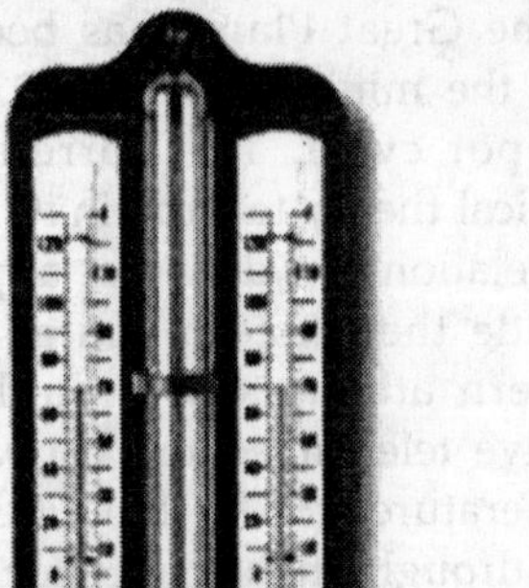

Fig. Dry bulb thermometre

Dry farming A type of farming practiced in semi-arid or dry grassland areas without irrigation using such approaches as fallowing, maintaining a finely broken surface, and growing drought-tolerant crops.

Dry line The boundary between the dry desert air mass of the Southwest and the moist air mass from the Gulf of Mexico. It usually lies north-south across the central and southern High Plains states during spring and summer. The passage of a dry line results in a sharp decrease in humidity, clearing skies, and a wind shift from southeasterly or south to southwesterly or west. Its presence influences severe weather development in the Great Plains.

Dry slot An area of dry, and usually cloud-free, air that wraps into the southern and eastern sections of a synoptic scale or mesoscale low pressure system. Best seen on a satellite picture, such as a water vapour image.

Dry-bulb thermometre Thermometre on a psychrometre used to determine current air temperature. This measurement and the reading from a wet-bulb thermometre are then used for the determination of relative humidity or dew point from a psychrometric table.

DTM Stands for digital terrain model, A method of transforming elevation data into a contoured surface or a three-dimensional display.

Dune 1. Stream bed deposit found streams whose channel is composed mainly of sand and silt. Dunes are about 10 or more centimetres in height and are spaced a metre or more apart and are common in streams with high velocities.

2. Terrestrial deposit of sand that resembles a mound or ridge that was formed from aeolian processes.

Dune field An extensive region covered by numerous sand dunes.

Dusk The period of waning light from the time of sunset to dark.

Dust A general term for solid particles of varying materials, sizes and origins, carried in suspension in the atmosphere. Most dust particles are considered to have diameters greater than 1 µm, but some volcanic dust may be finer than that. Larger particles remain in the troposphere, but volcanic eruptions sometimes push smaller particles into the stratosphere. Dust originates naturally as a result of volcanic activity, fires and the deflation of particles from deserts and other arid areas. Human activities such as mining, quarrying and various agricultural practices contribute dust directly, while activities which increase desiccation or desertification contribute indirectly by increasing the area susceptible to deflation or wind erosion. Dust in the atmosphere has a role in climate change through its ability to scatter radiation, and at the local level may present an environmental hazard.

Dust bowl An area of the Great Plains, stretching from Texas in the south to the Canadian prairies in the north, which suffered the effects of desertification in the 1930s. A combination of drought and inappropriate farming practices caused the destruction of the topsoil and allowed it to be carried away by the wind. The social and economic impacts of the Dustbowl helped to bring about a reassessment of agricultural development and land use in semi-arid areas such as the Plains.

Dust devil A small, rapidly rotating column of wind, made visible by the dust, dirt or debris it picks up. It usually occurs in arid or semi-arid areas and is most likely to develop on clear, dry, hot afternoons in response to surface heating. Also called a whirlwind.

Dust dome Dome of air that surrounds a city created from the urban heat island effect that traps pollutants like particulate matter.

Duststorm A dust storm (or sandstorm in some contexts) is a meteorological phenomenon common in dry, arid and semi-arid regions. Such a storm is usually the result of convection currents created by intense heating of the ground. The air over the sand becomes hot, and rises. This creates differences in air pressure and temperature, and the cooler winds begin to rush in. The wind is strong enough to move dunes, and it often interferes with travel, sometimes obliterating roads in flat, dry regions such as those of the western United States. For instance, on the 29th of November 1991 a convective dust storm formed over Interstate-5 in the San Joaquin Valley, California. The storm picked up dust in a suspension with winds at speeds of over 120km/h. Visibility was reduced to less than half a car's length, causing a pile-up accident involving 164 cars, killing 17 and injuring 151. The simoom or simoon (sîmûm, sîmûn) is the dust- and sand-laden desert wind of N Africa and Arabia that contributes largely to the atmospheric dust over Europe; evidence of the dust from simoon winds has also been found on the seafloor at considerable distances from shore. The haboob (hYbûb) is a sandstorm prevalent in the region of Sudan around Khartoum. Sandstorms, the leading edges of which often appear as solid walls of dust as much as 5,050 ft (1,525 m) high, also occur, although less frequently, in the SW United States. One that occurred near Tucson, Arizona, on July 16, 1971, was extensively documented by meteorologists. Similar duststorms from windborne particles are evident on the planet Mars and are thought to be seasonal. Severe dust storms can reduce visibility to zero, making travel impossible, and can blow away valuable topsoil, while depositing soil in places where it may not be wanted. Drought and wind contribute to the emergence of dust storms, as do poor farming and grazing practices. The dust picked up in such a storm can be carried thousands of kilometres: Sahara dust storms influence plankton growth in the western Atlantic Ocean and, according to some scientists, are an important source of scarce minerals for the plants of the

Amazon rainforest. Dust storms can often be observed from satellite photos.

Dyke Thin vertical veins of igneous rock that form when magma enters and cools in fractures found within the crust.

Dynamic equilibrium A dynamic equilibrium occurs when a system displays unrepeated average states through time.

Dynamic metamorphism Form of metamorphism that causes only the structural alteration of rock through pressure. The minerals in the altered rocks do not change chemically. The extreme pressures associated with mountain building can cause this type of metamorphism.

Dynamics A branch of mechanics that deals with forces and their relations to patterns of motion. In metorology, this relates especially to wind and water patterns.

Dyne A unit of force that creates an acceleration on a mass of 1 gram equal to 1 centimetre per second. 10^5 dynes equals one newton.

E

Earth albedo The reflectivity of the Earth's atmosphere and surface combined. Measurements indicate that the average Earth albedo is approximately 30 %.

Earth observing system (EOS) A series of small- to intermediate-sized spacecraft that is the centrepiece of NASAs Mission to Planet Earth (MTPE). Planned for launch beginning in 1998, each of the EDS spacecraft will carry a suite of instruments designed to study global climate change. MTPE will use space-, aircraft-, and ground-based measurements to study our environment as an integrated system. Designing and implementing the MTPE is, of necessity, an international effort. The MTPE program involves the cooperation of the U.S., the European Space Agency (ESA), and the Japanese National Space Development Agency (NASDA). The MTPE program is part of the U.S. interagency effort, the Global Change Research Program.

Earth Observing System Data & Information System (EOSDIS) The system that will manage a dataset of Earth science observations to be collected over a 15-year period. Existing data indicates that the Earth is changing, and that human activity increasingly contributes to this change. To monitor these changes, a baseline of "normal" performance characteristics must be obtained. For the Earth, these baseline characteristics must cover a global scale and a long enough period that the variation caused by seasonal changes and other cyclical or periodic events (e.g., El Nino and the solar cycle) may be included in the analyses. The baseline characteristics also must enable scientists to quantify processes that govern the Earth's system. Functionally EOSDIS will provide computing and networking facilities supporting EOS research activities, including data interpretation and modelling; processing, distribution, and archiving of EDS data; and command and control of EGS observatories.

Earth probes Discipline-specific satellites and instruments that will be used by NASA to obtain observations before the launch of EDS spacecraft. Generally smaller than the EGS satellites and instruments, Earth Probes are planned to complement the broad environmental measurements from EOS with highly focused studies in areas such as tropical rainfall (TRMM), ocean productivity, atmospheric ozone (TOMS), and ocean surface winds (NSCAT).

Earth Radiation Budget Experiment (ERBE) An experiment to obtain data to study the average radiation budget of the Earth and determine the energy transport gradient from the equator to the poles. Three satellites were flown in different orbits to obtain the data: the earth radiation budget satellite, ERBS (launched in October 1984), NOAA-9 (launched in December 1984), and NOAA-10 (launched in September 1986).

Earth revolution Refers to the orbit of the Earth around the sun. This celestial motion takes 365 1/4 days to complete one cycle. Further, the Earth's orbit around the sun is not circular, but elliptical.

Earth rotation Refers to the spinning of the Earth on its polar axis.

Earth sciences tradition Academic tradition in modern Geography that investigates natural phenomena from a spatial perspective.

Earth Station (aka ground station) Hardware necessary to acquire data directly from environmental satellites. The WEFAX Earth station diagram illustrates a basic ground station configuration for obtaining direct readout data from geostationary environmental (weather) satellites.

Earth system The Earth regarded as a unified system of interacting components, including geosphere (land), atmosphere (air), hydrosphere (water and ice), and biosphere (life).

Earth system science An integrated approach to the study of the Earth that stresses investigations of the interactions among the Earth's components in order to explain Earth dynamics, evolution, and global change.

Earthflow A rapid type of downslope mass movement that involves soil and other loose sediments. Usually triggered by water saturation from rainfall.

Earthlight (earthshine) The faint illumination of the dark part of the moon's disk produced by sunlight reflected onto the moon from the earth's surface and atmosphere.

Earthquake A series of earth movements brought about by a sudden release of energy during tectonic activity in the earth's crust. Earthquakes frequently accompany volcanic eruptions, but the most severe examples are associated with movements along fault lines. Movement along most active faults is infinitesimal and smooth. The rocks on either side of the fault line will move vertically or horizontally in relation to each other and the small amounts of energy involved will be dissipated into the environment. If for some reason no movement takes place, despite the addition of energy, the fault is said to be locked, and energy continues to accumulate in the fault system until it is sufficient to break the lock. The sudden release of accumulated energy causes the earthquake. Waves of energy - seismic waves - spread out in all directions from the origin or focus of the earthquake, but those which reach the surface immediately above the focus - the epicentre of the quake - and spread from there cause the most visible damage. Earthquakes are classified according to their magnitude and intensity. Magnitude is a measure of the energy released and is recorded on the Richter Scale, an open-ended logarithmic scale starting at zero, but with no upper limit, since there is no evidence that the values of past large earthquakes are the greatest possible. Maximum observed intensities range between 8.0 and 8.6 - the 1906 San Francisco earthquake, for example, had an estimated magnitude of 8.3 on the Richter Scale - but thousands of earthquakes of lesser magnitude occur every year. The intensity of an earthquake is reflected in the damage that it does to property and is measured on the Modified Mercalli Scale. The scale was developed in the early part of the twentieth century and modified in 1931. It includes twelve values ranging from Negligible (detected by instruments only) to Catastrophic (total destruction) based on the damage caused to property.

Earthquakes of the same magnitude and intensity may cause different levels of destruction depending upon local conditions. Structures built on a sand or clay substrate, or land reclaimed from the sea, for example, suffer greater damage than those built directly upon rock. Areas in which the buildings have been constructed or renovated to withstand earthquake shocks usually suffer less damage than those where the buildings have been constructed using local materials and techniques. Factors such as these often lie behind the great differences in the death toll associated with earthquakes of similar magnitude and intensity. In Tangshan, China in 1976, buildings of local brick using traditional methods simply disintegrated under the effects of an earthquake measuring 7.6 on the Richter Scale and more than 200,000 people died. Nearly 10 years later in Mexico City an earthquake of similar magnitude produced a much lower death-toll of between 9000 and 10,000, despite the city's high population density and the fact that it is built mainly on the bed of a dried-up lake. Most of the damage and the highest death-toll was in those areas where the sub-surface materials were unstable or the houses were of traditional construction. Earthquake proofed buildings and those on more stable ground had a much greater survival rate. Indirect effects of major earthquakes include fire - caused by the rupturing of gas lines and the availability of large amounts of combustible materials - which is difficult to fight because the water supply system has usually been ruined. The destruction of the water system may also allow disease to spread among the survivors of the earthquake. Earthquake occurrence is not evenly distributed around the earth, but is concentrated in those areas that are most tectonically active. Such areas include the boundary zones between tectonic plates and zones of sea floor spreading. As a result, the areas that experience most earthquakes are located in a belt around the Pacific. This circum-Pacific belt accounts for more than 80 per cent of the world's earthquakes, among them the most devastating in terms of both life and property damage. Japan, for example, has experienced seventeen earthquakes with a magnitude greater than 7 this century, resulting in a death toll of more than 165,000 and damage to nearly 720,000 buildings. On the other side of the Pacific in California, the San Andreas Fault is an earthquake zone which has the potential to devastate the large population centres of southern and central California. It has formed where the Pacific and North American tectonic plates grind together, and sudden movement along it was responsible for the 1906 San Francisco earthquake. Current concern is with the southern end of the fault which is locked and under great stress in a number of places. The sudden release of the energy represented by that stress would have a catastrophic effect on most of southern California. Other earthquake concentrations occur in a belt stretching through the Mediterranean to the Himalayas and then south-east through Indonesia to meet up with the circum-Pacific belt in the area of the Philippines. Mid-oceanic ridge development in the Atlantic Ocean and the Indian Ocean are also regions of increased earthquake activity. Outside these areas earthquakes may be less common, but they do occur in even the most stable parts of the earth's crust, such as the shield areas of Canada and Australia. Since they are unexpected, earthquakes in these areas often cause a disproportionate amount of damage.

Earthquakes cause major environmental change. They disrupt drainage patterns, both on the surface and underground; in mountainous areas they initiate landslides that change the geomorphology of these areas and alter vegetation patterns; coastal areas may be altered catastrophically by the tidal waves or tsunamis which follow major offshore earthquakes Attempts at earthquake prediction with the aim of reducing loss of life have met with little success. Potential precursors of damaging earthquakes have included swarms of minor earth tremors, emissions of gases from cracked bedrock or unusual animal behaviour. These may apply to individual earthquakes, but as yet there is no universal signal of an impending earthquake.

Earthquake focus Point of stress release in an earthquake.

Easterlies Usually applied to the broad patterns of persistent winds with an easterly component, such as the easterly trade winds.

Easterly wave An inverted, migratory wave-like disturbance or trough in the tropical region that moves from east to west, generally creating only a shift in winds and rain. The low level convergence and associated convective weather occur on the eastern side of the wave axis. Normally, it moves slower than the atmospheric current in which it is embedded and is considered a weak trough of low pressure. It is often associated with possible tropical cyclone development and is also known as a tropical wave.

Eastern hemisphere The Eastern Hemisphere consists of Africa, Asia, Australia, and Europe.

Easting First measurement of a grid reference used to specific the location of a point on a rectangular coordinate system. The distance measured eastward from the origin of a rectangular coordinate system.

Ebb tide Time during the tidal period when the tide is falling. Compare with flood tide.

Eccentricity One of six Keplerian elements, it describes the shape of an orbit. In the Keplerian orbit model, the satellite orbit is an ellipse, with eccentricity defining the "shape" of the ellipse. When e=0, the ellipse is a circle. When e is very near 1, the ellipse is very long and skinny.

Echo The energy return of a radar signal after it has hit the target. Also called a radar echo.

Eclipse The partial or total apparent darkening of the sun when the moon comes between the sun and the Earth (solar eclipse), or the darkening of the moon when the full moon is in the Earth's shadow (lunar eclipse).

Eclipse blindness Focus-point type of vision loss caused by looking at the sun for too long a time, which can burn a hole in the retina of the eye.

Ecliptic The sun's apparent path across the sky that tracks a circle through the celestial sphere.

Ecological diversity The variety of unique biological communities found on the Earth. A component of biodiversity.

Ecological niche Is all of the physical, chemical and biological conditions required by a species for survival, growth and reproduction. Two further abstractions of this concept are the fundamental niche and the realized niche.

Ecology Science dealing with the inter-relationships between living organisms and their environments.

Economies of agglomeration The economic advantages that accrue to an activity by locating close to other activities; benefits that follow from complementarity or shared public services.

Economies of scale Savings achieved in the cost of production by larger enterprises because the cost of initial investment can be defrayed across a greater number of producing units.

Ecosphere See biosphere.

Ecosystem Ecosystem A term coined in 1935 by the British ecologist A.G. Tansley to refer to a community of interdependent organisms and the physical environment they inhabit. Although there is no accepted ecosystem hierarchy, the ecosystem concept can be applied at a variety of scales, from the microscopic to the whole earth. The individual organisms interact with each other and with their environment in a series of relationships made possible by the flow of matter and energy within and through the system. The relationships are dynamic and routinely respond to change, without altering the basic characteristics of the ecosystem. Cyclical changes in animal populations, for example, are an integral part of most ecosystems. Despite the large fluctuations that may be involved, the functional relationships among organisms – such as those between predator and prey – continue, and ultimately some degree of balance is restored. Major environmental disruption such as that caused by climate change or fire, however, may alter a specific ecosystem irreversibly, and bring about its replacement by a system with different characteristics. Regular fires, for example, may create a grassland ecosystem where once a forest ecosystem existed. Natural ecosystems are theoretically self-sustaining, but increasing human interference is threatening their sustainability in many parts of the world, and various methods of ecosystem management have been introduced in an attempt to preserve and protect characteristic natural ecosystems.

Ecosystem diversity The variety of unique biological communities found on the Earth. A component of biodiversity.

Ecotone A transition zone between two ecosystems. Although usually most obvious through the changing distribution of vegetation, it also includes other organisms from the two communities. The transition zone may be many kilometres wide and include the characteristics of the two ecosystems, as in the parkland area between the forest and savannah grasslands of the sub-tropics, or it may be relatively narrow, with the change from one ecosystem to the other taking place within as little as a few hundred metres, as at some altitudinal tree lines. Where the zone is broad, the ecotone will include animals from both adjacent ecosystems and perhaps others that are characteristic of the ecotone itself, producing a community that may be more diverse than in the ecosystems. The location of particular ecotones is usually associated with changes in such environmental elements as climate, geology and soils, but the nature of the transition may also be influenced by such factors as fire, the number of grazing animals present in an area or the extent of the competition between the two communities it separates.

Eddy A small disturbance of wind in a large wind flow, which can produce turbulent conditions or turbulence. They can also be areas of warmer air north of the main westerlies or colder air south of the westerlies. In oceanic circulation, it is a circular movement of water usually formed where currents pass obstructions, between two adjacent currents flowing counter to each other, or along the edge of a permanent current.

Eddy diffusion Mixing of the atmosphere by chaotic air currents.

Edge matching Is an important part of the creation of a digital map or GIS database. One digital map may encompass many paper maps. But when the paper maps are laid edge-to-edge, features running across the boundaries of the map sheets are not always properly aligned. These misalignments occur for many reasons including survey error, cartographic generalization, and problems of map projection and compilation. Edge matching is a term applied to a variety of techniques employed to resolve these inconsistencies, including warping and rubber sheeting.

Edge wave A wave of water that moves parallel to the shore. This wave is usually a secondary wave of complex formation.

Effusive eruption Volcanic eruption where low-viscosity basaltic magma is released. This type of eruption is not explosive and tends to form shield volcanoes.

El Nino El Niño Southern Oscillation (ENSO) is a global coupled ocean-atmosphere phenomenon. The Pacific Ocean signatures, El Niño and La Niña (also written in English as El Nino and La Nina) are major temperature fluctuations in surface waters of the tropical Eastern Pacific Ocean. The names, from the Spanish for "the little boy" and "the little girl", refer to the Christ child, because the phenomenon is usually noticed around Christmas time in the Pacific Ocean off the west coast of South America. Their effect on climate in the southern hemisphere is profound. These effects were first described in 1923 by Sir Gilbert Thomas Walker from whom the Walker circulation, an important aspect of the Pacific ENSO phenomenon, takes its name. The atmospheric signature, the Southern Oscillation (SO) reflects the monthly or seasonal fluctuations in the air pressure difference between Tahiti and Darwin. El Nino affects Australia by drought.

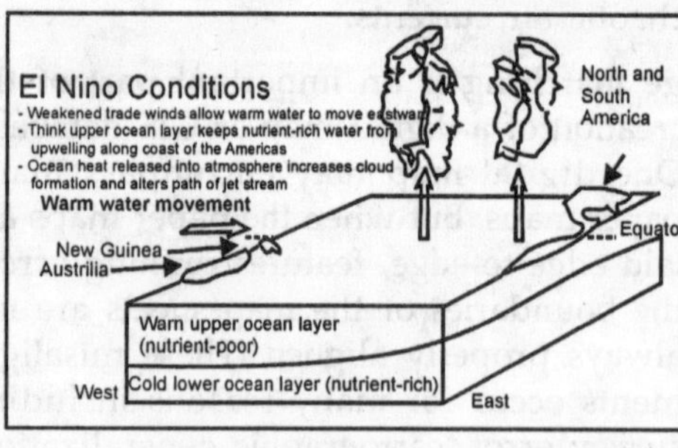

Fig. El Nino

ENSO is a set of interacting parts of a single global system of coupled ocean-atmosphere climate fluctuations that come about as a consequence of oceanic and atmospheric circulation. ENSO is the most prominent known source of inter-annual variability in weather and climate around the world (~3 to 8 years), though not all areas are affected. ENSO has signatures in the Pacific, Atlantic and Indian Oceans. In the Pacific, during major warm events El Niño warming extends over much of the tropical Pacific and becomes clearly linked to the SO intensity. While ENSO events are basically in phase between the Pacific and Indian Oceans, ENSO events in the Atlantic Ocean lag behind those in the Pacific by 12 to 18 months. Many of the countries most affected by ENSO events are developing countries within main continents (South America, Africa), with economies that are largely dependent upon their agricultural and fishery sectors as a major source of food supply, employment, and foreign exchange. New capabilities to predict the onset of ENSO events in the three oceans can have global socio-economical impacts. While ENSO is a global and natural part of the Earth's climate, whether its intensity or frequency may change as a result of global warming is an important concern. Low-frequency variability has been evidenced. Inter-decadal modulation of ENSO might exist.

El Niño and La Niña are officially defined as sustained sea surface temperature anomalies of magnitude greater than 0.5°C across the central tropical Pacific Ocean. El Niño is associated with a positive anomaly, and La Niña with a negative anomaly. When the condition is met for a period of less than five months, it is classified as El Niño or La Niña conditions; if the anomaly persists for five months or longer, it is classified as an El Niño or La Niña episode. Historically, it has occurred at irregular intervals of 2-7 years and has usually lasted one or two years.

El Niño's warm current of nutrient-poor tropical water, heated by its eastward passage in the Equatorial Current, replaces the cold, nutrient-rich surface water of the Humboldt Current, also known as the Peru Current, which support great populations of food fish. In most years the warming lasts only a few weeks or a month, after which the weather patterns return to normal and fishing improves.

In the Pacific, La Niña is characterized by unusually cold ocean temperatures in the eastern equatorial Pacific, compared to El Niño, which is characterized by unusually warm ocean temperatures in the same area. Atlantic tropical cyclone activity is generally enhanced during La Niña. The La Niña condition often follows the El Niño, especially when the latter is strong.

El Niño/Southern Oscillation- A shift in the normal relationship between the atmosphere and ocean in the tropical Pacific Ocean. Normally, strong winds (called trade winds because they aided sailing ships transporting goods) blow to the west in the Pacific, moving warmer surface water away from North and South America. Simultaneously, cold water from the ocean depths rises to the surface off the west coast of South America. This upwelling brings nutrients to the surface, supporting fisheries and ecosystems in the area. In an El Niño event, these trade winds die down, causing warmer surface water to accumulate off western North and South America. This leads to increased rainfall, storm activity, and flooding in the Americas (especially the southwestern United States and Peru) and drought conditions in Australia and other areas in the western Pacific.

Because El Niño's warm pool feeds thunderstorms above, it creates increased rainfall across the east-central and eastern Pacific Ocean. In South America, the effects of El Niño are direct and stronger than in North America. An El Niño is associated with warm and very wet summers (December-February) along the coasts of northern Peru and Ecuador, causing major flooding whenever the event is strong or extreme. The effects during the months of February, March and April may become critical. Southern Brazil and northern Argentina also experience wetter than normal conditions but mainly during the spring and early summer. Central Chile receives a mild winter with large rainfall, and the Peruvian-Bolivian Altiplano is sometimes exposed to unusual winter snowfall events. Drier and hotter weather occurs in parts of the Amazon River Basin, Colombia and Central America.

Direct effects of El Niño resulting in drier conditions occur in Indonesia, increasing forest fires, in the Philippines, and northern Australia. Drier than normal conditions are also generally observed in Queensland, inland Victoria, inland New South Wales and eastern Tasmania during June-August. West of the Antarctic Peninsula, the Ross, Bellingshausen, and Amundsen Sea sectors have more sea ice during El Niño. The latter two and the Weddell Sea also become warmer and have higher atmospheric pressure.

In North America, typically, winters are warmer than normal in the upper Midwest states and Canada, while central and southern California, northwest Mexico and the southeastern U.S., are wetter than normal. Summer is wetter in the intermountain regions of the U.S. The Pacific Northwest states, on the other hand, tend to be drier during an El Niño. During a La Niña, by contrast, the Midwestern U.S. tends to be drier than normal. El Niño is also associated with decreased hurricane activity in the Atlantic. Finally, East Africa, including Kenya, Tanzania and the White Nile basin, experiences in the long rains from March to May wetter than normal conditions. There also are drier than normal conditions from December to February in south-central Africa, mainly in Zambia, Zimbabwe, Mozambique and Botswana.

Elastic deformation Change in the shape of a material as the result of the force of compression or expansion. Upon release of the force, the material returns to its original shape. Also called plastic deformation.

Elastic limit Maximum level of elastic deformation of a material without rupture.

Elastic rebound theory Theory that describes how earthquakes arise from the horizontal movement of adjacent tectonic plates along a linear strike-slip fault. This theory suggests that the two plates moving in opposite directions become locked for some period of time because of friction. However, the accumulating stress overcomes the friction and causes the plate to suddenly move over a short time period which generates an earthquake.

Elastic wave An energy wave that causes elastic deformation in a material without its structure and shape being deformed.

Electrical energy Energy resulting from the force between two objects having the physical property of charge.

Electromagnetic energy Energy stored in electromagnetic waves or radiation. Energy is released when the waves are absorbed by a surface. Any object with a temperature above absolute zero (-273° Celsius) emits this type of energy. The intensity of energy released is a function of the temperature of the radiating surface. The higher the temperature the greater the quantity of energy released.

Electromagnetic radiation Energy propagated as time-varying electric and magnetic fields. These two fields are inextricably linked as a single entity since time-varying electric fields produce time-varying magnetic fields and vice versa. Light and radar are examples of electromagnetic radiation differing only in their wavelengths (or frequency). Electric and magnetic fields propagate through space at the speed of light.

Electromagnetic spectrum The entire range of radiant energies or wave frequencies from the longest to the shortest wavelengths—the categorization of solar radiation. Satellite sensors collect this energy, but what the detectors capture is only a small protion of the entire elctromagnetic spectrum. The spectrum usually is divided into seven sections: radio, microwave, infrared, visible, ultra-violet, x-ray, and gamma-ray radiation.

Electromagnetic wave Method of travel for radiant energy (all energy is both particles and waves), so called because radiant energy has both magnetic and electrical properties. Electromagnetic waves are produced when electric charges change their motion. Whether the frequency is high or low all electromagnetic waves travel at 300,000,000 metres per second.

Electromotive Producing an electric current through differences in potential. electromotive force The force that can alter the motion of electricity, measured in terms of the energy per unit charge imparted to electricity passing through the source of this force. Electromotive force causes current flow in a circuit.

Electron A sub-particle of an atom that contains a negative atomic charge.

Electronic distance measuring (EDM) device Instruments that measure the phase difference between transmitted and reflected or retransmitted electromagnetic waves of known frequency, or that measure the round-trip transit time of a pulsed signal, from which distance is computed.

Element A molecule composed of one type of atom. Chemists have recognized or created 112 different types of elements. Two or more different elements form a compound.

Element set Specific information used to define and locate a particular satellite. The set includes the catalog number; epoch year; day and fraction of day; period decay rate; argument of perigee; inclination; eccentricity; right ascension of the ascending node; mean anomaly; mean motion; revolution number at epoch; and element set number. This data is contained in the two-line orbital elements provided by NASA in the NASA Prediction Bulletin.

Elevation The measure of height with respect to a point on the earth's surface above mean sea level. Sometimes referred to as station elevation.

Elliptical orbits Bodies in space orbit in elliptical rather than circular orbits because of factors such as gravity and drag. The point where the orbiting satellite is closest to Earth is the perigee, sometimes called per-apsis or perifocus. The point where the satellite is farthest from Earth is called apogee apoapsis, or api-focus. A line drawn from perigee to apogee is the line-of-apsides, sometimes called the major-axis of the ellipse. It's simply a line drawn through the ellipse the long way.

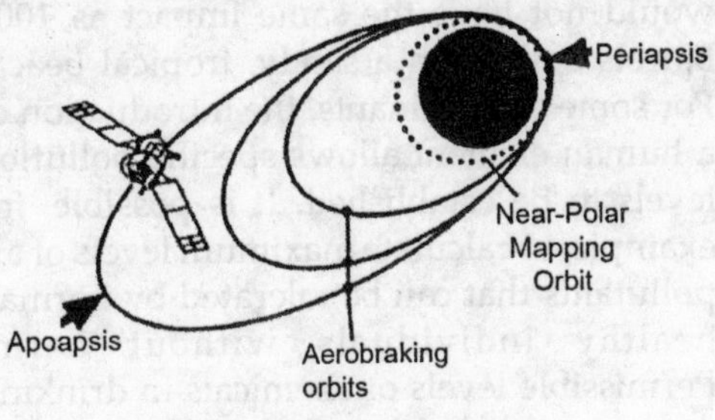

Fig. Elliptical orbits

Eluviation Movement of humus, chemical substances, and mineral particles from the upper layers of a soil to lower layers by the downward movement of water through the soil profile.

Emergent coastline A shoreline resulting from a rise in land surface elevation relative to sea level.

Emigration Migration of an organism out of an area for the purpose of changing its residence permanently.

Emissivity The ratio of total radiative output from a body per unit time per unit area at a specific temperature and wavelength to that of a black body under the same environmental conditions.

Enclave A tract or territory enclosed within another state or country.

Endangered species A species found in nature that has so few surviving individuals that the it could soon become extinct in all or most of its natural range.

Endogenic Refers to a system that is internal to the Earth.

Energy Is defined as the capacity for doing work. Energy can exist the following forms: radiation; kinetic energy; potential energy; chemical energy; atomic energy; electro-magnetic energy; electrical energy; and heat energy.

Energy budget A quantitative description of the energy exchange for a physical or ecological system. The budget includes terms for radiation, conduction, convection, latent heat, and for sources and sinks of energy.

Energy flux The rate of energy flow from, into, or through a substance.

ENSO (El Nino-southern oscillation) Interacting parts of a single global system of climate fluctuations. ENSO is the most prominent known source of interannual variability in weather and climate around the world, though not all areas are affected. The Southern Oscillation (SO) is a global-scale seesaw in atmospheric pressure between Indonesia /North Australia, and the southeast Pacific. In major warm events El Nino warming extends over much of the tropical Pacific and becomes clearly linked to the SO pattern. Many of the countries most affected by ENSO events are developing countries with economies that are largely dependent upon their agricultural and fishery sectors as a major source of food supply employment, and foreign exchange. New capabilities to predict the onset of ENSO events can have a global impact. While ENSO is a natural part of the Earth's climate, whether its intensity or frequency may change as a result of global warming is an important concern.

Entisols Soil order (type) of the United States Natural Resources Conservation Service Soil Classification System. Soil of recent development with no or poorly developed soil horizons.

Entrainment One of three distinct processes involved in erosion. Is the process of particle lifting by an agent of erosion.

Entropy Entropy is the measure of the disorder or randomness of energy and matter in a system.

Environment The sum total of all the external conditions that effect an organism, community, material , or energy.

Environmental gradient Spatial gradient where abiotic and biotic factors vary.

Environmental Lapse Rate (ELR) The rate of air temperature increase or decrease with altitude. The average ELR in the troposphere is an air temperature decrease of 6.5° Celsius per 1000 metres rise in elevation.

Environmental pollution The contamination of the physical and biological components of the earth/atmosphere system to such an extent that normal environmental processes are adversely affected. Although some writers and researchers regard pollution as anthropogenic in origin, natural sources can also provide sufficient extraneous material to disrupt normal environmental processes. Volcanic activity, for example, can cause major air pollution or water pollution and destroy flora and fauna by covering them with volcanic ash or lava. With time, the environment adjusts to take such changes into account, since the earth/atmosphere system includes mechanisms capable of dealing with almost any form of contamination or disruption. Larger air or water pollutants, for example, fall out of the air or water stream as a result of gravity, gases may be absorbed by plants, neutralized by oxidation or dissolved in water, and organic materials may be destroyed by bacteria. The capacity of the environment to deal with pollutants in this way is not unlimited, but it might be argued that pollution occurs only when the environment's capacity for dealing with additional material is surpassed. Such levels are difficult to establish, however, since the impact of a given amount of pollutant will vary with such factors as the physical nature of the environment, the age and mix of organisms and the timing or duration of the pollution event. Emissions from 100 cars in a restricted airshed would have the potential to cause a greater environmental impact than the same emissions in an open, well-ventilated location. Similarly, 1000 barrels of oil spilled on a rocky, sub-Arctic shoreline would not have the same impact as 1000 barrels spilled on a sandy, tropical beach. For some contaminants, the introduction of a human element allows specific pollution levels to be established. It is possible, for example, to calculate maximum levels of air pollutants that can be tolerated by normal, healthy individuals without harm. Permissible levels of chemicals in drinking water can be developed also. They are not ideal, however, since like the wider environment individual humans also react differently to the same levels of contamination. Young children, older adults or individuals with existing health problems commonly suffer more than healthy, mature adults. Furthermore, the tests used to estimate these levels are not necessarily conducted on human subjects, but on animals under laboratory conditions. Thus, attempts at establishing the levels of pollution that the environment or elements of the environment can withstand have met with only limited success. It is increasingly clear, however, that human activities are adding contaminants to the environment at rates which easily exceed its ability to cope. Pollution takes many forms, most of which started as local problems, but which are now global in extent. Even in the Arctic, far removed from industrial activity, air pollution is present in the form of Arctic Haze, and oil slicks can be found in the middle of the world's oceans. Some of the pollutants, such as human sewage or smoke, are obvious, whereas others, such as radioactive emissions, toxic gases or chemicals dissolved in water are invisible and may not evoke a response until they have caused major environmental damage. Noise pollution is similar, in that it is not visible. However, along with more traditional forms of pollution, it makes a serious contribution to the quality of the environment in cities and has begun to create problems in rural areas also as power boats, snow-machines and all-terrain vehicles increase in popularity. To be treated effectively, pollution must be controlled at source. There is little point, for example, in liming a lake to reduce the acidity of water, if industry continues to emit acid gases into the atmosphere. In the case of the nuclear industry and certain branches of the chemicals industry, potential pollutants may be so hazardous that there must be complete control, with a no-emissions policy and permanent or long-term management of waste products. Although the technology exists to handle most, if not all, pollution problems, socioeconomic or political factors may prevent its adoption. There have been some successes, such as the reduction in water pollution by detergent phosphates and control of the production and emission of CFCs, but other pollutants such as sewage and automobile exhaust gases remain serious and growing problems in some areas. Since the 1940s and 1950s, governments have introduced significant volumes of anti-pollution legislation and set standards for air and water quality. Success has been mixed, however, and as pollution problems become increasingly global in scope, international co-operation becomes essential. The successful implementation of the provisions of the Montreal Protocol shows that co-operation can work, but differences in national economic and political agendas create the potential for ongoing problems.

Environmental science Field of knowledge that studies of how humans and other species interact with one another and with the nonliving environment. It is both a physical and social science that integrates knowledge from a wide range of disciplines, including physics, chemistry, biology, geology, geography, economics, political science, sociology, psychology, and philosophy.

Environmental system A system where life interacts with the various abiotic components found in the atmosphere, hydrosphere, and lithosphere.

Enzyme Are types of proteins that are used to facilitate and regulate chemical reactions within cells.

Eolian Geomorphic process involving wind.

Eolian landform Is a landform formed from the erosion or deposition of weathered surface materials by wind. This includes landforms with some of the following geomorphic features: sand dunes, deflation hollows, and desert pavement.

Eon Longest geologic time unit.

EPA (Environmental Protection Agency) U.S. agency that ensures: Federal environmental laws are implemented and enforced effectively; U.S. policy—both foreign and domestic—fosters the integration of economic development and environmental protection so that economic growth can be sustained over the long term; public and private decisions affecting energy, transportation, agriculture, industry, international trade, and natural resources fully integrate considerations of environmental quality; national effort to reduce environmental risk are based on the best available scientific information communicated clearly to the public; everyone in our society recognizes the value of preventing pollution before it is created; people have the information and incentives they need to make environmentally-responsible choices in their daily lives; and schools and community institutions promote environmental stewardship as a national ethic.

Ephemeris A tabulation of a series of points that define the position and motion of a satellite.

Epicentre Surface location of an earthquake's focus.

Epiphyte Type of vegetation that gets its physical support from the branches of other plants. Commonly found in the tropical forests.

EPIRB Emergency position indicating radio beacon.

Epoch Geologic time unit that is shorter than a period.

Equator Zero degrees latitude. The Sun is directly overhead the equator at noon on the two equinoxes (March and Sept. 20 or 21). The equator divides the globe into the Northern and Southern hemispheres. The equator appears halfway between the North and South poles, at the widest circumference of the globe. It is 24,901.55 miles (40,075.16 km) long.

Equatorial rainforest One of the world's major biomes. Equatorial rainforests are located in a zone some 10°N and S of the equator, mainly in Amazonia, equatorial west and central Africa and south-east Asia, all of which experience heavy precipitation (1750–2500 mm) and high, fairly constant temperatures (25–28°C) through the year. Such hot, moist conditions encourage rapid and abundant plant growth, with the tallest trees exceeding 30 m in height, and supporting a canopy sufficiently thick that it allows little light to reach ground level. Many of the trees, shallow rooted in thin soils, are underpinned by buttresses. The trees in turn support a great variety of climbing plants and epiphytes that use the trees to reach the sunnier conditions in the upper levels of the forest. The rainforest vegetation is typically stratified, consisting of as many as five layers from the forest canopy to the forest floor. The greatest amount of biomass is present in the tree layers that have access to light, whereas, on the forest floor, where little light penetrates, the vegetation is poorly developed and widely spaced. With the absence of seasons, the trees do not shed their leaves at any particular time of year, and, as a result, the forest is considered to be evergreen. Because of the ideal growing conditions and the rapid recycling of nutrients through the system, the equatorial rainforest is biologically very productive and is claimed (although not without dispute) to have the greatest abundance and diversity of plants and animals of any of the terrestrial biomes. Many commercially important tree species such as ebony, teak, mahogany and rubber grow there, while forest plants also provide a wide range of raw materials that can be processed into pharmaceuticals. Like all tropical forests, the equatorial rainforest is under increasing pressure from development. In Asia, for example, high value hardwoods such as teak are being overharvested, while in Brazil the forest is being cleared at a rate of 10,000 km^2 per year to provide land for agricultural or industrial development. According to Friends of the Earth, such destruction is responsible for the extinction of more than 8000 rainforest species every year. Diverse as the rainforest flora and fauna may be, they cannot continue to undergo such stress without serious environmental effects. Additional impacts include changes to temperature and moisture regimes, reduced soil fertility and soil erosion which ensure that once the forest is destroyed it is very difficult to re-establish. On a global scale, rainforest destruction has been linked to global warming. Clearing the forest by burning introduces additional carbon dioxide (CO_2) into the atmosphere, while the removal of the vegetation reduces photosynthesis and therefore the recycling of the gas. As a result, destruction of the equatorial rainforest is the second largest cause of increased atmospheric CO_2, which contributes to the progressive enhancement of the greenhouse effect. Clearly the threat to the rainforest is serious and has implications that extend beyond the tropics, but attempts to improve the situation through such approaches as sustainable development, bans on logging and mining and the creation of forest reserves have as yet had only limited impact.

Equatorial trough The quasi-continuous area of low pressure between the subtropical high pressure areas in both the Northern and Southern Hemisphere.

Equilibrium Equilibrium describes the average condition of a system, as measured through one of its elements or attributes, over a specific period of time.

Equinox The point at which the ecliptic intersects the celestial equator. Days and

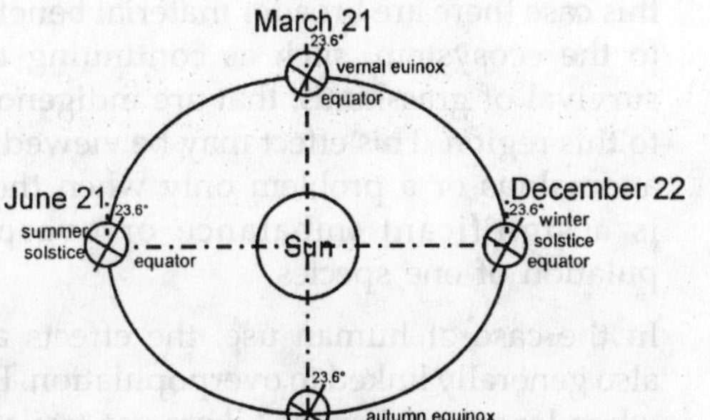

Fig. Equinox

nights are most nearly equal in duration. In the Northern Hemisphere, the vernal equinox falls on or about March 20 and the autumnal equinox on or about September 22.

Era Geologic time unit that is shorter than an eon but longer than a period.

Erg desert A region in a desert where sand is very abundant.

Erosion Erosion is the displacement of solids (soil, mud, rock and other particles) by the agents of wind, water or ice, by downward or down-slope movement in response to gravity or by living organisms (in the case of bioerosion). Erosion is to be distinguished from weathering, which is the decomposition of rock, although the two processes may be concurrent. Erosion is an intrinsic natural process but in many places it is increased by human land use. Poor land use practices include deforestation, overgrazing, unmanaged construction activity and road or trail building. However, improved land use practices can limit erosion, using techniques like terrace-building and tree planting. A certain amount of erosion is natural and, in fact, healthy for the ecosystem. For example, gravels continually move downstream in watercourses. Excessive erosion, however, can cause problems, such as receiving water sedimentation, ecosystem damage (including fish kills) and outright loss of soil.

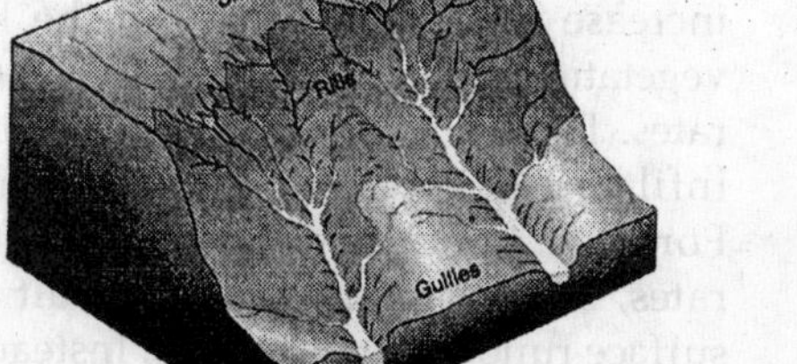

Fig. Erosion

What causes erosion to be severe in some areas and minor elsewhere? It is a combination of many factors, including the amount and intensity of precipitation, the texture of the soil, the gradient of the slope, ground cover (from vegetation, rocks, etc.) and land use. The first factor, rain, is the agent for erosion, but the degree of erosion is governed by other factors. The first three factors can remain fairly constant over time. In general, given the same kind of vegetative cover, you expect areas with high-intensity precipitation, sandy or silty soils and steep slopes to be the most erosive. Soils with a greater proportion of clay that receive less intense precipitation and are on gentle slopes tend to erode less.

The factor that is most subject to change is the amount and type of ground cover. When fires burn an area or when vegetation is removed as part of timber operations or building a house or a road, the susceptibility of the soil to erosion is greatly increased. Roads are especially likely to cause increased rates of erosion because, in addition to removing ground cover, they can significantly change drainage patterns. A road that has a lot of rock and one that is "hydrologically invisible" (that gets the water off the road as quickly as possible, mimicking natural drainage patterns) has the best chance of not causing increased erosion.

Understandably, many human activities remove vegetation from an area, making the soil easily eroded. Logging and heavy grazing can reduce vegetation enough to increase erosion. Changes in the kind of vegetation in an area can also affect erosion rates. Different kinds of vegetation effect infiltration rates of rain into the soil. Forested areas have higher infiltration rates, so precipitation will result in less surface runoff, which erodes. Instead much of the water will go in subsurface flows, which are generally less erosive. Leaf litter and low shrubs are an important part of the high infiltration rates of forested systems, the removal of which can increase erosion rates. Leaf litter also shelters the soil from the impact of falling raindrops, which is a significant agent of erosion. Vegetation can also change the speed of surface runoff flows, so grasses and shrubs can also be instrumental in this aspect.

One of the main causes of erosive soil loss in the year 2006 is the result of slash and burn treatment of tropical forests. When the total ground surface is stripped of vegetation and then seared of all living organisms, the upper soils are vulnerable to both wind and water erosion. In a number of regions of the earth, entire sectors of a country have been rendered unproductive. For example, on the Madagascar high central plateau, comprising approximately ten percent of that country's land area, virtually the entire landscape is sterile of vegetation, with gully erosive furrows typically in excess of 50 metres deep and one kilometre wide. Shifting cultivation is a farming system which sometimes incorporates the slash and burn method in some regions of the world.

When land is overused by animal activities (including humans), there can be mechanical erosion and also removal of vegetation leading to erosion. In the case of the animal kingdom, this effect would become material primarily with very large animal herds stampeding such as the Blue Wildebeest on the Serengeti plain. Even in this case there are broader material benefits to the ecosystem, such as continuing the survival of grasslands, that are indigenous to this region. This effect may be viewed as anomalous or a problem only when there is a significant imbalance or overpopulation of one species.

In the case of human use, the effects are also generally linked to overpopulation. For when large numbers of hikers use trails or extensive off road vehicle use occurs, erosive effects often follow, arising from vegetation removal and furrowing of foot traffic and off road vehicle tires. These effects can also accumulate from a variety of outdoor human activities, again simply arising from too many people using a finite land resource. One of the most serious and long-running water erosion problems worldwide is in China, on the middle reaches of the Yellow River and the upper reaches of the Yangtze River. From the Yellow River, over 1.6 billion tons of sediment flow each year into the ocean. The sediment originates primarily from water erosion in the Loess Plateau region of northwest China.

1. Gravity erosion-Mass wasting is the down-slope movement of rock and sediments, mainly due to the force of gravity. Mass wasting is an important part of the erosional process, as it moves material from higher elevations to lower elevations where transporting agents like streams and glaciers can then pick up the material and move it to even lower elevations. Mass-wasting processes are occurring continuously on all slopes; some mass-wasting processes act very slowly, others occur very suddenly, often with disastrous results. Any perceptible down-slope movement of rock or sediment is often referred to in general terms as a landslide. However, landslides can be classified in a much more detailed way that reflects the mechanisms responsible for the movement and the velocity at which the movement occurs. One of the visible topographical manifestations of a very slow form of such activity is a scree slope. Slumping happens on steep hillsides, occurring along distinct fracture zones, often within materials like clay that, once released, may move quite rapidly downhill. They will often show a spoon-shaped depression, within which the material has begun to slide downhill. In some cases, the slump is caused by water beneath the slope weakening it. In many cases it is simply the result of poor engineering along highways where it is a regular occurrence. Surface creep is the slow movement of soil and rock debris by gravity which is usually not perceptible except through extended observation. However, the term can also describe the rolling of dislodged soil particles 0.5 to 1.0 mm in diametre by wind along the soil surface.

2.Water erosion-Splash erosion is the detachment and airborne movement of small soil particles caused by the impact of raindrops on soil. Sheet erosion is the result of heavy rain on bare soil where water flows as a sheet down any gradient, carrying soil particles. Where precipitation rates exceed soil infiltration rates, runoff occurs. Surface runoff turbulence can often cause more erosion than the initial raindrop impact. Gully erosion results where water flows along a linear depression eroding a trench or gully. Valley or stream erosion occurs with continued water flow along a linear feature. The erosion is both downward, deepening the valley, and headward, extending the valley into the hillside. In the earliest stage of stream erosion, the erosive activity is dominantly vertical, the valleys have a typical V cross-section and the stream gradient is relatively steep. When some base level is reached the erosive activity switches to lateral erosion which widens the valley floor and creates a narrow floodplain. The stream gradient becomes nearly flat and lateral deposition of sediments becomes important as the stream meanders across the valley floor. In all stages of stream erosion by far the most erosion occurs during times of flood, when more and faster-moving water is available to carry a larger sediment load. In such processes, it is not the water alone that erodes, suspended abrasive particles, pebbles and boulders can also act erosively, as they traverse a surface.

3. Shoreline erosion-Shoreline erosion, on both exposed and sheltered coasts, primarily occurs through the action of

currents and waves but sea level (tidal) change can also play a role. Sediment is transported along the coast in the direction of the prevailing current (longshore drift). When the upcurrent amount of sediment is less than the amount being carried away, erosion occurs. When the upcurrent amount of sediment is greater, sand or gravel banks will tend to form. These banks may slowly migrate along the coast in the direction of the longshore drift, alternately protecting and exposing parts of the coastline.

4. Ice erosion-Ice erosion is caused by movement of ice, typically as glaciers. Glaciers can scrape down a slope and break up rock and then transport it, leaving moraines, drumlins and glacial erratics in their wake, typically at the terminus or during glacier retreat. Ice wedging is the weathering process in which water trapped in tiny rock cracks freezes and expands, breaking the rock. This can lead to gravity erosion on steep slopes. The scree which forms at the bottom of a steep mountainside is mostly formed from pieces of rock broken away by this means. It is a common engineering problem, wherever rock cliffs are alongside roads, because morning thaws can drop hazardous rock pieces onto the road.

5. Wind erosion-Wind erosion, also known as eolian erosion is the movement of rock and/or sediment by the wind. Windbreaks are often planted by farmers to reduce wind erosion. This includes the planting of trees, shrubs, or other vegetation, usually perpendicular or nearly so to the principal wind direction. The wind causes dust particles to be lifted and therefore moved to another region. Wind erosion generally occurs in areas with little or no vegetation, often in areas where there is insufficient rainfall to support vegetation. An example is the formation of sand dunes, on a beach or in a desert.

Erosional landform Is a landform formed from the removal of weathered and eroded surface materials by wind, water, glaciers, and gravity. This includes landforms with some of the following geomorphic features: river valleys, glacial valleys, and coastal cliffs..

Erratic A boulder that has been carried from its source by a glacier and deposited as the glacier melted. Thus, the boulder is often of a different rock type from surrounding types.

Escarpment A long cliff or steep slope separating two comparatively level or more gently sloping surfaces and resulting from erosion or faulting.

Esker Long twisting ridges of sand and gravel found on the Earth's surface. Created when the deposits of subsurface glacial streams are placed on the ground after glacial melting.

Establishment Subsequent growth and/or reproduction of a colonized species in a new territory.

Estimator An estimator is any value calculated from the sample data For example, the sample mean is an estimator of the population mean.

Estuary Somewhat enclosed coastal area at the mouth of a river where nutrient rich fresh water meets with salty ocean water.

Eukaryota All the organisms with a eukaryote cell type. This group includes animals, plants, fungi, and protists.

Eukaryote Organisms whose cells have a membrane-bound nucleus and many specialized structures located within their cell boundary. In these organisms, genetic material is organized into chromosomes that reside in the nucleus.

Eustacy Variations in sea-level that are related to changes in the volume of seawater in the oceans. The consideration of sea-level change on a worldwide scale. In recent geological time the major eustatic changes have been associated with the glaciations of the Quaternary era. These glacio-eustatic changes involved a lowering of sea level during glacial periods, as more water was retained on land in the form of ice, and a rising sea level during interglacials as the ice melted and the water was returned to the ocean basins. Estimates of the decline in sea level at various glacial maxima range from 85 m to 120 m below current sea level, whereas during the interglacials sea levels rose again, perhaps exceeding current levels by a few metres. In postglacial times, during the climatic optimum (hypsithermal), sea level was perhaps 6 m higher than at present. In northern latitudes, the question of eustatic sea-level changes during glaciations, interglacials and in postglacial times is complicated by isostatic movements of the earth's crust. It is difficult to obtain accurate values of eustatic change under such conditions, and most estimates have been obtained from observations in lower latitudes well away from the direct effects of ice. Although glacial variations have caused the greatest eustatic changes, other factors, such as tectonic changes, which alter the shape and capacity of the ocean basins, also have the potential to contribute. A change in temperature of sea water could also cause sea level to rise. An increase in 1°C of the mean temperature of the oceans would cause sufficient expansion of the water to bring about a rise in sea level of perhaps as much as 60 cm. Recent concern with global warming has drawn attention to this fact. Glacioeustatic change is also a potential consequence of global warming. The increase in the melting of glaciers and ice-caps following an estimated increase of 8°C in high latitudes has the potential to cause a measurable rise in sea level. In the extreme case of temperature increases sufficiently great to cause complete melting of the Greenland and Antarctic ice sheets, sea level might increase by some 60 m.

Eutrophic lake Lake that has an excessive supply of nutrients, mostly in the form of nitrates and phosphates.

Eutrophication The natural ageing of a water body characterized by increasing levels of dissolved nutrients in the water. These in turn encourage the growth of aquatic plants often in the form of algal blooms. Eutrophication also occurs in coastal sea areas, where 'red tides' are an indication of a rapid increase in local nutrient levels. When the plants begin to die, the increase in bacterial activity raises the biochemical oxygen demand (BOD), and so the water becomes deoxygenated. Eutrophication is a natural process, but it has been accelerated by human activities which add such products as sewage, agricultural fertilizers and detergents to the water bodies. The additional nitrogen (N) and phosphorus (P) from such effluents encourages plant growth. The introduction of phosphate-free detergents and the improvement of sewage treatment facilities have slowed eutrophication in some areas. In the late 1960s and 1970s, for example, Lake Erie in North America was in a state of advanced eutrophication that many environmentalists considered to be irreversible. In places it was covered by great mats of floating algae, and stretches of shoreline were contaminated by masses of rotting plant material. Since then, effluent control has reduced the problem significantly. Eutrophication remains a major concern in many areas, however, including enclosed seas such as the Mediterranean, which are more susceptible to the problem than the open ocean.

Evaporation The process of vaporization by which a liquid is converted into a gas or vapour. This change in state requires the absorption of energy normally supplied by the adjacent environment – for example, from a hot surface, or from the surrounding air. Once the change has taken place, the absorbed energy is retained by the vapour in the form of latent heat. When the process is reversed and the vapour is condensed back into a liquid, the latent heat is released into the environment in the form of sensible heat. Evaporation has a role in several major processes in the earth/ atmosphere system. It is an integral part of the hydrological cycle, for example, initiating the recycling process which involves both the physical redistribution and the cleansing of the system's water supply.

Evaporation also contributes to the earth's energy budget, by helping to redistribute energy within the atmosphere. The energy from solar radiation used to evaporate water in the tropics is retained by water vapour in the form of latent heat. If that vapour is carried into higher latitudes, it carries the latent heat with it, to be released as sensible heat if condensation takes place. Similar conditions also develop in frontal systems, and the net result is a redistribution of energy in the environment. The rate at which evaporation takes place depends upon moisture availability, temperature, humidity and air movement. If no moisture is available, no evaporation will take place. If moisture is available, however, a combination of high temperatures, low relative humidities and strong winds will produce maximum evaporation. Under calm conditions, the evaporation rate will decline with time, as the input of water vapour into the atmosphere increases the relative humidity. Air movement helps to maintain a high evaporation by replacing the saturated air with drier air that can absorb more moisture. The greatest annual evaporation rates occur over the tropical oceans. Rates are also high over the continents in tropical regions, where moisture is available. More evaporation takes place over the rainforest, for example, than over the tropical deserts. In the latter case, although high temperatures and low relative humidities encourage high evaporation rates there is insufficient moisture and actual evaporation is restricted. Rates are also high over the Gulf Stream in the western Atlantic Ocean and over the Kuro Siwo current in the Pacific, particularly when dry continental air moves out over the oceans. The accurate measurement of evaporation is not easy. Instruments such as evaporating pans, which provide evaporation rates from an open water surface, and atmometres, which provide evaporation rates from a saturated, porous surface, present information on potential rather than actual evaporation, since the supply of moisture is not limited. Comparisons with other variables such as precipitation and soil moisture availability allow actual evaporation rates to be calculated. Formulae have been developed for estimating evaporation from such variables as precipitation, drainage, runoff and soil moisture storage (water budget approach) or from duration of sunshine, air temperature, air humidity and wind speed (energy budget/ aerodynamic approach), but all contain potential sources of error, usually in the form of data reliability.

Evaporation fog Fog generated from cold air advection over warm water or warm or moist land. This type of fog is sometimes called steam fog or sea smoke.

Evaporation pan Meteorological instrument that is used to measure evaporation rates.

Evapourite Type of sedimentary rock that is formed from the concentration of dissolved salts through evaporation.

Evapotranspiration The total amount of water that is transferred from the earth's surface to the atmosphere. It is made up of the evaporation of liquid or solid water plus the transpiration from plants.

Evergreen vegetation Vegetation that keeps a majority of their leaves or needles throughout the year. Also see deciduous vegetation and succulent vegetation.

Evolution Is a process by which species come to possess genetic adaptations to their environment. Its mechanism is natural selection. It also requires genetic mutations.

Exfoliation dome A physical weathering feature associated with granite that is the result of the erosion of overburden material and pressure-release. With the release of pressure, layers of rock break off in sheets or shells leaving a dome-like bedrock feature.

Exogenic Refers to a system that is external to the Earth.

Exosphere This region is considered the very outer limits of the earth's atmosphere. Its lower boundary is often called the critical level of escape, where gas atoms are so widely spaced that they rarely collide with one another and have individual orbits. It is estimated to be some 400 plus miles (640 kilometres) above the surface.

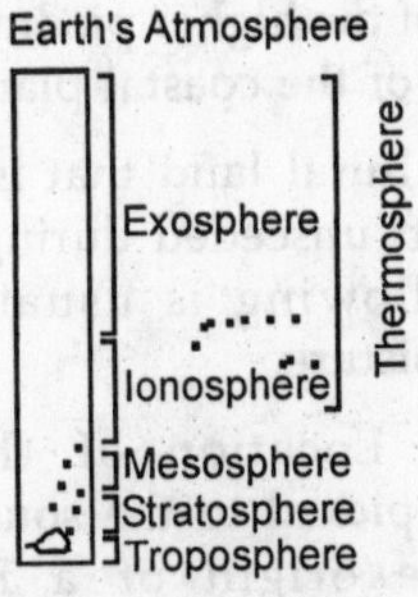

Fig. Exosphere

Exotic stream A stream that has a course that begins in a humid climate and end in an arid climate. Because of reductions in precipitation and increases in evaporation, the discharge of these streams deceases downslope. Examples of such streams are the Nile and Colorado Rivers.

Experiment A controlled investigation designed to evaluate the outcomes of causal manipulations on some system of interest.

Exploitation Form of competition where the indirect effects of the two or more species or individuals reduce the supply of the limiting resource or resources needed for survival.

Explosive eruption Volcanic eruption where high-viscosity granite-rich magma causes an explosion of ash and pyroclastic material. This type of eruption is common to composite and caldera volcanoes.

Extended family A family that includes three or more generations. Normally, that would include grandparents, their sons or daughters, and their children, as opposed to a "nuclear family," which is only a married couple and their offspring.

External forcing Influence on the Earth system (or one of its components) by an external agent such as solar radiation or the impact of extraterrestrial bodies such as meteorites.

Extinction Disappearance of a species from all or part of their geographic range.

Extratropical cyclone Any cyclone not of tropical origin. Generally considered to be a migratory frontal cyclone found in the middle and high latitudes. Also called an extratropical low or an extratropical storm.

Extrusive igneous rock Igneous rock that forms on the surface of the Earth. Also called volcanic igneous rock.

Exurb A region or district that lies outside a city and usually beyond its suburbs.

Eye The centre of a tropical storm or hurricane, characterized by a roughly circular area of light winds and rain-free skies. An eye will usually develop when the maximum sustained wind speeds exceed 78 mph. It can range in size from as small as 5 miles to up to 100 miles, but the average size is 20 miles. In general, when the eye begins to shrink in size, the storm is intensifying.

Eye wall An organized band of convection surrounding the eye, or centre, of a tropical cyclone. It contains cumulonimbus clouds, intense rainfall and very strong winds.

F

Facilitation model of succession This model of succession suggests that the change in plant species dominance over time is caused by modifications in the abiotic environment that are imposed by the developing community. Thus the entry and growth of the later species depends on earlier species preparing the ground.

Facilitation Modification of a system that makes subsequent modifications easier.

Fahrenheit temperature scale A temperature scale where water at sea level has a freezing point of +32 degrees F (Fahrenheit) and a boiling point of +212 degrees F. More commonly used in areas that observe the English system of measurement. Created in 1714 by Gabriel Daniel Fahrenheit (1696-1736), a German physicist, who also invented the alcohol and mercury thermometers.

Fair This is a subjective description. Considered as pleasant weather conditions with regard to the time of year and the physical location.

Fall Season between summer and winter. Astronomically it is the period from the autumnal equinox to the winter solstice in the Northern Hemisphere.

Fall line The physiographic border between the piedmont and coastal plain regions. The name derives from the river rapids and falls that occur as the water flows from hard rocks of the higher piedmont onto the softer rocks of the coastal plain.

Fallow Agricultural land that is plowed or tilled but left unseeded during a growing season. Fallowing is usually done to conserve moisture.

False origin Location of the starting coordinates picked to the south and west of the true origin of a rectangular coordinate system. False origins are used to avoid negative coordinates.

Falsification Falsification is a procedure used in science to test the validity of a hypothesis or theory. It involves stating some output from theory in specific and finding contrary cases among experiments or observations.

Far infrared Electromagnetic radiation, longer than the thermal infrared, with wavelengths between about 25 and 1000 micrometers.

Fathom The common unit of depth in the ocean for countries using the English system of measurement. It is six feet or 1.83 metres. It can also be used in expressing horizontal distance, since 120 fathoms is equal to one cable or nearly on tenth of a nautical mile.

Fault A fracture in the Earth's crust accompanied by a displacement of one side of the fracture.

Fault block mountain A mountain mass created either by the uplift of land between faults or the subsidence of land outside the faults.

Fault plane The plane that represents the fracture surface of a fault.

Fault scarp The section of the fault plane exposed in a fault. Also called an escarpment.

Fault zone A fracture in the Earth's crust along which movement has occurred. The movement may be in any direction and involve material on either or both sides of the fracture. A "fault zone" is an area of numerous fractures.

Feature separation Process of preparing a separate drawing, engraving, or negative for selected types of data in the preparation of a map or chart.

Feedback loop Process where the output of a system causes positive or negative changes to some measured component of the system.

Feeder bands In tropical parlance, the lines or bands of thunderstorms that spiral into and around the centre of a tropical system. Also known as outer convective bands, a typical hurricane may have three or more of these bands. They occur in advance of the main rain shield and are usually 40 to 80 miles apart. In thunderstorm development, they are the lines or bands of low level clouds that move or feed into the updraft region of a thunderstorm.

Feedhorn A metallic cylinder closed at one end, used to obtain and direct radio frequency (RF) energy reflected from a satellite dish. It acts as a wave guide at microwave frequencies. RE energy inside the horn is picked up by a small probe; once inside the horn, the wavelength (energy) of the microwave radiation changes to a guided wave.

Feldspar A group of common aluminium silicate minerals that contains potassium, sodium, or calcium.

Felsic magma Magma that is relatively rich in silica, sodium, aluminium, and potassium. This type of magma solidifies to form rocks relatively rich in silica, sodium, aluminium, and potassium.

Fen A habitat composed of woodland and swamp.

Feng Yun Chinese geostationary environmental satellite that was destroyed by an explosion before launch in April 1994. The name Feng Yun, meaning Wind and Cloud, was originally applied to the Chinese polar-orbiting environmental satellite launched in September 1991 (Feng Yun 1-2), which offered direct readout services. The Chinese polar-orbiter program has since been abandoned.

Feral animal A wild or untamed animal, especially one having reverted to such a state from domestication.

Fermentation Decomposition and breakdown of organic matter by anaerobic means.

Fern A group of about 11,000 species of vascular seedless plants that belong to the division Pterophyta. About 75 percent of the various species of ferns are found in the tropics. Some ferns grow on the branches of trees as epiphytes.

Ferrel cell Three-dimensional atmospheric circulation cell located at roughly 30 to 60° North and South of the equator.

Ferricretes Sedimentary rock created by the chemical precipitation of iron.

Fertilizer Substance that adds inorganic or organic nutrients to soil for the purpose of increasing the growth of crops, trees, or other vegetation.

Fetch An area of the water surface over which waves are generated by a wind having a constant direction and speed. Also, the name given to the length of the fetch area, measured in the direction of the wind from which the seas are generated.

Few The amount of sky cover for a cloud layer between 1/8th and 2/8ths, based on the summation layer amount for that layer.

Field The set of influences (electricity magnetism, gravity) that extend throughout space.

Field capacity The water remaining in a soil after the complete draining of the soil's gravitational water.

Field of view The range of angles that are scanned or sensed by a system or instrument, measured in degrees of arc.

Filling Used in describing the history of a low pressure system or an area of cyclonic circulation, it means an increase in the central pressure of the system. Although it usually describes the action of a pressure system on a constant pressure chart, it also means a surface low is decreasing in cyclonic circulation and losing its characteristics. The opposite of deepening.

Filter Device that while selectively passing desired frequencies removes undesired ones.

Firewhirl A tornado-like rotating column of fire and smoke created by intense heat from a forest fire or volcanic eruption.

Firn Névé on a glacier that survives the year's ablation season. With time much of the firn is transformed into glacial ice.

Firn limit The lower boundary of the zone of accumulation on a glacier where snow accumulates on an annual basis. Also called the Firn Line.

First gust Another name for the initial wind surge observed at the surface as the result of downdrafts forming the leading edge or gust front of a thunderstorm.

Fish Group of vertebrate animals that inhabit aquatic habitats.

Fish ladder A series of shallow steps down which water is allowed to flow; designed to permit salmon to circumvent artificial barriers such as power dams as the salmon swim upstream to spawn.

Fission (nuclear) Process where the mass of an atomic nucleus is made smaller by the removal of subatomic particles. This process releases atomic energy in the form of heat and electromagnetic radiation.

Fissionable isotope Isotope that can undergo nuclear fission when hit by a neutron at the right speed. Examples include uranium-235 and plutonium-239.

Fissure Opening or crack in the Earth's crust.

Fitness A measure of the health of a species in terms of physiology and future reproductive success.

Fixed energy A process, like photosynthesis, where organisms repackage inorganic energy into organic energy.

Fjord A glacial valley or glacial trough found along the coast that is now filled with a mixture of fresh water and seawater.

Flanking line A line of attached cumulus or towering cumulus clouds of descending height, appearing as stair steps (usually on the southwest side) of the most active part of a supercell.

Flash flood A flood that rises and falls quite rapidly with little or no advance warning, usually as the result of intense rainfall over a relatively small area. Flash floods can be caused by situations such as a sudden excessive rainfall, the failure of a dam, or the thaw of an ice jam.

Flocculation Chemical processes where salt causes the aggregation of minute clay particles into larger masses that are too heavy to remain suspended water.

Flood The inundation of normally dry land by water. Flooding causes millions of worth of property damage and takes hundreds of lives every year. It is most common in river valleys or along the coastal areas of lakes, seas and oceans. River floods are caused when a river channel is incapable of carrying the volume of water added to it, and the excess spills over on to the adjacent floodplain. Heavy and prolonged precipitation, snowmelt, channel constrictions, dam failures and alterations to drainage basins may produce or contribute to flooding either singly or in combination. Intense, prolonged precipitation will cause flooding, particularly if the ground is already saturated, but the flooding may be aggravated by urban development or deforestation within a drainage basin which leads to an increase in the rate of runoff. Similarly, in many areas snowmelt causes annual spring flooding, but the extent and duration of the flooding can be increased when ice accumulation in the river channel constricts the flow. This is common in Canada and Siberia, where snow in the headwaters of north-flowing rivers begins to melt, while the lower reaches remain icebound and unable to cope with the increased flow. Some degree of coastal flooding frequently accompanies high tide levels, but the area affected may be increased significantly when high tides are accompanied by river floods or by storm surges in which strong onshore winds drive the water ashore. The 1953 North Sea floods were caused by a combination of high spring tides and strong north winds that pushed large volumes of water into the southern end of the basin to inundate the low-lying coastal areas of the Netherlands and eastern England. Similar storm surges accompany the hurricanes that pass through the Caribbean or the cyclones in the Bay of Bengal that regularly devastate Bangladesh. Seismic sea waves or tsunamis are less common, but following earthquakes or volcanic eruptions, they too can cause very rapid and destructive flooding in coastal areas. Global warming through the increased melting of glaciers and ice sheets, and the subsequent rise in sea level, has the potential to increase the frequency and extent of coastal flooding.

Floods are a natural part of the hydrological cycle and contribute to both erosion and deposition. In human terms, however, they are seen as a serious hazard to life and property, and human responses to flooding reflect that. The most obvious response is to provide protection against it by building embankments or barriers, such as the dykes which protect the coast of the Netherlands, or the levées which line major rivers such as the Mississippi. The diversion of rivers and the straightening or deepening of their channels allow them to carry more water and therefore reduce the amount that spills over on to the floodplain. At the other end of the scale, the response may include acceptance of flooding plus some form of adjustment to minimize the impact. Zoning by-laws, for example, may prevent the use of a floodprone area for residential development, or require floodproofing of buildings or structures so that flood damage is minimized. Acceptance of flooding in this way is usually combined with flood frequency analysis and emergency measures procedures to warn people and secure structures when flooding is imminent. A modern approach to flood problems is to consider the integrated nature of the entire drainage basin rather than only the area prone to flooding. This recognizes that activities allowed in one part of the basin may have serious consequences elsewhere in the system. The channelization of one section of a river, for example, might cause flooding downstream where the existing channel may be unable to accommodate the new flow regime. Similarly, a change in land use in the upper reaches of a drainage basin may well have consequences far downstream. Deforestation of the upper reaches of the Ganges and Bhramaputra rivers in the foothills of the Himalayas has been blamed by some researchers for major flooding problems in Bangladesh, where the rivers combine to flow into the Bay of Bengal. Such circumstances suggest that flood prevention requires the management of an entire drainage basin rather than just the areas that are obviously prone to flooding.

Flood plain An area of limited relief bordering a river inundated when the river overflows its banks during a *flood*. Floodplains are generally low and flat, but possess some relief in the form of the levées which border the main channel and the abandoned channels which indicate the former course of the river as it meandered across the floodplain. Poor drainage on the floodplain encourages the persistence of wetlands, such as the backswamp areas found towards the landward edge of the plain. The floodplain surface is composed of fine sediments deposited during flooding. Regular deposition of these sediments maintains the fertility of the floodplain soils and makes them attractive to settlement, despite the hazards posed by regular flooding. The agrarian civilizations of Egypt and Mesopotamia depended upon this regular renewal of soil quality, and the large populations that inhabit the floodplains of Bangladesh and China attest to their continued attraction. The availability of flat land and proximity to the river providing water for industry and transportation also made floodplains attractive for urban development. As a result, many of the world's largest cities, such as London, Paris, New Orleans, Shanghai and Calcutta, have grown up on floodplains or on deltas where the floodplains reach the sea. Modern environmental planners, however, tend to regard all floodplains as hazard land, best suited for recreational and certain types of agricultural land use rather than for urban residential or industrial development.

Flood stage The level of a river or stream where overflow onto surrounding areas can occur.

Flood tide Time during the tidal period when the tide is rising. Compare with ebb tide.

Fluid drag Reduction in the flow velocity of a fluid by the frictional effects of a surface.

Fluid Substance, gas or liquid, that has the property of flow.

Fluvial Involving running water. Usually pertaining to stream processes.

Focal length 1. In optics, the distance—usually expressed in millimetres—from the principal point of a lens or concave mirror to its focal point.

2. The distance, measured from the centre of the surface of a parabolic or spherical reflector (e.g., satellite dish) where RF energy is brought to essential point focus.

Focal point The area where weak signals collected by a satellite dish, concentrated into a smaller receiving area, converge.

Focality The characteristic of a place that follows from its interconnections with more than one other place. When interaction within a region comes together at a place (i.e., when the movement focuses on that location), the place is said to possess "focality.".

Foehn A warm dry wind on the lee side of a mountain range, whose temperature is increased as the wind descends down the slope. It is created when air flows downhill from a high elevation, raising the temperature by adiabatic compression. Examples include the Chinook wind and the Santa Ana wind. Classified as a katabatic wind.

Fog A visible aggregate of minute water droplets suspended in the atmosphere at or near the surface of the earth, reducing horizontal visibility to less than 5/8 statute miles. It is created when the temperature and the dew point of the air have become the same, or nearly the same, and sufficient condensation nuclei are present. It is reported as "FG" in an observation and on the METAR.

Fog bank A fairly well-defined mass of fog observed in the distance. Most commonly seen at sea, over a lake, or along coastal areas.

Fogbow A whitish semicircular arc seen opposite the sun in fog. The outer margin has a reddish tinge, its inner margin has a bluish tinge, and the middle of the band is white. An additional bow with reversed colours sometimes appears inside the first.

Föhn wind European equivalent of Chinook wind.

Fold Wavelike layers in rock strata that are the result of compression.

Folding The deformation of rock layers because of compressive forces to form folds.

Foliar leaching Process in which water from precipitation removes plant nutrients from the surface of leaves.

Foliation Process where once randomly distributed platy minerals in a rock become reoriented, because of metamorphism, in a parallel manner.

Food chain A group of organisms linked to each other through their production and consumption of food and energy. Most food chains are short (perhaps four or five links) and linear, with some form of green plant at one end and a carnivore or omnivore at the other.

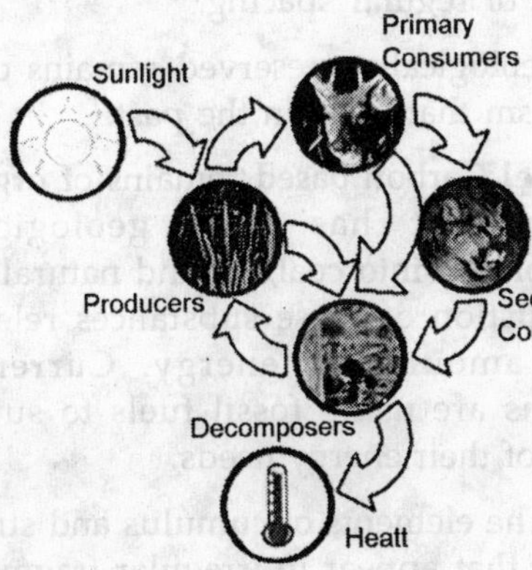

Fig. Food chain

The grass produces the food and energy through photosynthesis. It is then consumed by the antelope, which in turn is consumed by the lion. In this way food and energy are passed along the chain, with each stage referred to as a trophic level. This is an example of a grazing food chain. Ultimately the energy reaches the decomposers. They convert dead organic matter into its constituent parts, releasing nutrients and initiating a detrital food chain, which includes a variety of detrivores and other detritus consumers. The conversion process in any food chain is relatively inefficient, with as much as 90 per cent of the useful energy being lost during the conversion from one level to another, usually in the form of heat. Thus only 1 per cent of the energy available from the grass would be stored in the body of the lion. Aquatic food chains tend to be longer than terrestrial chains, and as a result the amount of the original solar energy that reaches the ultimate consumer is even less. The numbers of organisms in a food chain also tend to decline as the levels increase, creating the so-called pyramid of numbers. Large herds of antelope are required to support only a few lions, for example. Linkages among the elements in the chain are so strong that disruption at one level will be felt along the entire chain. Removal of the producer, for example, will reduce the food supply for the primary consumers, perhaps causing a decline in their numbers by starvation, which in turn reduces the food supply for the predators. At the other end of the chain, a reduction in the numbers of predators will allow the population of the grazing primary consumers to increase, thus placing stress on their food supply. The introduction of rabbits into Australia, for example, created a short food chain with only a limited number of predators. This allowed the rabbit population to increase to such an extent that it competed with other primary consumers, both natural and domesticated, for food. In some places, the abundance of rabbits caused the complete destruction of the vegetation and the initiation of soil erosion. The introduction of the viral disease myxomatosis in the 1950s was so effective in killing off the rabbits that predators such as foxes and eagles turned to other native species, but eventually their populations also slumped. A similar disruption of the food chain occurred following the accidental escape of the experimental rabbit calicivirus in 1995.

Humans are at the upper end of many food chains, and can adapt to the relative inefficiency of the chains. In the developed nations, for example, farmers grow grain to be fed to cattle or pigs. The beef or pork is then eaten by humans. Being omnivorous, however, they could take a more energy-efficient approach, and eat the grain before it is processed by the animals. The resulting food savings could then be made available to the less developed nations where under-nourishment and even malnutrition are major problems. Harvesting the lower levels of a food chain has also been considered in the aquatic environment, where experimental fishing of large zooplankton, such as Antarctic krill, has been carried out. Lack of knowledge of the impact of such harvesting on the oceanic food chains, which depend upon krill, plus negative economic factors have ensured that it has not gone beyond the experimental stage.

Food web A model describing the organisms found in a food chain. Food webs describe the complex patterns of energy flow in an ecosystem by modelling who consumes who.

Foot wall The bottommost surface of an inclined fault.

Force Process that changes the state of rest or motion of a body.

Force of acceleration Force resulting in the speed of a moving body to increase.

Forecast A statement of expected future occurrences. Weather forecasting includes the use of objective models based on certain atmospheric parameters, along with the skill and experience of a meteorologist. Also called a prediction.

Foreset bed Deltaic deposit of alluvial sediment that is angled 5 to 25° from horizontal. Most of the delta is made up of these deposits.

Foreshock Small earth tremors that occur seconds to weeks before a significant earthquake event.

Forest Ecosystem dominated by trees. Major forest biomes include tropical evergreen forest, tropical savannah, deciduous forest, and boreal forest.

Forestry map A forestry map is a map that notes the density, kind, size, and value of the trees in an area.

Forminifera Microscopic organisms of the group protozoa that are found living mainly in marine environments. These organisms produce shells rich in calcium carbonate. Sedimentation and lithification of these shells produces the sedimentary rock chalk.

Formlines Lines, resembling contour lines, drawn to present a conception of the shape of the terrain without regard to a true datum or regular spacing.

Fossil Geologically preserved remains of an organism that lived in the past.

Fossil fuel Carbon based remains of organic matter that has been geologically transformed into coal, oil and natural gas. Combustion of these substances releases large amounts of energy. Currently, humans are using fossil fuels to supply much of their energy needs.

Fractus The elements of cumulus and stratus clouds that appear in irregular fragments, as if they had been shred or torn. Never appears in cirrus clouds. Also known as scud.

Frame A single image or picture. A single complete vertical scan of the cathode ray tube (CRT).

Free radicals Atomic or molecular species with unpaired electrons or an otherwise open shell configuration, usually very reactive. Specific to atmospheric chemistry free radicals are: short-lived, highly reactive, intermediate species produced by dissociation of the source molecules by solar ultraviolet radiation or by reactions with other stratospheric constituents. Free radicals are the key to intermediate species in many important stratospheric chain reactions in which an ozone molecule is destroyed and the radical is regenerated.

Freeze-thaw action Processes associated with daily and seasonal cycles of freezing and melting.

Freezing The change in state of matter from liquid to solid that occurs with cooling. Usually used in meteorology when discussing the formation of ice from liquid water.

Freezing drizzle Drizzle, falling as a liquid, but freezing on impact with the colder ground or other exposed surfaces. It is reported as "FZDZ" in an observation and on the METAR.

Freezing fog Used to describe the phenomena when fog is present and the air temperature is below 0 degrees Celsius. It is reported as "FZFG" in an observation and on the METAR.

Freezing point/freeze The process of changing a liquid to a solid. The temperature at which a liquid solidifies under any given set of conditions. Pure water under atmospheric pressure freezes at 0 degrees Celsius or 32 degrees Fahrenheit. It is the opposite of fusion. In oceanography, the freezing point of water is depressed with increasing salinity.

Freezing precipitation Precipitation that is liquid, but freezes upon impact with a solid surface, such as the ground or other exposed surfaces. Examples include freezing rain and freezing drizzle.

Freezing rain Type of precipitation. Occurs when rain hits a cold surface and freeze. For this to occur a surface temperature inversion is required. In such an inversion, the surface must have a temperature below freezing, while the temperature of the atmosphere where the precipitation forms is above freezing. Surface inversions may develop from a variety of causes, but typically they occur near the leading edge of cold air from the north as it pushes southward.

Frequency (F) Number of cycles and parts of cycles completed per second. $F=1/T$, where T is the length of one cycle in seconds.

Frequency division multiplexing The combining of a number of signals to share a medium by dividing it into different frequency bands for each signal.

Frequency Modulation (FM) The instantaneous variation of the frequency of a carrier wave in response to changes in the amplitude of a modulating signal. As applied to APT the radio signal from the satellite is broadcast on an FM transmitter and received on the ground on an FM radio receiver.

Fresh water Water found rivers, lakes, and rain, that is distinguished from salt water by its appreciable lack of salinity.

Friction In meteorology, it is the turbulent resistance of the earth on the atmosphere. Considered as the resistance of fluids (air and water) to the relative motion of a solid body. The amount is dependent on the size and shape of the body.

Friction layer The thin layer of atmosphere adjacent to the earth's surface. Surface friction is effective in slowing down wind up to approximately 1,500 to 3,000 feet above the ground. Above this level, air tends to flow parallel to the isobars. Wind distribution within this layer is determined by vertical temperature gradient and the physical contours of the underlying surface features. Also known as a surface boundary layer or boundary layer.

Frictional force Force acting on wind near the Earth's surface due to frictional roughness. Causes the deceleration of wind.

Front A boundary between two different air masses. The difference between two air masses sometimes is unnoticeable. But when the colliding air masses have very different temperatures and amounts of water in them, turbulent weather can erupt.

A cold front occurs when a cold air mass moves into an area occupied by a warmer air mass. Moving at an average speed of about 20 mph, the heavier cold air moves in a wedge shape along the ground. Cold fronts bring lower temperatures and can create narrow bands of violent thunderstorms. In North America, cold fronts form on the eastern edges of high pressure systems.

A warm front occurs when a warm air mass moves into an area occupied by a colder air mass. The warm air is lighter; 50 it flows up the slope of the cold air below it. Warm fronts usually form on the eastern sides of low pressure systems, create wide areas of clouds and rain, and move at an average speed of 15 mph.

When a cold front follows and then overtakes a warm front (warm fronts move more slowly than cold fronts) lifting the warm air off the ground, an occluded front forms.

Frontal fog Fog produced when weather fronts, especially warm fronts, pass through an area. Precipitation falling into the colder air ahead of the warm front may evaporate enough water to cause the formation of small droplets as fog near the ground.

Frontal lifting Lifting of a warmer or less dense air mass by a colder or more dense air mass at a frontal transitional zone.

Frontal passage It is considered as the passage of a front over a specific point on the surface. It is reflected by the change in dew point and temperature, the shift in wind direction, and the change in atmospheric pressure. Accompanying a passage may be precipitation and clouds.

Frontogenesis The birth or creation of a front. This occurs when two adjacent air masses exhibiting different densities and temperatures are brought together by prevailing winds, creating a front. It could happen when either air mass, or both, move over a surface which strengthens their original properties. However, it occurs most often along the eastern coasts of North America and Asia, when the air mass moving out over the ocean has a weak or no distinct boundary. The opposite of frontolysis.

Frontolysis The destruction or dying of a front where the transition zone is losing its contrasting properties. The opposite of frontogenesis.

Frost Deposition of ice at the Earth's surface because of atmospheric cooling.

Frost creep Slow mass movement of soil downslope that is initiated by freeze-thaw action. Occurs where the stresses on the slope material are too small to create a rapid failure.

Frost point Is the temperature at which water vapour saturates from an air mass into solid usually forming snow or frost. Frost point normally occurs when a mass of air has a relative humidity of 100 %.

Frost wedging A process of physical weathering in which water freezes in a crack and exerts force on the rock causing further rupture.

Frozen precipitation Precipitation that reaches the ground in a frozen state.

Examples include snow, snow pellets, snow grains, ice crystals, ice pellets, and hail.

Fujita tornado intensity scale Tornado classification system developed by T. Theodore Fujita. This system six levels from F0 to F5. These levels are based on the estimated speed of the tornado's winds from proxy information like property damage.

Fujita Pearson scale A scale that classifies the severity of wind damage intensity based on the degree of destruction as it relates to the wind speed as well as path length and path width of the event. It is normally used to identify the most instense damage exhibited by a tornado. Developed by T. Theodore Fujita and Allen Pearson.

Functional diversity The characteristic of a place where a variety of different activities (economic, political, social) occur; most often associated with urban places.

Fundamental niche Describes the total range of environmental conditions that are suitable for a species existence without the effects of interspecific competition and predation from other species.

Fungi Group Multicellular organisms that have a eukaryotic cell type, mitochondria, and a cell wall composed of chitin and other noncellulose polysaccharides.

Funnel cloud A tornado which is beginning its descent from the base of a cumulonimbus cloud. This severe weather event may or may not reach the ground surface.

Fusion The change of state from a solid to a liquid at the same temperature. The heat of fusion is the number of gram calories of heat necessary to change one gram of a substance from the solid to the liquid state. It is the opposite of freezing.

G

Gabbro An intrusive igneous rock that develops from mafic magma and whose mineral crystals are coarse. Mineralogically this rock is identical to basalt.

Gaia hypothesis The Gaia hypothesis states that the temperature and composition of the Earth's surface are actively controlled by life on the planet. It suggests that if changes in the gas composition, temperature or oxidation state of the Earth are induced by astronomical, biological, lithological, or other perturbations, life responds to these changes by growth and metabolism.

Gain The increase in signal power produced by an amplifier; usually expressed in decibels as the ratio of the output to the input. A measure of the effectiveness of a directional antenna as compared to a non-directional antenna.

Gala hypothesis The hypothesis that the Earth's atmosphere, biosphere, and its living organisms behave as a single system striving to maintain a stability that is conducive to the existence of life.

Galaxy An assemblage of millions to hundreds of billions of stars.

Gale On the Beaufort Wind Scale, a wind with speeds from 28 to 55 knots (32 to 63 miles per hour). For marine interests, it can be categorized as a moderate gale (28 to 33 knots), a fresh gale (34 to 40 knots), a strong gale (41 to 47 knots), or a whole gale (48 to 55 knots). In 1964, the World Meteorological Organization defined the categories as near gale (28 to 33 knots), gale (34 to 40 knots), strong gale (41 to 47 knots), and storm (48 to 55 knots).

Gale warning A warning for marine interests for impending winds from 28 to 47 knots (32 to 54 miles per hour).

Gall-Peters projection Map projection system that reduces the area distortion found in Mercator projections.

Gamete A haploid reproductive cell.

Gamma radiation A type of ionizing, electromagnetic radiation that readily penetrates the body tissues of organisms. Has a wavelength less than 0.03 nanometres.

Gap A spatial opening in a plant community. Can be caused by natural death or by some other abiotic or biotic disturbance.

Gas A state of matter where molecules are free to move in any direction they like. The state of matter where the substance completely fills any container that it occupies.

Gelifluction Form of mass movement in periglacial environment where a permafrost layer exists. It is characterized by the movement of soil material over the permafrost layer and the formation of lobe-shaped features.

Gelisols Soil order (type) of the United States Natural Resources Conservation Service Soil Classification System. This soil is common to high latitude tundra environments. The main identifying feature of this soil is a layer of permafrost within one metre of the soil surface.

Gene(s) Organic material that allows organisms to pass on the inheritance of adaptations or traits. In most organisms these adaptations are coded through the organic molecule DNA. New adaptations appear by way of mutations.

Gene frequency Frequency of alleles at an individual or population level.

Gene pool Sum total of all the genes found in the individuals of the population of a particular species.

General Circulation Models (GCMS) Climate models take various forms, and involve various levels of complexity depending upon the application for which they are designed. GCMs are threedimensional models that provide full spatial analysis of the atmosphere. They incorporate major atmospheric processes plus local climate features predicted through the process of parametreization. The simulations of current and future climates provided by these models require powerful computers capable of processing as many as 200,000 equations at tens of thousands of points in a three-dimensional grid covering the earths surface, and reaching through two to fifteen levels as high as 30 km in to the atmosphere. In addition to these gridpoint models, spectral models have been developed. In these, the emphasis is on the representation of atmospheric disturbances or waves by a finite number of mathematical functions. Many of the more advanced models incorporate this approach. GCMs can be programmed to recognize the role of land and sea in the development of global climates and climate change. Their complex representation of atmospheric processes allows the inclusion of important feedback mechanisms, and they can deal with the progressive change set in motion when one or more of the components of the atmosphere is altered. In an attempt to emulate the integrated nature of the earth/ atmosphere system, atmospheric GCMs have been coupled with other environmental models. Recognizing the major contribution of the oceans to world climatology, the most common coupling is with ocean models. In theory, such models combining the atmospheric and oceanic circulations should provide a more accurate representation of the earths climate. This is not always the case, however. The coupling of the models leads to the coupling of any errors included in the individual models. The so-called model drift which occurs can be treated, but it remains a constraint for coupled models. Another major problem is the difference in time-scales over which atmospheric and oceanic phenomena develop and respond to change. The atmosphere generally responds within days, weeks or months, while parts of the oceans for example, the ocean deeps may take centuries or even millennia to respond. As a result, running a completely interactive coupled ocean-atmosphere model, until all elements reach equilibrium, is time-consuming and costly. Because of this, the oceanic element in most coupled models is much less comprehensive than the atmospheric element. The ocean is commonly modelled as a slab which represents only the uppermost layer of water in which the temperature is relatively uniform with depth. Oceanic heat storage is calculated only from the chosen depth of the layer and other elements such as oceanic heat transport and exchanges with the deeper parts of the ocean are neglected or calculated only indirectly. Thus the accuracy of the results is limited. Sea-ice models, carbon cycle models and chemical models have also been recognized as having the portential to contribute to climate simulation when coupled to existing GCMs. Carbon cycle models, particularly important in studies of global warming, have already been coupled to ocean models, and chemical models have been developed to investigate the influence of other trace gases on the general circulation of the atmosphere. Most of the effort in the development of GCMs has gone into producing equilibrium models. In these, change is introduced into a model which represents existing climate conditions and the model is allowed to run until a new equilibrium is reached. The new model climate can then be compared with the original to establish the overall impact of the change. Most of the numerous GCMs used to study the impact of a doubling of atmospheric carbon dioxide (CO_2) on world climates have been of this type. They make no attempt to estimate changing conditions during the transient phase of the model run, although these conditions may well have important environmental impacts long before equilibrium is reached. The development of transient or time-dependent models which would provide the interim information lagged behind that of equilibrium models, but that discrepancy is now being addressed. In the IPCC second assessment of climate change compiled in 1995, for example, ten experiments were run using transient models compared with only one available for the first assessment five years earlier. Despite the growing sophistication of general circulation models, problems remain, often associated with their inability to deal adequately with elements that are integral to the functioning of the earth/ atmosphere system. These include the roles of clouds, oceans and feedback mechanisms. Clouds are important because of the significant influence which they exert on the earths heat budget through the reflection and absorption of radiation, but they are difficult to simulate, in part because they develop at the regional level, whereas GCMs are global in scale. Parametreization provides only a partial solution, but progress is being made in the provision of regional scale conditions to GCMs for example, through nested models which may improve the representation of clouds. Despite lengthy experimentation with coupled ocean-atmosphere models, representation of the oceans remains a source of uncertainty. Part of the problem is the paucity of observational data, and this reflected to some extent in the relatively low resolution of the ocean components in coupled models. Attempts must be made to address the inadequacy of the data through the development of an oceanic observing network, if the ocean component in models is to be improved. Both clouds and oceans are also involved in feedback mechanisms. The representation of feedbacks is a highly complex but important element in the consideration of climate-related processes. Feedbacks are incorporated in some form in most GCMs, with their number and complexity varying from model to model. The IPCC investigators in 1995 identified feedbacks involving clouds, surface radiation budgets and the carbon cycle in the terrestrial biosphere as particularly important sources of uncertainty. The constraints that this imposes on GCMs must be recognized and appropriate allowances made, when analysing their predictions or incorporating them in policy-making decisions. All models represent a compromise between the complexities of the earth/ atmosphere system and the technical constraints of computer hardware and software. Despite the tens of thousands of hours of computing time that went into the production of the IPCC 1995 reports, for example, the results are far from definitive. Progress in the development of GCMs since 1990 is impressive, however, and with further development and refinement, the courses now being pursued by modellers have considerable potential for narrowing the gap between climate simulation and reality.

Generalist species Species that can survive and tolerate a broad range of environmental conditions.

Genetic adaptation Changes in the genetic makeup of organisms of a species due to mutations that allow the species to reproduce and gain a competitive advantage under changed environmental conditions.

Genetic diversity Genetic variability found in a population of a species or all of the populations of a species.

Genus A group in the classification of organisms. Classification level above the species group. It consists of similar species. Similar genera (plural form of genus) are grouped into a family.

Geocoding The conversion of features found on an analog map into a computer-digital form. In this process, the spatial location of the various features is referenced geographically to a coordinate system used in the computer's software system.

Geodesy Science concerned with the measurement and mathematical description of the size and shape of the earth and its gravitational fields. Geodesy also includes the large-scale, extended surveys for determining positions and elevations of points, in which the size and shape of the earth must be taken into account.

Geographic coordinates Latitude and longitude lines form an imaginary grid over the Earth's surface. By combining longitude and latitude measurements, any location on earth can be determined. The units of measurement for geographic coordinates are degrees (°), minutes (′), and seconds (″). Like a circle, the Earth has 360 degrees. Each degree is divided into 60 minutes, which in turn is divided into 60 seconds. Latitude and longitude coordinates also include cardinal directions: north or south of the equator for latitude, and east or west of the prime meridian for longitude. The geographic coordinates of New York City, for example, are 40° N, 74° W, meaning that it is located 40 degrees north latitude and 74 degrees west longitude. Using minutes and seconds as well as degrees, the coordinates for New York would be 40°42′51" N, 74°0′23" W. (Latitude is always listed first.) A less common format for listing coordinates is in decimal degrees. The Tropic of Cancer, for example, can be expressed in degrees and minutes (23°30' N) or in decimal degrees (23.5° N).

Geographic cycle Theory developed by William Morris Davis that models the formation of river-eroded landscapes. This theory suggests that landscapes go through three stages of development (youth, maturity, and old age) and argues that the rejuvenation of landscapes arises from tectonic uplift of the land.

Geographic Information System (GIS) A system for archiving, retrieving, and manipulating data that has been stored and indexed according to the geographic coordinates of its elements. The system generally can utilize a variety of data types, such as imagery maps, tables, etc.

Geographic range Spatial distribution of a species. The geographic ranges of species often fluctuate over time.

Geographical coordinate system A geographical coordinate system is a system that uses latitude and longitude to describe points on the spherical surface of the globe.

Geography The study natural and human constructed phenomena relative to a spatial dimension.

Geoid True shape of the Earth, which deviates from a perfect sphere because of a slight bulge at the equator.

Geologic map A geologic map is a map that notes the structure and composition of geologic features, like the presence of minerals, rock types, earthquake faults, underground water, and landslide areas.

Geologic time scale 1. Scale used to measure time relative to events of geological significance.

2. Time scale that occurs over millions and billions of years.

Geology The field of knowledge that studies the origin, structure, chemical composition, and history of the Earth and other planets.

Geomorphic threshold The amount of slow accumulated change a landform can take before it suddenly moves into an accelerated rate of change that takes it to a new system state.

Geomorphology The study of the arrangement and form of the Earth's crust and of the relationship between these physical features and the geologic structures beneath.

Geophysics The study of the physics or nature of earth and its environment. It deals with the composition and physical phenomena of the earth and its liquid and gaseous envelopes. Areas of studies include the atmospheric sciences and meteorology, geology, seismology, and volcanology, and oceanography and related marine sciences, such as hydrology. By extension, it includes astronomy and the related astro-sciences.

Geosphere Considered the solid portions of the earth, including the hydrosphere and the lithosphere, as opposed to the atmosphere, which lies above it. At their conjunction is the biosphere.

Geostationary Describes an orbit in which a satellite is always in the same position (appears stationary) with respect to the rotating Earth. The satellite travels around the Earth in the same direction, at an altitude of approximately 35,790km (22,240 statute miles) because that produces an orbital period equal to the period of rotation of the Earth (actually 23 hours, 56 minutes, 04.09 seconds). A worldwide network of operational geostationary meteorological satellites provides visible and infrared images of Earths' surface and atmosphere. The satellite systems include the U.S. GOES, METEOSAT (launched by the European Space Agency and operated by the European Weather Satellite Organization-summerset), the Japanese GMS, and most commercial, telecommunications satellites.

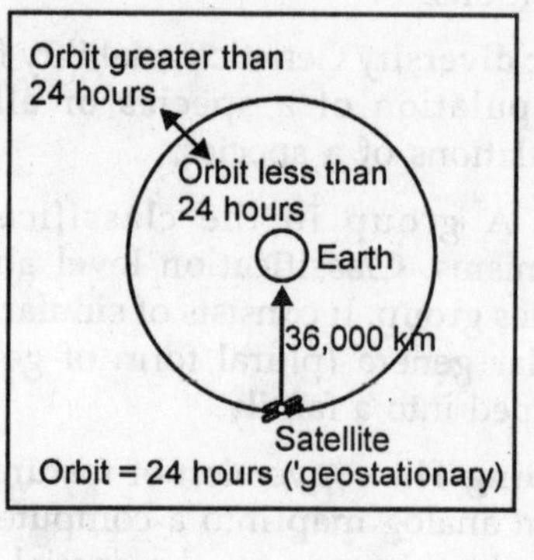

Fig. Geostationary

Geostationary Operational Environmental Satellite (GOES) NASA-developed, NOAA-operated series of satellites that: provide continuous day and night weather observations;

Monitor severe weather events such as hurricanes, thunderstorms, and flash floods;

Relay environmental data from surface collection platforms to a processing centre;

Perform facsimile transmissions of processed weather data to low-cost receiving stations;

Monitor the Earths' magnetic field, the energetic particle flux in the satellite's vicinity and x-ray emissions from the sun;

Detect distress signals from downed aircraft and ships.

GOES observes the U.S. and adjacent ocean areas from vantage points 35,790 km (22,240 miles) above the equator at 75o west and 35o west. GOES satellites have an equatorial, Earth-synchronous orbit with a 24-hour period, a resolution of 8 km, an IR resolution of 4 km, and a scan rate of 1864 statute miles in about three minutes. The transmission of processed weather data (both visible and infrared) by GOES is called weather facsimile (WEFAX). GOES WEFAX transmits at 1691 + MHz and is accessible via a ground station with a satellite dish antenna.

GOES carries the following five major sensor systems: 1. The imager is a multispectral instrument capable of sweeping simultaneously one visible and four infrared channels in a north-to-south swath across an east-to-west path, providing full disk imagery once every thirty minutes.

2. The sounder has more spectral bands than the imager for producing high-quality atmospheric profiles of temperature and moisture. It is capable of stepping one visible and eighteen infrared channels in a north-to-south swath across an east-to-west path.

3. The Space Environment Monitor (SEM) measures the condition of the Earth's magnetic field, the solar activity and radiation around the spacecraft, and transmits these data to a central processing facility.

4. The Data Collection System (DCS) receives transmitted meteorological data from remotely located platforms and relays the data to the end users.

5. The Search and Rescue Transponder can relay distress signals at all times, but cannot locate them. While only the polar-orbiting satellite can locate distress signals, the two types of satellites work together to create a comprehensive search and rescue system.

Geostationary orbit Satellite that has an orbit that keeps it over the same point on the Earth at all times. This is accomplished by having the satellite travel in space at the same angular velocity as the Earth.

Geostationary satellite An orbiting weather satellite that maintains the same position over the equator during the earth's rotation. Also known as GOES, an acronym for Geostationary Operational Environmental Satellite. Compare with a polar-orbiting satellite.

Geostrophic wind A steady horizontal motion of air along straight, parallel isobars or contours in an unchanging pressure or contour field. It is assumed that there is no friction, that the flow is straight with no curvature and there is no divergence or convergence with no vertical acceleration.

Geosynchronous (GEO) Synchronous with respect to the rotation of the Earth.

Geothermal energy Energy available in the molten and semi-molten rocks beneath the earths crust. The high temperatures that this creates in adjacent solid rocks in certain areas causes sub-surface water to be superheated or converted into steam, which can be used for direct space heating or converted into electricity in a conventional power plant. Most existing geothermal plants depend upon the production of naturally heated water, but the expansion of production is technically possible through the injection of cold surface water into hot, dry rocks 3 to 5 km beneath the earths surface. The injected water would be under pressure and become superheated. When returned to the surface and normal pressure it would convert instantly into steam, which would provide the power for the generation of electricity. After condensation the water would be returned again beneath the surface for reheating. Despite having the advantage of supplying apparently free and renewable energy, geothermal zones are being developed only slowly. Development and production costs remain sufficiently high that geothermal energy cannot readily compete with conventional energy sources. Technical problems also exist. For example, the rapid extraction of energy from the geothermal zone may allow the rocks to cool down and reduce the efficiency of the process. In addition, the dissolved gases and minerals present in geothermal fluids can lead to rapid corrosion of the system. The environmental impacts of geothermal power plants are much less than those of conventional plants, but they include the primary problems of noise, odour and groundwater contamination, plus secondary concerns such as the disruption of local hot springs or geysers and even induced seismic activity. Geothermal fields in the USA, Iceland, Italy and New Zealand already contribute to power production, but their impact is largely local and the contribution of geothermal zones to global energy requirements is limited.

Germination The beginning of vegetative growth of a plant from a seed.

Geyser A geyser is a natural hot spring that occasionally sprays water and steam above the ground.

Ghetto Originally, the section of a European city to which Jews were restricted. Today, commonly defined as a section of a city occupied by members of a minority group who live there because of social restrictions on their residential choice.

GIS Is the abbreviation for geographic information system. GIS are special-purpose digital databases in which a common spatial coordinate system is the primary means of reference. GIS contain subsystems for: 1. data input;

2. data storage, retrieval, and representation;

3. data management, transformation, and analysis; and

4. data reporting and product generation. It is useful to view GIS as a process rather than a thing. A GIS supports data collection, analysis, and decision making and is far more than a software or hardware product. Other terms for GIS, and special-purpose GIS, include: Land-Base Information System, Land Record System, Land Information System, Land Management System, Multipurpose Cadastre, and AM/FM System.

Glacial (Glaciation) 1. Period of time during an ice age when glaciers advance because of colder temperatures.

2. Involving glaciers and moving ice. Usually pertaining to processes associated with glaciers.

Glacial drift A generic term applied to all glacial and glaciofluvial deposits.

Glacial ice A very dense form frozen water that is much harder than snow, névé, or firn.

Glacial lake A natural impoundment of meltwater at the front of a glacier.

Glacial milk Term used to describe glacial meltwater which has a light coloured or cloudy appearance because of clay-sized sediment held in suspension.

Glacial polish The abrasion of bedrock surfaces by materials carried on the bottom of a glacier. This process leaves these surfaces smooth and shiny.

Glacial retreat The backwards movement of the snout of a glacier.

Glacial surge A rapid forward movement of the snout of a glacier.

Glacial till The mass of rocks and finely ground material carried by a glacier, then deposited when the ice melted. Creates an unstratified material of varying composition.

Glacial trough A deep U-shaped valley with steep valley walls that was formed from glacial erosion. At the base of many of these valleys are cirques.

Glacial uplift Upward movement of the Earth's crust following isostatic depression from the weight of the continental glaciers.

Glacial valley Valley that was influenced by the presence of glaciers. The cross-section of such valleys tends to be U-shaped because of glacial erosion. Similar to glacial trough.

Glaciation Having been covered with a glacier or subject to glacial epochs.

Glacier A body of snow and ice lying wholly or mostly on land and capable of movement. Glaciers occur when the accumulation of winter snow exceeds summer melting or ablation. When the accumulation is sufficiently great, the base of the snowpack is converted into ice through a process of compaction and recrystallization. The amount of snow that must collect before ice is formed depends upon such factors as local temperature and the density of the snow, and may vary between 13 and 80 m, but once formed, it is the deformation of the basal ice that causes glaciers to move. Once moving they are capable of erosion, transportation and deposition, producing characteristic glacial landforms. Glaciers are commonly classified according to their morphological (e.g. valley glaciers, ice sheets), dynamic (e.g. active fast moving, passive slow moving) or thermal (e.g. temperate, polar) characteristics. Currently, glaciers cover some 10 per cent of the earths land surface, with the ice sheets of Greenland and Antarctica accounting for most of that total, but during the Pleistocene, ice sheets covered much of the northern hemisphere, reaching as far south as the Great Lakes in North America, and southern Britain, Germany and Poland in Europe. The ice retreated from these areas some 10- or 12,000 years ago but they retain typically glaciated landscapes formed at that time.

Glacier winds Air flow that descends from glaciers, occasionally at a high rate of speed. Caused by the temperature difference between the air in contact with the glacier and the air at the same altitude, it reaches maximum intensity in the early afternoon. A form of katabatic wind.

Glaciofluvial Geomorphic feature whose origin is related to the processes associated with glacial meltwater.

Glaze A smooth clear icy coating of supercooled water droplets that spread out and freeze onto objects on contact. A storm that produces the accretion of glaze is called an ice storm.

Gleization A soil formation process that occurs in poorly drained environments. Results in the development of extensive soil organic layer over a layer of chemically reduced clay that takes on a blue colour.

Gleysol soil Soil order (type) of the Canadian System of Soil Classification. This soil type is found in habitats that are frequently flooded or permanently waterlogged. Its soil horizons show the chemical signs of oxidation and reduction.

Global measurement All of the activities required to specify a global variable, such as ozone. These activities range from data acquisition to the generation of a data-analysis product, and include estimates of the uncertainties in that product. A global measurement often will consist of a combination of observations from a spacecraft instrument (required for global coverage) and measurements in situ (needed to provide reference points for long-term accuracy).

Global Positioning System (GPS) System used to determine latitude, longitude, and elevation anywhere on or above the Earth's surface. This system involves the transmission of radio signals from a number of specialized satellites to a hand held receiving unit. The receiving unit uses triangulation to calculate altitude and spatial position on the Earth's surface.

Global variables Functions of space and time that describe the large scale state and evolution of the Earth system. The Earth system's geosphere, hydrosphere, atmosphere, and biosphere and their components are, or potentially are, global variables.

Global warming Since the beginning of the century, there has been a rise in global mean temperatures of about 0.5°C. The basic cause of this warming is seen as the enhancement of the greenhouse effect over the same period brought on by rising levels of anthropogenically produced greenhouse gases. Carbon dioxide (CO_2) emissions have received most attention but methane (CH_4), nitrous oxide (N_2O) and CFCs have also contributed through their ability to retain terrestrial radiation in the atmosphere, and thus produce warming. The observed warming is what might be expected from the rising level of greenhouse gases, but the change has not been consistent. The main increase took place between 1910 and 1940 and again after 1975. Between 1940 and 1975, despite rising greenhouse levels, global temperatures declined. Such variations are not uncommon in the climate record, and the changes that have taken place are well within the range of normal natural variations in global temperatures. However, the warming between the 1960s and 1980s was more rapid than that between the 1880s and 1940, and James Hansen of the Goddard Institute for Space Studies (GISS) claimed in 1988 that the global greenhouse signal is sufficiently strong for a cause-and-effect relationship between the anthropogenically produced carbon dioxide increase and global warming to be inferred. His conclusion was not widely accepted at the time, but in 1996 the IPCC also claimed to recognize a human influence in the current global warming. Estimates of warming are commonly obtained by employing general circulation models, set to provide information on the temperature impact of a set increase in carbon dioxide. By the mid-1980s it was estimated that a doubling of CO_2 would 1990, the IPCC assessment produced values cause an average warming of 1.34C. In of 1.54.5C with a best estimate of 2.5°C. By the time of their second assessment in 1995, their best estimate had been reduced to 2°C with a maximum of 3.5°C, to be reached by 2100. Although the estimated temperatures are not particularly high mainly because they are global averages evidence from past world temperature changes indicates that they are of a magnitude which could lead to significant changes in climate and climate-related activities. During the Climatic Optimum some 5000 to 7000 years ago, temperatures in North America and Europe were only 23°C higher than the present average, but they produced major environmental changes. Evidence from that time period and from another warm spell in the early Middle Ages 800 to 1000 years ago also suggests that the greatest impact of any change will be felt in mid- to high latitudes in the northern hemisphere. Northern Canada, for example, might experience an increase of 812C in mean temperatures in the winter, whereas in lower latitudes in Southern Europe, the Sahel and south-east Asia the predicted increases generally fall between 12°C. Because the various elements in the atmospheric environment are closely interrelated, it is only to be expected that if temperature increases, changes in other elements will also occur. Moisture patterns are likely to be altered, for example. Semiarid grassland areas would likely experience a reduction in precipitation, but some projections indicate that more rainfall is possible in parts of Africa and south-east Asia. Confidence in the ability of computerized models to predict precipitation changes is low, however.

Given the range of possibilities presented in the estimates of future global warming it is difficult to predict its environmental and socioeconomic impacts. However, using a combination of investigative techniques ranging from laboratory experiments with plants to the creation of computer-generated models of the atmosphere and the analysis of past climate anomalies researchers have produced results which provide a general indication of what the consequences might be in certain key sectors. Warming would bring about changes in the length and intensity of the growing season which, coupled with new moisture regimes would disrupt existing vegetation patterns, both natural and cultivated. Across the northern regions of Canada, Scandinavia and Russia, the trees of the boreal forest would begin to colonize the tundra as they have done during warmer spells in the past. Along its southern margin, however, they would come under pressure from the pole ward spread of deciduous trees and grassland. Higher temperatures might threaten the forests indirectly through an increase in the frequency of forest fires and insect infestations, and if the rate of change takes place more rapidly than the forest can respond, rapid die-off of large numbers of trees is a possibility. Whatever the final outcome, it appears likely that global warming in the northern forest would disrupt the northern ecosystems, and in turn have a significant impact on those countries, such as Canada, Sweden, Finland and Russia, where national and regional economies depend very much on the harvesting of softwoods from the boreal forest. In lower latitudes where the temperature element is less dominant, the impact of global warming would be felt through changes in the amount and distribution of moisture. In northern Australia, for example, the increased pole ward penetration of monsoon rains would allow the expansion of tropical and subtropical vegetation. In China, agricultural output is expected to rise as a result of the northward spread of crops such as rice, corn and cotton, and the increase in the area in which tropical and subtropical fruits can be cultivated. Of major concern to many agriculturalists is the expected decrease in precipitation in the worlds grain-producing areas, which along with higher temperatures, could increase the frequency and severity of drought in the grain belts of North America, Russia and Ukraine, leading to more frequent crop failures. Martin Parry a leading analyst of the agricultural implications of global warming has predicted a decline in grain yields of 1520 per cent in Africa, tropical Latin America and much of India and south-east Asia by the middle of the twenty-first century. Another major concern associated with water is a rising sea level brought about by the thermal expansion of the oceans and the additional water added to the sea by the melting of snow and ice as warming progresses. The IPCC has predicted a rise of 3855 cm by the year 2100, but other estimates suggest that an increase of as much as a metre is possible. Even with the lower estimates, coastal regions would experience increased flooding, greater coastal erosion and the disruption of economic activities. For some nations occupying islands that are currently only a few metres above sea level (the Alliance of Small Island States, for example), the predicted rise would be disastrous. Some, such as the Maldive Islands in the Indian Ocean, would become uninhabitable, particularly if the increased storminess predicted by some models came to pass. The range of potential impacts is great, but in many cases the reality of the situation may only become apparent when the changes have occurred, for there are many variables involved in the predictions. The human factors, as always, are particularly unpredictable. Technology, politics, socioeconomic conditions and even demography can influence warming, for example, through their contribution to changes in greenhouse gas concentrations, yet the nature and magnitude of the variations in such elements is almost impossible to forecast. Despite such uncertainties and a certain degree of dissension among scientists and politicians, the international view as expressed by the Framework Convention on Climate Change is to see global warming as a serious threat to the environment and society in the future, which must be dealt with now. Attempts to arrest continued warming have centred mainly on the control and reduction of greenhouse gas emissions, but the degree of success is as yet limited.

Globe A true-to-scale map of the Earth that duplicates its round shape and correctly represents areas, relative size and shape of physical features, distances, and directions.

Glucose Simple six-carbon sugar. The chemical formula for glucose is $C_6H_{12}O_6$.

Gneiss A metamorphosed coarse grained igneous rock. In this rock you get the recrystallization of quartz, feldspars, micas and amphiboles into bands.

GOES (Geostationary Operational Environmental Satellite) Series of geostationary meteorological satellites launched by the United States starting in 1968. The main purpose behind these satellites was to use a variety of remote sensing devices for weather forecasting and environmental monitoring.

GPS Is an acronym for global positioning system. Developed for the military for navigation and surveying, the GPS relies on satellites (and ground stations) for precise determination of location. Although GPS can be used to determine location very precisely (within centimetres given the correct controls and proper use), it does not solve all the problems of locational determination in GIS databases.

Graben fault This fault is produced when tensional stresses result in the subsidence of a block of rock. On a large scale these features are known as Rift Valleys.

Graded stream A stream that has a long profile that is in equilibrium with the general slope of the landscape. A graded profile is concave and smooth. Stream's maintain their grade through a balance between erosion, transportation, and deposition. Erosion removes material from bumps in the profile and deposition fills in dips.

Gradient The steepness of a slope as measured in degrees, percentage, or as a distance ratio (rise/run).

Gradient wind Horizontal wind in the upper atmosphere that moves parallel to curved isobars. Results from a balance between pressure gradient force, Coriolis force, and centripetal force.

Granite Medium to coarse grained igneous rock that is rich in quartz and potassium feldspar. Derived from felsic magma.

Granitic magma Felsic magma that generates mainly granite rocks.

Graphic scale Way of expressing the scale of a map with a graphic.

Grass Type of plant that has long slender leaves that extend from a short stem or the soil surface.

Grasslands Regions in which the climax vegetation is dominated by grasses, but may contain a variety of annual and perennial plants. Grasslands form one of the earths major biomes. They occur in both temperate and subtropical regions, and are naturally associated with areas where the available moisture is insufficient to support trees. Even where sufficient moisture may be available for the growth of woody plants, for example, close to the grassland margins, the presence of grazing animals or natural fires tends to favour the survival of the grasses. Large grazing animals are the main fauna in grassland areas, from the bison of North America to the various kinds of antelope in Africa and the kangaroos in Australia. Associated with these are the predators such as lions, cheetahs, wolves and hyenas which prey on the grazing animals. Smaller burrowing mammals such as rabbits and prairie dogs, which can easily work the grassland soil are also characteristic of the worlds grasslands. In both temperate and subtropical grassland the return of nutrients and humus to the soil when the grass dies and the enrichment provided by the droppings of grazing animals tends to produce fertile soils. Given the low levels of moisture characteristic of such areas, however, overgrazing or cultivation can make them susceptible to soil erosion and even desertification. Most grassland areas are no longer natural. Native grasses have been replaced by cultivated species to improve grazing conditions, or by grains such as wheat or maize. In many areas, grasslands have been created by clearing forests or improving wasteland for example, draining swamps where the potential agricultural return warrants it.

Graticule, geographic System of coordinates of latitude and longitude used to define the position of a point on the surface of the Earth with respect to the reference spheroid.

Graupel A form of frozen precipitation consisting of snowflakes or ice crystals and supercooled water droplets frozen together. Also known as snow pellets.

Gravel A term used to describe unconsolidated sediments composed of rock fragments. These rock fragments have a size that is greater than 2 millimetres.

Gravitation The mutual attraction between two masses of matter. The rotation of the earth and the atmosphere modifes this attraction to produce the field of gravity.

Gravitational water Water that moves through soil due to gravitational forces. Soil water in excess of hygroscopic water and capillary water.

Gravity Is the process where any body of mass found in the universe attracts other bodies with a force proportional to the product of their masses and inversely proportional to the distance that separates them. First proposed by Sir Issac Newton in 1686.

Grayscale Environmental satellite scanners, rather than photographing a scene, scan a scene line-byline measuring light or heat levels and transmitting this information as a video image via an amplitude modulated (AM) subcarrier contained in the satellite's FM signal. The video image-a 2400 Hz tone-is amplitude modulated to correspond to the light and dark areas sensed, with the louder portion of the tone representing the lighter areas.

Grazing food chain Model describing the trophic flow of organic energy in a community or ecosystem.

Great circle An imaginary circle drawn on the Earth's surface that has its centre synchronize to the centre of the planet. The equator is a great circle.

Great circle route The shortest distance between two places on the Earth's surface. The route follows a line described by the intersection of the surface with an imaginary plane passing through the Earth's centre.

Green flash A brilliant green colouration of the upper edge of the sun, occasionally seen as the sun's apparent disk is about to set below a clear horizon.

Greenhouse effect Process by which significant changes in the chemistry of Earth's atmosphere may enhance the natural process that warms our planet and elevates temperatures. If the effect is intensified and Earth's average temperatures change, a number of plant and animal species could be threatened with extinction.

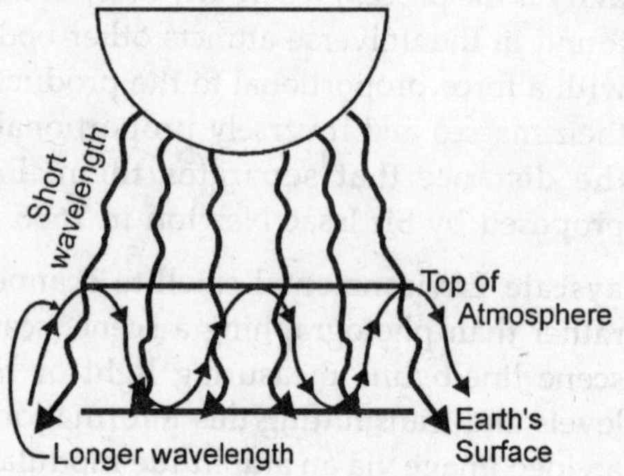

Fig. Greenhouse effect

Certain gaseous components of the atmosphere, called greenhouse gases, transmit the visible portion of solar radiation but absorb specific spectral bands of thermal radiation emitted by the Earth. The theory is that terrain absorbs radiation, heats up, and emits longer wavelength thermal radiation that is prevented from escaping into space by the blanket of carbon dioxide and other greenhouse gases in the atmosphere. As a result, the climate warms.

Because atmospheric and oceanic circulations play a central role in the climate .of the Earth, improving our knowledge about their interaction becomes essential.

Greenhouse gas A gaseous component of the atmosphere contributing to the greenhouse effect. Greenhouse gases are transparent to certain wavlengths of the suns' radiant energy allowing them to penetrate deep into the atmosphere or all the way into the Earths' surface. Greenhouse gases and clouds prevent some of infrared radiation from escaping, trapping the heat near the Earths' surface where it warms the lower atmosphere. Alteration of this natural barrier of atmospheric gases can raise or lower the mean global temperature of the Earth.

Greenhouse gases include carbon dioxide. Methane, nitrous oxide, chlorofluorocarbons, and water vapour Carbon dioxide, methane, and nitrous oxide have significant natural and human sources while only industries produce chlorofluorocarbons. Water vapour has the largest greenhouse effect, but its concentration in the troposphere is determined within the climate system. Water vapour will increase in response to global warming, which in turn may further enhance global warming.

Greenwich Mean Time (GMT) Former standard world time as measured at Greenwich, England (location of the Prime Meridian). Replace in 1928 with Universal Time (UT).

Grid A pattern of lines on a chart or map, such as those representing latitude and longitude, which helps determine absolute location.

Grid north The direction north as measured on the Universal Transverse Mercator grid system.

Grid south The direction south as measured on the Universal Transverse Mercator grid system.

Gross feature map Map that displays geographic characteristics rather than political boundaries.

Gross primary productivity Total amount of chemical energy fixed by the processes of photosynthesis.

Gross secondary productivity Total amount of chemical energy assimilated by consumer organisms.

Gross sediment transport The total amount of sediment transported along a shoreline in a specific time period.

Ground clutter A pattern of radar echoes reflecting off fixed ground targets such as buildings or hills near the radar. This may hide or confuse the proper return echo signifying actual precipitation.

Ground control (points) Identifiable points on the ground whose locations on the surface of the Earth are accurately known for use as geodetic references in mapping, charting, and other related menstruation applications.

Ground fog Fog created when radiational cooling at the earth's surface lowers the temperature of the air near the ground to or below its initial dew point. Primarily takes place at night or early morning. Also known as radiation fog.

Ground frost Frost that penetrates the soil surface in response to freezing temperatures.

Ground ice General term used to describe all bodies of ice in the ground surface of the permafrost layer. Also called anchor ice. Some forms of ground ice include: pore ice, needle ice, ice wedge, segregated ice, sand wedge, and ice lenses.

Ground moraine A thick layer of till deposited by a melting glacier.

Ground track The inciination of a satellite, together with its orbital altitude and the period of its orbit, creaies a track defined by an imaginary line connecting the satellite and the Earth's centre; The intersection on the line with the Earth's surface is the subsatellite point. As the Earth turns on its axis and the satellite orbits overhead, a line is created by the satellite's apparent path over the ground (the series of subsatellite points connected). A geostationary satellite has an inclination of essentially zero, and, because its orbital period exactly matches the Earth's rotation, its ground track is reduced to an apparent stationary point on the equator.

Groundwater The water that accumulates in the pore spaces and cracks in rocks beneath the earths surface. It originates as precipitation and percolates down into sub-surface aquifers. The upper limit of groundwater saturation is the water table. Groundwater moves under the influence of gravity, although usually only slowly, and may return to the surface naturally for example, through springs. Increasingly it is pumped from wells and boreholes for human use. The rate of withdrawal commonly exceeds the rate of recharge, and in many areas the groundwater supply is declining rapidly. In the United States, most of the major aquifers are not being recharged by precipitation as rapidly as they are being exploited and consequently are overdrawn. The withdrawal rate from the Ogallala Aquifer beneath the Great Plains in the United States is estimated to be 100 times the recharge rate, and the rapidly dropping water table means that water is being pumped to the surface from depths in excess of 1800 m. Similar problems of aquifer depletion are occurring in other areas such as Saudi Arabia, China and India, where surface water is in short supply, and groundwater is essential to meet local needs. In Mexico City, the demand is so high that the water-table is declining at a rate of more than 3 m per year. Another threat to groundwater supplies comes from contamination. Petroleum and chemical spills frequently seep into groundwater systems, and use of excess fertilizers and pesticides also contributes to pollution in some areas.

Groundwater flow Underground topographic flow of groundwater because of gravity.

Groundwater recharge The replenishment of groundwater with surface water.

Growing season Considered the period of the year during which the temperature of cultivated vegetation remains sufficiently high enough to allow plant growth. Usually considered the time period between the last killing frost in the spring and the first killing frost of the autumn. The frost-free growing season is between the first and last occurrence of 32 degrees Fahrenheit temperatures in spring and autumn.

Growing season The period from the average date of the last frost (in the United States, this occurs in the spring) to the first frost in the fall.

Guided wave Electromagnetic or acoustic wave that is constrained within certain boundaries, as in a wave guide (transmission line).

Gulf A gulf is a part of the ocean (or sea) that is partly surrounded by land (it is usually larger than a bay).

Gulf stream The warm, well-defined, swift, relatively narrow ocean current which exists off the east coast of the United States, beginning near Cape Hatteras. The term also applies to the oceanic system of currents that dominate the western and northern Atlantic Ocean: the Florida current, which flows through the Florida Straits between the Florida Keys and Cuba and northwards; the Gulf Stream, which begins around Cape Hatteras and flows northeasterly off the continental slope into the North Atlantic; and the North Atlantic current, which begins around the Grand Banks off Newfoundland and continues east-northeastwards towards the British Isles.

Gully washer A heavy rain shower that occurs suddenly, possibly creating a flash flood.

Gust A sudden significant increase in or rapid fluctuations of wind speed. Peak wind must reach at least 16 knots (18 miles per hour) and the variation between peaks and lulls is at least 10 knots (11.5 miles per hour). The duration is usually less twenty seconds.

Gust front A boundary found ahead of a thunderstorm that separates cold storm downdrafts from warm humid surface air. Winds in this phenomenon are strong and fast.

Gustnado A weak, and usually short-lived, tornado that forms along the gust front of a thunderstorm, appearing as a temporary dust whirl or debris cloud.

Gymnosperm Plant that bears naked seeds. Representatives of this group include the conifers.

Gypsum Sedimentary rock created by the chemical precipitation of calcium, sulphur, and oxygen.

Gyre Arrangement of surface ocean currents into a large macro-scale circular pattern of flow.

H

Habitat The specific environment in which an organism lives. Although the term may be linked with a particular species for example, the habitat best suited for elephants any given environment will be shared by a variety of organisms that have requirements in common, or depend upon other organisms in the habitat. Habitats range in scale, from worldwide to continental, to regional or local and even microscopic. The nature of any habitat is determined by a large number of variables, but most can be grouped into climatic, topographic, edaphic and biotic factors. Of these, the climatic factors for example, light, heat, moisture and wind are generally considered to be of most importance, particularly at larger scales. The others become more important at a regional or local scale, sometimes in combination with climate. The combination of topography and climate, for example, leads to the zonation of vegetation with altitude, with a consequent zonation in habitats. Edaphic factors include elements of soil structure, texture, chemistry and moisture characteristics. Biotic factors include the impact of animal grazing and, increasingly, human interference. The combination of all these elements determines the resource availability in a specific habitat and therefore its carrying capacity. Habitats tend to include a natural patchiness, brought about by local variations in topographical or edaphic factors, for example, and they change naturally as a result of normal environmental dynamics. However, the response to change may differ from one part of a habitat to another and as a result a specific habitat such as a forest may include a mosaic of areas in different stages of development. Changes may be minor or temporary for example, the result of a forest fire or much more dramatic and long lasting for example, following the major climate change which produced the ice ages. The impact on organisms also varies, from minor to catastrophic, with the complete loss of habitat perhaps bringing about the extinction of a particular species. Increasingly, habitat change or loss is being caused by human activities such as deforestation, development of agriculture, industrialization and urbanization. In many parts of the world the nature of the local or regional habitat is anthropogenically determined, and even in areas generally considered natural, such as the Arctic, the human imprint is present.

Haboob Sudanese name for duststorm or sandstorm with strong winds that carry small particles of dirt or sand into the air, particularly severe in areas of drought.

Hachure Any series of lines used on a map to indicate the general direction and steepness of slopes. The lines are short, heavy, and close together for steep slopes; longer, lighter, and more widely spaced for gentle slopes.

Hadean Geologic eon that occurred from 3800 to 4600 million years ago. The Earth's oldest rocks date to the end of this time period.

Hadley cell Three-dimensional atmospheric circulation cell located at roughly 0 to 30° North and South of the equator. The Hadley cell consists of rising air (intertropical convergence zone) at the equator and descending air (subtropical highs) at 30° North and South.

Hail Hail is a destructive form of precipitation that is 5 to 190 millimetres in diametre. The large downdrafts in mature thunderstorm clouds provide the mechanism for hail formation. Hailstones normally have concentric shells of ice alternating between those with a milky appearance and those that are clear. The milky white shells, containing bubbles and partially melted snowflakes, correspond to a period of rapid freezing, while the clear shells develop as the liquid water freezes much more slowly.

Hair hygrometre Hygrometre that uses the expansion and contraction of hair to determine atmospheric humidity.

Half-life Time required for one half of the nuclei in a radioisotope to emit its radiation. Half-lifes for radioisotopes range from a few millionths of a second to several billion years.

Halftone A picture in which the gradations of light are obtained by the relative darkness and density of tiny dots produced by photographing the subject through a fine screen.

Halite Sedimentary rock created by the chemical precipitation of sodium and chlorine.

Halo The ring of light that seems to encircle the sun or moon when veiled by cirrus clouds. To produce this phenomena, the ice crystals must be in a heterogeneous arrangement to refract the sunlight. The most commonly observed is a halo that forms at a 22 degrees radius, although another one at 46 degrees radius may also be seen.

Hamada A very flat desert area of exposed bedrock.

Hanging valley A secondary valley that enters a main valley at an elevation well above the main valley's floor. These features are result of past erosion caused by alpine glaciers. Hanging valleys are often the site of spectacular waterfalls.

Hanging wall The topmost surface of an inclined fault.

Haploid Cell that contains only one set of chromosomes.

Hardpan Impervious layer found within the soil. It can result from the precipitation of iron, illuviation of clay or the cementing of sand and gravel by calcium carbonate precipitates.

Hardware The electrical and mechanical components of a system, as opposed to software.

Hazard Phenomenon which can cause loss of life, injury, disease, economic loss, or environmental damage.

Haze A suspension of fine dust and/or smoke particles in the air. Invisible to the naked eye, the particles reduce visibility by being sufficiently numerous to give the air an opalescent appearance. It is reported as "HZ" in an observation and on the METAR.

Headlands A strip of land that juts seaward from the coastline. This feature normally bordered by a cliff.

Headwaters Upper portion of stream's drainage system.

Heat A form of energy transferred between two systems by virtue of a difference in temperature. The first law of thermodynamics demonstrated that the heat absorbed by a system may be used by the system to do work or to raise its internal energy.

Heat balance The equilibrium which exists on the average between the radiation received by the earth and atmosphere from the sun and that emitted by the earth and atmosphere. The balance between heat loss (long wave radiation from the earth back into the atmosphere) and heat gain (incoming solar radiation).

Heat capacity Is the ratio of the amount of heat energy absorbed by a substance compared to its corresponding temperature rise.

Heat Capacity Mapping Mission (HCMM) A two-channel radiometer launched by NASA to measure the thermal properties of the terrestrial surface. It had an application to identify and locate rocks and minerals. One radiometer channel was in the visible to near infrared (0.5-1.1 micrometers), and the other in the thermal infrared (10.5-12.5 micrometers). The instantaneous field of view (IEOV) was about 600 metres.

Heat energy A form of energy representing aggregated internal energy of motions of atoms and molecules in a body.

Heat exhaustion Introduced to the body by overexposure to high temperatures, particularly when accompanied by high humidity. The body has difficulty in cooling the body. Signs of heat exhaustion include a general weakness, heavy sweating and clammy skin, dizziness and/or fainting, and muscle cramps.

Heat index The combination of air temperature and humidity that gives a description of how the temperature feels. This is not the actual air temperature. For an example, check out the heat index chart.

Heat island The dome of relatively warm air which develops over the centre of urbanized areas.

Heat lightning Lightning that appears as a glowing flash on the horizon. It is actually lightning occurring in distant thunderstorms, just over the horizon and too far away for thunder to be heard.

Heat stroke Introduced to the body by overexposure to high temperatures, particularly when accompanied by high humidity. The signs of heat stroke include when an individual's body temperature is greater than 105 degrees Fahrenheit, the skin is hot and dry, there is a rapid and irregular pulse, perspiration has stopped, and one has lost consciousness. May be called a sun-stroke when caused by direct exposure to the sun.

Heat wave A period of abnormally and uncomfortably hot and usually humid weather. It could last from several days to several weeks. The Weather Channel uses the following criteria for a heat wave: a minimum of ten states with 90 degree plus temperatures and the temperatures must be at least five degrees above normal in parts of that area for at least two days or more.

Heating degree day One heating degree day is given for each degree that the daily mean temperature is below 65 degrees Fahrenheit. It is used as an indication of fuel consumption. Refer to degree day or cooling degree day.

Heavy industry Manufacturing activities engaged in the conversion of large volumes of raw materials and partially processed materials into products of higher value; hallmarks of this form of industry are considerable capital investment in large machinery, heavy energy consumption, and final products of relatively low value per unit weight.

Helical flow Movement of water within a stream that occurs as spiral flows.

Helicity A property of a moving fluid, such as air, representing the potential for helical flow (flow that follows a corkscrew pattern). Computed from the vertical wind profile of the lower atmosphere and measured relative to the motion as a storm, it is used to forecast the formation of mesocyclones.

Hemisphere A hemisphere is half the Earth's surface. The four hemispheres are the Northern and Southern hemispheres, divided by the equator (0° latitude), and the Eastern and Western hemispheres, divided by the prime meridian (0° longitude) and the International Date Line (180°).

Herb A nonwoody angiosperm whose above ground vegetation dies off seasonally.

Herbivore Heterotrophic organism that consumes plants for nutrition. Also known as a primary consumer.

Heredity The transmission of behavioural, physiological and morphological characteristics from parent to offspring.

Hertz The international unit of frequency equal to one cycle per second. Radio frequencies are usually expressed in kilohertz/kHz (1,000 cycles per second) or megahertz/MHz (1,000,000 cycles per second).

Hertzian waves Radio waves or other electromagnetic radiation resulting from the oscillations of electricity in a conductor.

Heterogeneity State of being dissimilar or diverse.

Heterosphere The upper layer in a two part classification of the atmosphere based on the general homogeneity of chemical composition. In this layer, oxygen atoms and nitrogen molecules dominate and remain constant in their relative quantities. The heterosphere extends upward from a height of 80 to 100 kilometres depending on latitude. Below this layer is the homosphere.

Heterotroph Organism that must consume energy rich organic molecules for survival. Energy is released from these molecules through the chemical process respiration.

Hierarchical database stores related information in terms of pre-defined categorical relationships in a "tree-like" fashion. Information is traced from a major group, to a subgroup, and to further subgroups. Much like tracing a family tree, data can be traced through parents along paths through the hierarchy. Users must keep track of the hierarchical structure in order to make use of the data. The relational database provides an alternative means of organizing datasets.

High A digital logic state corresponding to a binary "1.".

High clouds A term used to signify cirriform clouds that are composed of ice crystals and generally have bases above 20,000 feet. The main types of high clouds are cirrus, cirrocumulus, and cirrostratus. This altitude applies to the temperate zone. In the polar regions, these clouds may be found at lower altitudes. In the tropics, the defining altitudes for cloud types are generally higher.

High latitudes The latitude belt roughly between 60 and 90 degrees North and South. Also referred to as the polar region.

High pressure An area of atmospheric pressure within the Earth's atmosphere that is above average. If this system is on the Earth's surface and contains circular wind flow and enclosed isobars it is called an anticyclone.

High pressure system An area of relative pressure maximum that has diverging winds and a rotation opposite to the earth's rotation. This is clockwise the in Northern Hemisphere and counter clockwise in the Southern Hemisphere. Also known as an anticyclone, it is the opposite of an area of low pressure or a cyclone.

High water Maximum height reached by a rising tide. The height may be due solely to the periodic tidal forces or it may have superimposed upon it the effects of prevailing meteorological conditions. Use of the "high tide" is discouraged.

High water line Intersection of the land with the water surface at an elevation of high water.

High water mark Line or mark left upon tidal flats, beach, or along shore objects indicating the elevation or the intrusion of high water.

Hill A hill is a raised area or mound of land.

Hinterland The area tributary to a place and linked to that place through lines of exchange, or interaction.

Histosols Soil order (type) of the United States Natural Resources Conservation Service Soil Classification System. This soil is composed more than 30 % organic matter as a result of saturated environmental conditions.

Hoarfrost Another name for frost. A deposit of hoarfrost occurs when air with a dew point below freezing is brought to saturation by cooling.

Holistic Concerned with complete system.

Holocene epoch Period of time from about 10,000 years ago to today. During this period glaciers retreated because of a warmer global climate. Time of modern humans.

Homeostatic (Homeostasis) A constant or non-changing state of equilibrium in a system despite changes in external conditions.

Homosphere The lower layer in a two part classification of the atmosphere based on the general homogeneity of chemical composition. In this layer, nitrogen, oxygen, argon, carbon dioxide, and the trace gases dominate and remain constant in their relative proportions. The homosphere extends from the Earth's surface to a height of 80 to 100 kilometres depending on latitude. Above this layer is the heterosphere.

Hook echo A radar reflectivity pattern observed in a thunderstorm, appearing like a fish hook and indicating favourable conditions for tornado development. However, hook echoes and tornadoes do not always accompany each other.

Horizon One of several lines or planes used as reference for observation and measurement relative to a given location on the surface of the earth. The geographic horizon, also called the apparent horizon, is the distant line along which earth and sky appear to meet. This is the usual concept of horizon and is used in weather observing. The local horizon is the actual lower boundary of the observed sky or the upper outline of terrestrial objects including nearby natural obstructions, such as mountains.

Horn Pyramidal peak that forms when several cirques erode a mountain from three or more sides.

Horse latitudes Located between 30 degrees North and South in the vicinity of the equator, this area typically has calm or light and variable winds. Another name for the equatorial trough, the Intertropical Convergence Zone (ITCZ), or the doldrums.

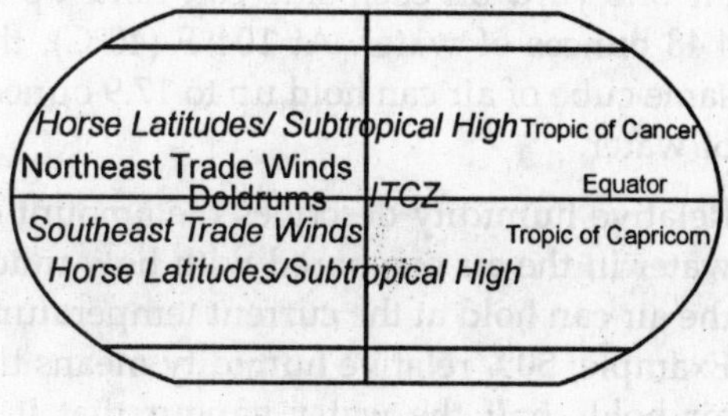

Fig. Horse latitudes

Horst fault A fault that is produced when two reverse faults cause a block of rock to be push up.

Host Organism that develops disease from a pathogen or is being feed on by a parasite.

Hot spot A volcanic area on the surface of the Earth created by a rising plume of magma.

Hudson bay low An area of low pressure over or near the Hudson Bay area of Canada that often introduces cold air to the north central and northeast United States.

Human geography Field of knowledge that studies human-made features and phenomena on the Earth from a spatial perspective. Subdiscipline of Geography.

Human land tradition Academic tradition in modern Geography that investigates human interactions with the environment.

Humbolt current Also known as the Peru Current, this ocean current flows northward along the western side of South America, offshore Chile and Peru. There is considerable upwelling of the colder subsurface waters due to the prevailing southerly winds. Dominate weather in this area includes coastal fog and low clouds. The presence or lack of this current is a vital part of the weather pattern known as El Nino.

Humidity The amount of water vapour in the air. The higher the temperature, the greater the number of water molecules the air can hold. For example: at 60ºF (15ºC), a cube of air one yard on each side can hold up to 4.48 ounces of water. At 104ºF (40ºC), the same cube of air can hold up to 17.9 ounces of water.

Relative humidity describes the amount of water in the air compared with how much the air can hold at the current temperature. Example: 50% relative humidity means the air holds half the water vapour that it is capable of holding; 100% relative humidity means the air holds all the water vapour it can. At 100% humidity no more evaporation can occur until the temperature rises, or until the water vapour leaves the air through condensation. Absolute humidity is the ratio of the mass of water vapour present in a system of moist air to the volume occupied by the mixture, that is, the density of water vapour.

Humus Dark coloured semi-soluble organic substance formed from decomposition of soil organic matter.

Hurricane Severe tropical storms whose winds exceed 74 mph. Hurricanes originate over the tropical and subtropical North Atlantic and North Pacific oceans, where there is high humidity and light wind. These conditions prevail mostly in the summer and early tall. Since hurricanes can take days or even weeks to form, time is usually available for preventive or protective measures. From space, hurricanes look like giant pinwheels, their winds circulating around an eye that is between 5 and 25 miles in diameter. The eye remains calm with light winds and often a clear sky. Hurricanes may move as fast as 50 mph, and can become incredibly destructive when they hit land. Although hurricanes lose power rapidly as soon as they leave the ocean, they can cause high waves and tides up to 25 feet above normal. Waves and heavy flooding cause the most deaths during a hurricane. The strongest hurricanes can cause tornadoes.

Hurricane warning A formal advisory issued by forecasters at the National Hurricane Centre when they have determined that a tropical storm or hurricane may threaten a coastal area or group of islands within a 24 hour period. A warning is used to inform the public and marine interests of the storm's location, intensity, and movement.

Hurricane watch A formal advisory issued by forecasters at the National Hurricane Centre when they have determined that a tropical storm or hurricane may threaten a coastal area or group of islands within a 24 to 36 hour period. A watch is used to inform the public and marine interests of the storm's location, intensity, and movement.

Hydration A form of chemical weathering that involves the rigid attachment of H^+ and OH^- ions to the atoms and molecules of a mineral.

Hydraulic gradient The slope of the water table or aquifer. The hydraulic gradient influences the direction and rate of groundwater flow.

Hydrocarbon Organic compound composed primarily of hydrogen and carbon atoms. An example of a hydrocarbon is methane (CH_4)

Hydrochloroflourocarbon (HCFC) One of a class of compounds used primarily as a CFC substitute. Work on CFC alternatives began in the late 1970s after the first warnings of CFC damage to stratospheric ozone. By adding hydrogen to the chemical formulation, chemists made CFCs less stable in the lower atmosphere enabling them to break down before reaching the ozone layer. However, HCFCs do release chlorine and have contributed more to atmospheric chlorine build-up than originally predicted. Development of non-chlorine based chemical compounds as a substitute for CFCs and HCFCs continues.

Hydrograph A graph describing stream discharge over time.

Hydrographic survey Survey of water area, with particular reference to submarine relief, and any adjacent land.

Hydrography Science that deals with the measurement and description of the physical features of the oceans, seas, lakes, rivers, and their adjoining coastal areas, with particular reference to their use for navigation.

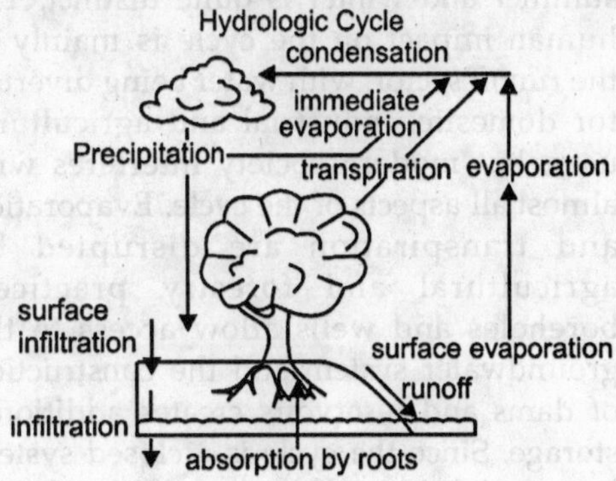

Fig. Study of water systems

Hydrologic cycle A complex group of processes by which water in its various forms is circulated through the earth/ atmosphere system. It is powered by solar radiation which provides the energy to maintain the flow through such processes as evaporation, transpiration, precipitation and runoff. Short- and long-term storage of

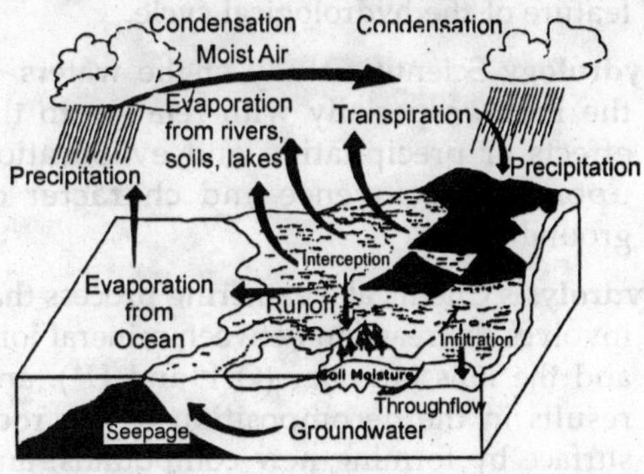

Fig. Hydrological Cycle

water in lakes, oceans, ice sheets and the groundwater reservoir is also part of the cycle. Although it is usual to consider the hydrological cycle as one all-encompassing system, there are in fact many regional and seasonal variations that may have environmental or societal implications. The relationship between the various elements in the cycle in a hot, semi-arid area, for example, is very different from that in a cool, moist location. Similarly, in higher latitudes the difference in the cycle between summer and winter is quite distinct. The human impact on the cycle is mainly in the runoff sector, with water being diverted for domestic, industrial and agricultural uses, but modern society interferes with almost all aspects of the cycle. Evaporation and transpiration are disrupted by agricultural and forestry practices, boreholes and wells allow access to the groundwater system and the construction of dams and reservoirs creates additional storage. Since the cycle is a closed system in material terms, human activities do not deplete the entire system, but excess withdrawal from the runoff or groundwater sectors can create local shortages of water. Most human uses involve only short-term withdrawal from the system, but when the water is returned its quality is often much impaired by a variety of pollutants. The cleansing of such polluted water, mainly through evaporation and precipitation, is an important but less well-acknowledged feature of the hydrological cycle.

Hydrology Scientific study of the waters of the Earth, especially with relation to the effects of precipitation and evaporation upon the occurrence and character of ground water.

Hydrolysis Chemical weathering process that involves the reaction between mineral ions and the ions of water (OH^- and H^+), and results in the decomposition of the rock surface by forming new compounds, and by increasing the pH of the solution involve through the release of the hydroxide ions.

Hydrometeor Any any form of atmospheric water vapour, including those blown by the wind off the earth's surface. Liquid or solid water formation that is suspended in the air includes clouds, fog, ice fog, and mist. Drizzle and rain are example of liquid precipitation, while freezing drizzle and freezing rain are examples of freezing precipitation. Solid or frozen precipitation includes ice pellets, hail, snow, snow pellets, snow grains, and ice crystals. Water vapour that evaporates before reaching the ground is virga. Examples of liquid or solid water particles that are lifted off the earth's surface by the wind includes drifting and blowing snow and blowing spray. Dew, frost, rime, and glaze are examples of liquid or solid water deposits on exposed objects.

Hydroponics The growing of plants, especially vegetables, in water containing essential mineral nutrients rather than in soil.

Hydrosphere The hydrosphere describes the waters of the Earth. Water exists on the Earth in various stores, including the: atmosphere, oceans, lakes, rivers, glaciers, snowfields and groundwater. Water moves from one store to another by way of: evaporation, condensation, precipitation, deposition, runoff, infiltration, sublimation, transpiration, and groundwater flow.

Hydrostatic pressure Force caused by water under pressure.

Hygrograph An instrument that records the hygrometer's measure of water vapour.

Hygrometer Instrument that measures water vapour content in the air and communicates changes in humidity visibly and immediately through a graph or a dial. There are three types of hygrometers: The hair hygrometer uses a human hair as the sensing instrument. The hair lengthens when the air is moist and contracts when the air is dry but remains unaffected by air temperature. However; the hair hygrometer cannot respond to rapid fluctuations in humidity.

An electric hygrometer uses a plate coated with carbon. Electrical resistance of the carbon coating changes as the moisture content of the air changes-changes that translate into relative humidity. This type of hygrometer is used frequently in the radiosonde.

An infrared hygrometer uses a beam of light containing two separate wave lengths to gauge atmospheric humidity. One of the wavelengths is absorbed by water vapour; the other is unaffected, providing an extremely accurate index of water vapour for paths of a few inches or thousands of feet.

Hygroscopic Substances that have the ability to absorb water and therefore accelerate the condensation process.

Hygroscopic coefficient Maximum limit of hygroscopic water around the surface of a soil particle.

Hygroscopic water Water held within 0.0002 millimetres of the surface of a soil particle. This water is essentially non-mobile and can only be removed from the soil through heating.

Hyphae Thread like structures found on a fungus.

Hypothermia Occurs when the core temperature of one's body falls below normal. It is the failure of the body to maintain adequate production of heat under conditions of extreme cold.

Hypothesis A tentative assumption that is made for the purpose of empirical scientific testing. A hypothesis becomes a theory when repeated testing and evidence suggests the hypothesis has a strong chance of being correct.

Hypothesis testing Process where an alternative and a null hypothesis are statistically tested for the purpose of falsifying a hypothesis.

Hypsography Topography referred to the national geodetic vertical datum of 1929. The science or art of describing heights of land surfaces with reference to this datum.

Hypsometry Science or art of determining terrain relief, by any method.

I

Ice Frozen form of water. Ice has a specific gravity (0.9166) that is slightly less than water. This difference in specific gravity causes ice to float on water.

Ice age Period of time when glaciers dominate the landscape of the Earth. The last major Ice Age was during the Pleistocene epoch.

Ice cap Large dome-shaped glacier found covering a large expanse of land. Smaller than an ice sheet.

Ice crystals Precipitation in the form of slowly falling, singular or unbranched ice needles, columns, or plates. They make up cirriform clouds, frost, and ice fog. Also, they produce optical phenomena such as halos, coronas, and sun pillars. May be called "diamond dust". It is reported as "IC" in an observation and on the METAR.

Ice fall An area of crevassed ice on a glacier. Caused when the base of the glacier flows over steep topography.

Ice field Large level area of glacial ice found covering a large expanse of land. Similar in size to an ice cap but does not have a dome-shape.

Ice fog Fog composed of suspended ice crystals. Common is Arctic locations when temperatures are below -30° Celsius and a abundant supply of water vapour exists.

Ice jam An accumulation of broken river ice caught in a narrow channel, frequently producing local flooding. Primarily occurs during a thaw in the late winter or early spring.

Ice lense Horizontal accumulation of permanently frozen ground ice.

Ice pellets Precipitation in the form of transparent or translucent pellets of ice, which is round or irregular in shape. They have a diameter of 0.2 inches (5 mm) or less. They are classified into two types: hard grains of ice consisting of frozen rain drops or largely melted and refrozen snowflakes; pellets of snow encased in a thin layer of ice which have formed from the freezing of droplets intercepted by pellets or water resulting from the partial melting of pellets. It is reported as "PE" in an observation and on the METAR. Also known as sleet.

Ice sheet A dome-shaped glacier covering an area greater than 50,000 square kilometres. Greenland and Antarctica are considered ice sheets. During the glacial advances of the Pleistocene ice sheets covered large areas of North America, Europe, and Asia. Larger than an ice cap.

Ice shelf A thick mass of ice extending from a polar shore. The seaward edge is afloat and sometimes extends hundreds of miles into the sea.

Ice storm A severe weather condition characterized by falling freezing precipitation. Such a storm forms a glaze on objects, creating hazardous travel conditions and utility problems.

Ice wedge Wedge-shaped, ice body composed of vertically oriented ground ice that extends into the top of a permafrost layer. These features are approximately 2 to 3 meters wide at their top and extend into the soil about 8 to 10 meters. Form in cracks that develop in the soil during winter because of thermal contraction. In the spring, these cracks fill with liquid water from melting snow which subsequently refreezes. The freezing process causes the water to expand in volume increasing the size and depth of the crack. The now large crack fills with more liquid water and again it freezes causing the crack to enlarge. his process continues for many cycles until the ice wedge reaches its maximum size.

Iceberg A mass of ice found floating in the ocean or a lake. Often icebergs form when ice calves from land-based glaciers into the water body. Icebergs can be dangerous to shipping in high and mid-latitude regions of the ocean because 90 percent of their mass lies below the ocean surface.

Icelandic low A semi-permanent, subpolar area of low pressure in the North Atlantic Ocean. Because of its broad area and range of central pressure, it is an area where migratory lows tend to slow down and deepen. It is strongest during a Northern Hemisphere winter and early spring, centred over Iceland and southern Greenland, and is the dominate weather feature in the area. During the summer, it is weaker, less intense, and might divide into two parts, one west of Iceland, the other over the Davis Strait between Greenland and Baffin Island. Then the Azores or Bermuda High becomes the dominate weather feature in the North Atlantic.

Icicle Ice that forms in the shape of a narrow cone hanging point down. It usually forms when liquid water from a sheltered or heated source comes in contact with below-freezing air and freezes more or less rapidly as it flows.

Icing The forming or depositing of ice on an object.

Ideal gas law This law describes the physical relationships that exist between pressure, temperature, volume, and density for gases. Two mathematical equations are commonly used to describe this law: Pressure × Volume = Constant×Temperature and

Pressure = Density × Constant × Temperature.

Igneous rock Rocks formed by solidification of molten magma either beneath (intrusive igneous rock) or at (extrusive igneous rocks) the Earth's surface.

Illuviation Deposition of humus, chemical substances, and fine mineral particles in the lower layers of a soil from upper layers because of the downward movement of water through the soil profile.

Image Pictorial representation of data acquired by satellite systems, such as direct readout images from environmental satellites. An image is not a photograph. An image is composed of two-dimensional grids of individual picture elements (pixels). Each pixel has a numeric value that corresponds to the radiance or temperature of the specific ground area it depicts.

Image-res The area represented by each pixel of a satellite image. The smaller the area represented by a pixel, the more accurate and detailed the image. For example, if a U.S. map and a world map are printed on identically sized sheets of paper, one square inch on the U.S. map will represent far less area and provide for more detail than one square inch on the world map. In this example the U.S. map has higher resolution. APT has a resolution of 4 km,

HRPT has a resolution of 1.1 km and WEFAX resolution is 8 km.

Imager A satellite instrument that measures and maps the Earth and its atmosphere. Imager data are converted by computer into pictures.

Imagery Visible representation of objects and (or) phenomena as sensed or detected by cameras, infrared and multispectral scanners, radar, and photometers. Recording may be on photographic emulsion (directly as in a camera or indirectly after being first recorded on magnetic tape as an electrical signal) or on magnetic tape for subsequent conversion and display on a cathode ray tube.

Immigrant species Species that migrate into an ecosystem or that are deliberately or accidentally introduced into an ecosystem by humans. Some of these species are beneficial, whereas others can take over and eliminate many native species. Compare with indicator species, keystone species, and native species.

Immigration Migration of an organism into an area for the purpose of changing its residence permanently.

In situ Latin for "in original place." Refers to measurements made at the actual location of the object or material measured. Compare with remote sensing.

Inceptisols These soils have horizons in the early stages of pedogenesis. Many of these soils are found on glacial drift deposits.

Inches of mercury (Hg) The name comes from the use of mercurial barometers which equate the height of a column of mercury with air pressure. One inch of mercury is equivalent to 33.86 millibars or 25.40 millimetres. First devised in 1644 by Evangelista Torricelli (1608-1647), an Italian physicist and mathematician, to explain the fundamental principles of hydromechanics.

Inclination One of the six Keplerian elements, it indicates the angle of the orbit plane to the central body's equator.

The elliptical path of a satellite orbit lies in a plane known as the orbital plane. The orbital plane always goes through the centre of the Earth but may be tilted at any angle relative to the equator. Inclination is the angle between the equatorial plane and the orbital plane measured counter-clockwise at the ascending node.

A satellite in an orbit that exactly matches the equator has an inclination of 0o; whereas one whose orbit crosses the Earth's poles has an inclination of 90o. Because the angle is measured in a counter clockwise direction, it is quite possible for a satellite to have an inclination of more than 90o. An inclination of 180o would mean the satellite is orbiting the equator; but in the opposite direction of the Earths' rotation. Some sun-synchronous satellites that maintain the same ground track throughout the year have inclinations of as much as 98o. U.S. scientific satellites that study the sun are placed in orbits closer to the equator; frequently at 28o inclination. Most weather satellites are placed in high-inclination orbits so they can oversee weather conditions worldwide.

Indentured labour Work performed according to a binding contract between two parties. During the early colonial period in America, this often involved long periods of time and a total work commitment.

Independent variable Variable in a statistical test that is thought to be controlling through cause and effect the value of observations in another dependent variable modelled in the test.

Index contour Contour line that is accentuated in thickness and is often labelled with the appropriate measure of elevation. Index contours occur every four or fifth contour interval and help the map user read elevations on a map.

Indian summer A period of abnormally warm weather in mid to late autumn with clear skies and cool nights. A first frost normally precedes this warm spell.

Indicator species Species that can be used as a early indicator of environmental degradation to a community or an ecosystem. Compare with immigrant species, keystone species, and native species.

Indigo A plant that yields a blue vat dye.

Induction Inference of a generalized conclusion from particular instances. In a science like Physical Geography, inductive reasoning would involve the development of a theory to explain previously collected facts or observed phenomenon.

Industrial revolution Major change in the economy and society of humans brought on by the use of machines. This period in human history began in England in the late 18th century.

Industrial smog Form of air pollution that develops in urban areas. This type of air pollution consists of a combination of sulphur dioxide, suspended droplets of sulphuric acid, and a variety of suspended solid particles.

Inertia costs of location Costs borne by an activity because it remains located at its original site, even though the distributions of supply and demand have changed.

Inferential statistics Statistical test that makes generalizations about a population based of the numeric information obtained from a sample based on the laws of probability.

Infiltration The absorption and downward movement of water into the soil layer.

Infiltration capacity The ability of a soil to absorb surface water.

Infiltration rate Rate of absorption and downward movement of water into the soil layer.

Information system All of the means and mechanisms for data receipt, processing, storage, retrieval, and analysis. Information systems can be designed for storage and dissemination of a variety of data products-including primary data sets and both intermediate and final analyses and for an interface providing connections to external computers, external data banks, and system users. To be effective, the design and operation of an information system must be carried out in close association with the primary producers of the data sets, as well as other groups producing integrated analyses or intermediate products.

Infrared The long wave, electromagnetic radiation of radiant heat emitted by all hot objects. On the electromagnetic spectrum, it can be found between microwave radiation and visible light. Water vapour, ozone, and carbon dioxide are capable of absorbing or transmitting infrared radiation. Also referred to as IR.

Infrared radiation (IR) Infrared is electromagnetic radiation whose wavelength spans the region from about 0.7 to 1000 micrometers (longer than visible radiation, shorter than microwave radiation). Remote-sensing instruments work by sensing radiation that is naturally emitted or reflected by the Earth's surface or from the atmosphere, or by sensing signals transmitted from a satellite and reflected back to it. In the visible and near-infrared regions, surface chemical composition, vegetation cover, and biological properties of surface matter can be measured. In the mid-infrared region, geological formations can be detected due to the absorption properties related to the structure of silicates. In the far infrared, emissions from the Earth's atmosphere and surface offer information about atmospheric and surface temperatures and water vapour and other trace constituents in the atmosphere. Since IR data are based on temperatures rather than visible radiation, the data may be obtained day or night.

Infrared scanner (thermal mapper) Instrument that detects infrared radiation and converts the detected energy to an electrical signal for recording on photographic film or magnetic tape.

Infrastructure Refers to the roads, cables, wires, pipes, bridges, canals, reservoirs, and sewers that support economy and society. Infrastructure development and management are important in urban and rural areas at the local, state, regional, national, or international levels.

Inhibition model of succession This model of succession suggests that the change in plant species dominance over time is caused by death and small scale disturbances and variations is plant species longevity and ability to disperse. Over time species turnover favors plant species with longer life spans.

Inner core Inner region of the Earth's core. It is thought to be solid iron and nickel with a density of about 13 grams per cubic centimeter. It also has a diameter of about 1220 kilometres.

Inorganic Non-living thing. Usually refers to the physical and chemical components of an organism's environment. Some times called abiotic.

Inosilicate Subclass of the silicate class of minerals. Inosilicates have two distinct forms: single and double chain silicates.

Input Addition of matter, energy, or information to a system.

INSAT Indian national satellite.

Insect Relatively small and simple animals that have a rigid external skeleton, three body sections, three pairs of legs, and antennae. These organisms are the most abundant group of eukaryotes on the Earth.

Inselberg A German term used to describe a steep-sided hill composed of rock that rises from a pediplain.

Insolation Solar radiation or heating received at the earth's surface The name is derived from incoming solar radiation.

Insolation weathering Form of physical weathering. Involves the physical breakdown of minerals and rock due to thermal expansion and contraction.

Instability Atmospheric condition where a parcel of air is warmer that the surrounding air in the immediate environment. This condition causes the parcel to rise in the atmosphere.

Instantaneous field of view (IFOV) The field of view of a scanning detector system at a given instant. The range of angles scanned by the system is then called the field of view, or swath width.

Instrument flight rules (IFR) Refers to the general weather conditions pilots can expect at the surface and applies to the weather situations at an airport during which a pilot must use instruments to assist take off and landing. IFR conditions for fixed wing aircraft means the minimum cloud ceiling is greater than 500 feet and less than 1,000 feet and/or visibility is greater than 1 mile and less than 3 miles.

Instrument shelter A boxlike structure designed to protect temperature measuring instruments from exposure to direct sunshine, precipitation, and condensation, while at the same time time providing adequate ventilation.

Insular Either of an island, or suggestive of the isolated condition of an island.

Interaction (biological) Mutual or reciprocal influence between two or more similar organisms or individuals of different species. Major biotic interactions are: competition, mutualism, predation, parasitism, amensalism, and commensialism.

Interception Is the capture of precipitation by the plant canopy and its subsequent return to the atmosphere through evaporation or sublimation. The amount of precipitation intercepted by plants varies with leaf type, canopy architecture, wind speed, available radiation, temperature, and the humidity of the atmosphere.

Interference Form of competition where an individual directly prevents the physical establishment of another individual in a portion of a habitat.

Interglacial Period of time during an ice age when glaciers retreated because of milder temperatures.

Intermittent stream A stream that flows only for short periods over a year. Flow events are usually initiated by rainfall.

Intermountain high An area of high pressure that occurs during the winter between the Rocky Mountains and the Sierra-Cascade ranges. It blocks the eastward movement of Pacific cyclones. Also called Plateau High or Great Basin High.

International date line A line drawn almost parallel to the 180 degree longitude meridian that marks the location where each day officially begins. The location of the International Date Line was decided upon, by international agreement.

International date line Located at 180° longitude (180° E and 180° W are the same meridian). Regions to the east of the International Date Line are counted as being one calendar day earlier than the regions to the west. Although the International Date Line generally follows the 180° meridian (most of which lies in the Pacific Ocean), it does diverge in places. Since 180° runs through several countries, it would divide those countries not simply into two different time zones, but into two different calendar days. To avoid such unnecessary confusion, the date line dips and bends around countries to permit them to share the same time.

International designator An internationally agreed-upon naming convention for satellites. The designator contains the last two digits of the launch year; the launch number of the year; and the part of the launch, i.e., "A" indicates payload, "B" the rocket booster; or second payload, etc.

International geophysical year (IGY) (1957-58) The IGY was organized by the scientific community through the International Council of Scientific Unions (ICSU) . It was highlighted by international cooperation in the exploration of world-wide geophysical phenomena and by the inauguration of the space age through the launching of the first satellites (USSR's Sputnik I and US Explorer 1) to study the upper atmosphere and Earth's nearby environment.

International space year (ISY) (1992. Designated the first international celebration of humanity's future in the space age. Themes included the global perspective of the space age, discovery exploration, and scientific inquiry. An important ISY scientific focus was Mission to Planet Earth. A wide range of educational programs and public events emphasized ISY's global perspective. 1992 also commemorated the 500th anniversary of Columbus' voyage to the New World and the 35th anniversary of the International Geophysical Year.

International system of units (SI) The International System of Units prescribes the symbols and prefixes shown in the table to form decimal multiples and submultiples of SI units.

The following examples illustrate the use of these prefixes.

0.000,001 meters = 10^{-6} meters = 1 micrometer 1pm

1000 meters = 10^{3} meters = 1 kilometers = 1km

1,000,000 cycles per second = 10^{6} hertz = 1 megahertz = 1 MHz.

Interspecific Interaction between different species.

Intertropical convergence zone (ITCZ) An area where the Northern and Southern Hemispheric trade winds converge, usually located between 10 degrees North and South of the equator. It is a broad area of low pressure where both the Coriolis force and the low-level pressure gradient are weak, occasionally allowing tropical disturbances to form. It fluctuates in location, following the sun's rays, so that during the Northern Hemisphere summer, the ITCZ moves northward over the southern North Atlantic and southern Asia.

Intervening opportunity The existence of a closer, less expensive opportunity for obtaining a good or service, or for a migration destination. Such opportunities lessen the attractiveness of more distant places.

Intracoastal waterway system A waterway channel, maintained through dredging and sheltered for the most part by a series of linear offshore islands, that extends from New York City to Florida's southern tip and from Brownsville, Texas, to the eastern end of Florida's panhandle.

Intraspecific Interaction within a single species.

Intrusive igneous rock A mass of igneous rock that forms when magma from the mantle migrates upward and cools and crystallizes near, but not at, the Earth's surface. Also called plutonic igneous rock.

Inverse square law This law suggests that the amount of radiation passing through a specific area is inversely proportional to the square of the distance of that area from the energy source. Mathematically, the Inverse Square Law is described by the equation: Intensity = I/d^2

where I is the intensity of the radiation at 1 unit distance and d is the distance travelled in those units.

Inversely proportional Cause and effect relationship between two variables where a positive or negative change in the quantity of one causes a predictable opposite change in quantity in the other.

Inversion A departure from the usual increase or decrease of an atmospheric property with altitude. It usually refers to an increase in temperature with increasing altitude, which is a departure from the usual decrease of temperature with height.

Invertebrate Animal that does not have a backbone.

Ion An atom, molecule or compound that carries either a positive (cation) or negative (anion) electrical charge.

Ionizing radiation The emission of alpha or beta particles or gamma rays from radioisotopes. These emitted particles can dislodge one or more electrons from atoms they strike. The free electrons can form charged ions in living tissue that can react with and damage cells.

Ionosphere A region in the atmosphere above 50 kilometres from the surface where relatively large concentrations of ions and free electrons exist. The ionosphere is important for human communications because it re-directs AM radio transmissions. This process extends the distance that radio transmissions can travel.

IPS Inches per second.

Isallobar The line of equal change in atmospheric pressure during a certain time period. It marks the change in pressure tendency.

Island An island is a piece of land that is surrounded by water.

Island arc A line of volcanic islands found of the ocean that have been created by the convergence of two tectonic plates and the subsequent subduction of one of the plates beneath the other. Subduction cause magma plumes to rise to the Earth's surface creating the volcanic islands.

Isobar The line drawn on a weather map connecting points of equal barometric pressure.

Isodrosotherm The line drawn on a weather map connecting points of equal dew point.

Isogonic chart Chart showing isogonic lines properly labelled with their magnetic declination.

Isogonic line Line joining points on the Earth's surface having equal magnetic declination as of a given date.

Isohel A line drawn through geographic points having equal duration of sunshine or another form of solar radiation during a specified time period.

Isohyet The line drawn through geographic points recording equal amounts of rainfall during a given time or for a given of storm.

Isolated system A system that has no interactions beyond its boundary layer. Many controlled laboratory experiments are this type of system.

Isoline Lines on a map joining points of equal value.

Isopleth A line connecting equal points of value. Also called an isoline.

Isostacy The buoyant condition of the Earth's crust floating in the asthenosphere. The greater the weight of the crust the deeper it floats into the asthenosphere. When weight is remove the crust rises higher.

Isostatic depression Large scale sinking of the crust into the asthenosphere because of an increase in weight on the crustal surface. Common in areas of continental glaciation where the crust was depressed by the weight of the ice.

Isostatic rebound The upward movement of the Earth's crust following isostatic depression.

Isotach A line connecting equal wind speeds.

Isotherm The line of equal or constant air temperature. If something is isothermal, it is of equal or constant temperature with respect to either time or space.

Isothermal Of or indicating equality of temperature.

Isothermal layer Vertical layer in the atmosphere where temperature remains unchanged. In the Earth's atmosphere, three isothermal layers are found in the lower regions of the stratosphere, mesosphere, and the thermosphere.

Isotherms Lines connecting points of equal temperature on a weather map.

Isotope Form of an element where the number of neutrons in its atomic nucleus is different than the number of protons.

Isotopic dating Dating technique used to determine the age of rock and mineral through the decay of radioactive elements.

Isthmus An isthmus is a narrow strip of land connecting two larger landmasses. An isthmus has water on two sides.

Isthmus Narrow strip of land located between two bodies of water; connecting two larger land areas.

J

Jet stream An area of strong winds that are concentrated in a relatively narrow band in the upper troposphere of the middle latitudes and subtropical regions of the Northern and Southern Hemispheres. Flowing in a semi-continuous band around the globe from west to east, it is caused by the changes in air temperature where the cold polar air moving towards the equator meets the warmer equatorial air moving polarward. It is marked by a concentration of isotherms and strong vertical shear. Various types include the arctic, the low level, the polar, and the subtropical jets.

Joint A fracture in a rock where no movement has taken place or where no movement has taken place perpendicular to the surface of the fracture. Important in rock weathering because it increases the exposed surface area.

Joule Unit for measuring energy. One joule is the energy used by a force of one Newton in moving its point of application in the direction of the force one meter.

Jurassic Geologic period that occurred roughly 144 to 208 million years ago. During this period, the first birds and mammals appear and large areas of the continents are covered by shallow seas.

Jurisdiction The right and power to apply the law; the territorial range of legal authority or control.

K

K index The measure of thunderstorm potential based on the vertical temperature lapse rate, the moisture content of the lower atmosphere and the vertical extent of the moist layer.

K selected species (Logistic Strategy) Long-lived species that produces only a few, often fairly large progeny.

Kame terrace A long flat ridge composed of glaciofluvial sediment. This feature forms along the margin of a valley glacier where the glacial ice meets the valley's slope. Sediment is deposited by laterally flowing meltwater streams.

Kaolinite A type of clay that is widespread in tropical and subtropical regions. It is produced from the weathering of granite.

Karst Landform type with limestone bedrock and dominated by geomorphic features created from solution chemical weathering.

Katabatic wind A wind that is created by air flowing downhill. When this air is warm, it may be called a foehn wind, and regionally it may be known as a Chinook or Santa Ana. When this air is cold or cool, it is called a drainage wind, and regionally it may be known as a mountain breeze or glacier wind. The opposite of an anabatic wind.

Katafront A front where the warm air descends the frontal surface (except in the low layers of the atmosphere).

Kelsh plotter Double-projection plotting instrument utilizing swinging lamps to transmit light through contact- size diapositives (positive transparencies).

Kelvin temperature scale A temperature scale with the freezing point of -273 degrees K (Kelvin) and the boiling point of +273 degrees K. It is used primarily for scientific purposes. Also known as the Absolute Temperature Scale. Proposed in 1848 by William T. Kelvin, 1st Baron of Largs (1824-1907), Irish-born Scottish physicist and mathematician.

Keplerian elements (satellite orbital elements) The set of six independent constants which define an orbit-named for Johannes Kepler (1571-1630). The constants define the shape of an ellipse or hyperbola, orient it around its central body and define the position of a satellite on the orbit. The classical orbital elements are: A: semi-major axis, gives the size of the orbit,

e: eccentricity, gives the shape of the orbit,

i: inclination angle, gives the angle of the orbit plane to the central body's equator

: right ascension of the ascending node, which gives the rotation of the orbit plane from reference axis,

: argument of perigee is the angle from the ascending nodes to perigee point, measured along the orbit in the direction of the satellites motion

: true anomaly gives the location of the satellite on the orbit.

Kepler's three laws of motion Any spacecraft launched into orbit obeys the same laws that govern the motions of the planets around our sun, and the moon around the Earth. Johannes Kepler formulated three laws that describe these motions: 1. Each planet revolves around the sun in an orbit that is an ellipse with the sun as its focus or primary body Kepler postulated the lack of circular orbits-only elliptical ones-determined by gravitational perturbations and other factors. Gravitational pulls, according to Newton, extend to infinity although their forces weaken with distance and eventually become impossible to detect. Spacecraft orbiting the Earth are primarily influenced by the Earth's gravity and anomalies in its composition, but they also are influenced by the moon and sun and possibly other planets.

2. The radius vector—such as the line from the centre of the sun to the centre of a planet, from the centre of Earth to the centre of the moon, or from the centre of Earth to the centre of gravity of a satellite—sweeps out equal areas in equal periods of time.

3. The square of a planet's orbital period is equal to the cube of its mean distance from the sun times a constant. As extended and generalized, this means that a satellite's orbital period increases with its mean distance from the planet.

Kettle hole Depression found in glacial deposits. Created when a piece of ice from a retreating glacier becomes embedded in soft glacial till or glacial drift deposits. Many are filled with water to form a small lake or pond.

Kettle moraine An area of glaciofluvial influenced moraine deposits pitted with kames and kettle holes.

Key The key of a map (also called the legend) is a small table accompanying the map that explains the symbols that are used on the map.

Keystone species Species that interacts with a large number of other species in a community. Because of the interactions, the removal of this species can cause widespread changes to community structure. Compare with immigrant species, indicator species, and native species.

Kilocalorie (KCAL) Unit of energy equal to 1,000 calories.

Kilohertz (kHz) One thousand hertz, i.e., one thousand cycles per second.

Kilometre (km) Metric unit of distance equal to 3,280.8 feet or .621 statute miles.

Kilopascal (KPA) A unit measurements for quantifying force. Used to measure atmospheric pressure. Equivalent to 10,000 dynes per square centimetre.

Kilowatt (KW) Unit of electrical power equal to 1,000 Watts.

Kinetic energy The energy due to motion.

Kingdom Top most level of the common system used to classify life. Generally, five kingdoms are recognized: Monera, Protista, Fungi, Animalia, and Plantae.

Kirchoff's law This law suggests that good emitters of radiation are also good absorbers of radiation at specific electromagnetic radiation wavelength bands. It also suggests that poor emitters of radiation are also poor absorbers of radiation at specific wavelength bands.

Knot A nautical unit of speed equal to the velocity at which one nautical mile is travelled in one hour. Used primarily by marine interests and in weather observations. A knot is equivalent to 1.151 statute miles per hour or 1.852 kilometres per hour.

Köppen climate classification System that uses monthly precipitation and temperature data and total annual precipitation data to classify a location's climate into one of five main categories: Tropical Moist Climates; Dry Climates; Moist Mid-latitude Climates with Mild Winters; Moist Mid-Latitude Climates with Cold Winters; and Polar Climates. These categories are further divided into number of subcategories. First developed in 1918 by German biologist W. Köppen, this system has undergone a number of modifications.

Kudzu A vine, native to China and Japan but imported into the United States; originally planted for decoration, for forage, or as a ground cover to control erosion. It now grows wild in many parts of the southeastern United States.

L

La Nina Condition opposite of an El Nino. In a La Nina, the tropical Pacific trade winds become very strong and an abnormal accumulation of cold water occurs in the central and eastern Pacific Ocean.

Lacustrine plain A nearly level land area that was formed as a lake bed.

Lagoon 1. A body of seawater that is almost completely cut off from the ocean by a barrier beach.

2. The body of seawater that is enclosed by an atoll.

Lahar A very rapid type of down slope mass movement that involving mudflows from volcanic ash.

Lake A body standing water found on the Earth's continental land masses. The water in a lake is normally fresh.

Lake effect snow Snow showers that are created when cold dry air passes over a large warmer lake, such as one of the Great Lakes, and picks up moisture and heat.

Laminar flow Movement of water within a stream that occurs as uninterrupted parallel flows. Laminar flow generally occurs in areas where friction is low.

Land base Includes the natural and man-made environmental features within which infrastructure is developed or by which natural resources are indexed and analyzed.

Land breeze A diurnal coastal breeze that blows offshore, from the land to the sea. It is caused by the temperature difference when the sea surface is warmer than the adjacent land. Predominate during the night, it reaches its maximum about dawn. It blows in the opposite direction of a sea breeze.

Land use classification system Coding system of categories and subcategories designed for use on a map to designate land or water use.

Landfall The coastline location where a tropical storm or hurricane moves from ocean onto land.

Landmark Monument of material mark or fixed object used to designate a land boundary on the ground: any prominent object on land that may be used to determine a location or a direction in navigation or surveying.

Landsat Land Remote-Sensing Satellite, operated by the U.S. earth observation satellite company (EOSAT). Commercialized under the Land Remote-Sensing Commercialization Act of 1984, Landsat is a series of satellites (formerly called ERTS) designed to gather data on the Earth's resources in a regular and systematic manner. Objectives of the mission are: land use inventory, geological/mineralogical exploration, crop and forestry assessment, and cartography. Landsat has a spatial resolution of 28.5 meters.

Restructured Federal agency responsibilities for the Landsat program are effective for the acquisition and operation of Landsat 7. New operating policy specifies that NOAA will be responsible for satellites after they are placed in orbit, NASA will be responsible for the development and launch of Landsat 7, and that the U.S. government will provide unenhanced data to users at no cost beyond the cost of fulfilling their data request.

Landslide Term used to describe the downslope movement of soil, rock, and other weathered materials because of gravity.

Landspout A small, weak tornado, which is not formed by a storm-scale rotation. It is generally weaker than a supercell tornado and is not associated with a wall cloud or mesocyclone. It may be observed beneath cumulonimbus or towering cumulus clouds and is the land equivalent of a waterspout.

Landward Positioned or located away from a water body but towards the land.

Langley Unit of the intensity of radiation measured per minute and equal to one calorie.

Lapse rate The decrease of an atmospheric variable, usually temperature, with height. A steep lapse rate implies a rapid decrease in temperature with height and is a sign of instability.

Laser (Light amplification by stimulated emission of radiation) Active instrument that produces discretely coherent pulses of light (light waves with no phase differences, or with predictable phases differences, are said to be coherent).

Laser ranging The use of lasers to measure distances.

Latent heat Is the energy required to change a substance to a higher state of matter (solid > liquid > gas). This same energy is released from the substance when the change of state is reversed (gas > liquid > solid).

Latent heat flux Latent heat flux is the global movement of latent heat energy through circulations of air and water. Atmospheric circulation moves latent heat energy vertically and horizontally to cooler locations where it is condensed as rain or is deposited as snow releasing the heat energy stored within it. Large quantities of radiation energy are transferred into the Earth's tropical oceans. Radiant energy enters these water bodies at the surface when absorbed radiation is converted into heat energy. The warmed surface water is then transferred downward into the water column by conduction and convection. Horizontal transfer of this heat energy from the equator to the poles is accomplished by ocean currents.

Latent heat of condensation The amount of heat energy release to the environment when a gas changes its state to a liquid. For one gram of water, the amount of heat energy released is 540 calories at a temperature of 100° Celsius.

Latent heat of vapourization The amount of heat energy required from the environment to change the state of a liquid to a gas. For one gram of water, the amount of heat energy required is 540 calories at a temperature of 100° Celsius.

Lateral moraine Moraine that is found along the sides of a glacier. Commonly found on glaciers that occupy a valley.

Laterite Hard subsurface deposit of oxides of aluminium and iron found in tropical soils where the water table fluctuates with seasonal changes in precipitation.

Laterization Soil forming process that creates a laterite layer.

Latitude Latitude is a north-south measurement of position on the Earth. It is defined by the angle measured from a horizontal plane located at the Earth's centre that is perpendicular to the polar axis. A line connecting all places of the same latitude is termed a parallel. Latitude is measured in degrees, minutes, and seconds. Measurements of latitude range from equator (0°) to 90° North and South from this point.

Latitude lines Imaginary lines running horizontally around the globe. Also called parallels, latitude lines are equidistant from each other. Each degree of latitude is about 69 miles (110 km) apart. Zero degrees (0°) latitude is the equator, the widest circumference of the globe. Latitude is measured from 0° to 90° north and 0° to 90° south—90° north is the North Pole and 90° south is the South Pole.

Latosol Soil order (type) of the United States Natural Resources Conservation Service Soil Classification System. This soil is characterized by a thin 0 horizon, the présence of a laterite layer, and a deeply weathered profile.

Laurasia Northern section of Pangaea.

Lava Molten magma released from a volcanic vent or fissure.

Lava flow Stream of lava flowing from a volcanic vent.

Law of basin areas Morphometric relationship observed in the mean basin area size of stream segments of a particular classification order in stream channel branching. Proposed by R.E. Horton.

Law of conservation of energy This law states that energy can be transferred from one system to another in many forms, however, it can not be created nor destroyed. Thus, the total amount of energy available in the universe is constant.

Law of stream lengths Morphometric relationship observed in the cumulative size of stream segment lengths in stream channel branching. Propòsed by R.E. Horton.

Law of stream number Morphometric relationship observed in the number of stream segments of a particular classification order in stream channel branching. Proposed by R.E. Horton.

Law of the minimum This biological law suggests that organisms are normally limited by only one single physical factor that is in shortest supply relative to demand.

Layer Is a subdivision of a CAD or GIS database containing related data. Layers can be visualized as "transparencies" which allow the user to view and analyze information selectively by theme. Some GIS build their databases as a series of layers covering a single area. Layers are fundamental to overlay analysis.

LDC See less developed country.

Leachate Solution containing material leached from a soil.

Leaching A process of soil nutrient removal through the erosive movement and chemical action of water.

Lead line Line weighted with lead for making depth soundings in water.

Leaf drip The rain water that fall to the ground surface from plant leaves after it has been intercepted by these structures.

Lee Side of a slope that is opposite to the direction of flow of ice, wind, or water. Opposite of stoss.

Lee/leeside/leeward The side of an object or obstacle, such as a ship's sail, a mountain, or a hill, furthest away from the wind, and therefore, protected from the direct force of the wind. The opposite of windward.

Legend The legend of a map (also called the key) is a small table accompanying the map that explains the symbols that are used on the map.

Legume Angiosperm plant species that is a member of the Fabaceae (Pea or Bean) family. These plants form symbiotic relationships with specific bacteria species for the purpose of acquiring nitrogen for growth.

Lenticular cloud A cloud species which has elements resembling smooth lenses or almonds and more or less isolated. These clouds are caused by a wave wind pattern created by the mountains. They are also indicative of down-stream turbulence on the leeward side of a barrier.

Less Developed Country (LDC) Country characterized by minimal industrialization, low technological development, low per capita income, and high population growth rates. Many of these countries are found in Asia, Africa, and Central and South America.

Levee Artificial bank confining a stream channel or limiting adjacent areas subject to flooding; an embankment bordering a submarine canyon or channel, usually occurring along the outer edge of a curve.

Level of Free Convection (LFC) The level at which a parcel of saturated air becomes warmer than the surrounding air and begins to rise freely. This occurs most readily in a conditionally unstable atmosphere.

Level surface Surface which at every point is perpendicular to the plumbline or the direction in which gravity acts.

Levelling Surveying operation in which heights of objects and points are determined relative to a specified datum.

Liana Species of plant that uses the support of wood plants to elevate its leaves above the forest canopy.

Libraries Contain frequently used symbols, subroutines, and commands. Rather than recreating a symbol everytime it is required, a user can call the symbol from a library using a unique identification code. The same is true of subroutines, sometimes referred to as macros.

Lichen Organism that consists of a symbiotic joining of a species of fungi and a species of algae.

Life cycle stage A period of uneven length in which the relative dependence of an individual on others helps define a complex of basic social relations that remains relatively consistent throughout the period.

Lifted Index (LI) A measure of atmospheric instability that is obtained by computing the temperature that the air near the ground would have if it were lifted to a higher level and comparing it to the actual temperature at that altitude. Positive values indicate more stable air and negative values indicate instability.

Lifting Condensation Level (LCL) The height at which a parcel of moist air becomes saturated when it is lifted dry adiabatically.

Light 1. Form of radiant energy that acts upon the retina of the eye, optic nerve, etc., making sight possible. This energy is transmitted at a velocity of about 186,000 miles per second by wavelike or vibrational motion.

2. A form of radiant energy similar to this, but not acting on the normal retina, such as ultraviolet and infrared radiation.

Interplay between light rays and the atmosphere cause us to see the sky as blue, and can result in such phenomena as glows, halos, arcs, flashes, and streamers.

Light industry Manufacturing activities that use moderate amounts of partially processed materials to produce items of relatively high value per unit weight.

Light waves That part of the electromagnetic spectrum that contains visible light. The colours, from longest wave length to shortest, are red, orange, yellow, green, blue, indigo, and violet (ROY G. BIV).

Light year Distance that light travels in the vacuum of space in one year. Approximately 9.7 trillion kilometres.

Lightning A discharge of atmospheric electricity accompanied by a vivid flash of light. During thunderstorms, static electricity builds up within the clouds. A positive charge builds in the upper part of the cloud, while a large negative charge builds in the lower portion. When the difference between the positive and negative charges becomes great, the electrical charge jumps from one area to another; creating a lightning bolt. Most lightning bolts strike from one cloud to another; but they also can strike the ground. These bolts occur when positive charges build up on the ground. A negative charge called the "faintly luminous streamer" or "leader" flows from the cloud toward the ground. Then a positively charged leader; called the return stroke, leaves the ground and runs into the cloud. What is seen as a lightning bolt is actually a series of downward-striking leaders and upward-striking return strokes, all taking place in less than a second.

Lightning bolts can heat the air to temperatures hotter than the surface of the sun. This burst of heat makes the air around the bolt expand explosively producing the sound we hear as thunder Since light travels a million times faster than sound, we see lightning bolts before we hear their thunderclaps. By counting the seconds between a flash of lightning and the thunderclap and dividing by five, we can determine the approximate number of miles to the lightning stroke.

Lignite A low-grade brownish coal of relatively poor heat-generating capacity.

Limb viewing (occultation) The process of viewing the atmosphere at a tangent to the Earth's surface. The viewing signal, from a star or another satellite, is occulted or obscured by the intervening atmosphere. The absorption of light from the sun or star provides information on the properties of the atmosphere at different heights. Limb viewing instruments can also sense infrared or microwave-emitted radiation from the atmosphere.

Limestone Sedimentary rock composed of carbonate minerals, especially calcium carbonate. Limestone can be created by clastic and non-clastic processes. Clastic limestones are formed from the break up and deposition of shells, coral and other marine organisms by wave-action and ocean currents. Non-clastic limestones can be formed either as a precipitate or by the lithification of coral reefs, marine organism shells, or marine organism skeletons.

Limiting factor Abiotic condition that most controls the growth of a species. For most terrestrial plants this condition is the supply of the nutrient nitrogen in the soil.

Line Copy (line drawing) Map copy suitable for reproduction without the use of a screen; a drawing composed of lines as distinguished from continuous- tone copy.

Line Echo Wave Pattern (LEWP) A bulge in a line of thunderstorms, producing a wave-shaped in the line. It is seen as a scalloped radar echo signature and is often associated with severe weather.

Line of apsides (major-axis of the ellipse) The straight line drawn from the perigee (point of orbit closest to Earth) to the apogee (point of orbit farthest from Earth) is the line-of-apsides.

Line of nodes The line created by the intersection of the equatorial plane and the orbital plane.

Line of sight Area within which visible contact can be made. For example, NOAA polar-orbiting satellites continuously transmit the APT signal. Radio reception of the APT signal is possible only when the satellite is above the horizon of a particular location (not obstructed by the Earth's surface), with a line-of-sight contact with the satellite.

Lipid Is an organic compound composed of carbon atoms that have two hydrogen atoms attached. Lipids are commonly known as fats and oils, and belong to the family of molecules known as hydrocarbons.

Liquefaction Temporary transformation of a soil mass of soil or sediment into a fluid mass. Occurs when the cohesion of particles in the soil or sediment is lost. Often triggered by seismic waves from an earthquake. For this condition to take place the pore spaces between soil particles must be at or near saturation.

Liquid A state of matter where molecules have the ability to flow and the surface of this mass displays the property of surface tension.

LIS The acronym for land information system. LIS and GIS are virtually synonymous, although it could be maintained that an LIS is but one component of a GIS.

Lithification Process by which sediments are consolidated into sedimentary rock.

Lithometeor Atmospheric phenomena which affect the state of the atmosphere. They constitute dry particles that hang suspended in the atmosphere, such as dust, smoke, sand, and haze.

Lithosphere Is the solid inorganic portion of the Earth (composed of rocks, minerals, and elements). It can be regarded as the outer surface and interior of the solid Earth.

Litter Accumulation of leaves, twigs and other forms of organic matter on the soil surface. In most soils, the surface layer of litter is at various stages of decomposition.

Litterfall Movement of leaves, twigs and other forms of organic matter from the biosphere to the litter layer found in soil.

Little climatic optimum Time period from 900 - 1200 AD. Warmest period since the Climatic Optimum.

Little ice age Time period from 1550 to 1850 AD. During this period, global temperatures were at their coldest since the beginning of the Holocene.

Littoral drift The sediment that is transported by waves and currents through beach drift and longshore drift along coastal areas.

Littoral transport The process of sediment moving along a coastline. This process has two components: longshore transport and onshore-offshore transport.

Littoral zone The zone along a coastline that is between the high and low-water spring tide marks.

Loam A soil that contains a roughly equal mixture of clay, sand, and silt. Good for growing most crops.

Lobe A tongue-like extension of some material. For example, the ice lobe of an alpine glacier.

Location A term used in geography that deals with the relative and absolution spatial position of natural and human-made phenomena.

Loess Deposits of silt laid down by aeolian processes over extensive areas of the mid-latitudes during glacial and postglacial times.

Logarithm Exponent of the power to which it is necessary to raise a fixed number (the base) to produce the given number. For example, the logarithm of 100 (base 10) is 2 because $10^2 = 100$.

Logarithmic scale Measurement scale based on logarithms. Values increase on this scale exponentially.

Long wave A large wave in the polar jet stream and the westerlies that extends from the middle to the upper troposphere. Often associated with the formation of a mid-latitude cyclone at the ground surface. Contrasts with short waves. Also called Rossby waves.

Long wave trough A wave in the prevailing westerly flow aloft which is characterized by a large length and amplitude. A long wave moves slowly and is persistent. Being associated with major cold troughs and warm ridges, their position and intensity govern weather patterns over a period of day or weeks.

Longitude lines Imaginary lines, also called meridians, running vertically around the globe. Unlike latitude lines, longitude lines are not parallel. Meridians meet at the poles and are widest apart at the equator. Zero degrees longitude (0°) is called the prime meridian. The degrees of longitude run 180° east and 180° west from the prime meridian.

Longshore current A water current that moves parallel to the shoreline.

Longshore drift The movement and deposition of coastal sediments because of longshore currents.

Longshore transport The transport of sediment in water parallel to a shoreline.

Longwave radiation See infrared radiation.

Loss of Signal (LOS) The inability to receive a satellite signal because the satellite's orbital path has taken it below the antenna's horizon. This term is relevant to all satellites except geostationary.

Lovelock, James E British scientist and naturalist who is most famous for his development of the Gaia Hypothesis. This theory suggests that life on the Earth functions like super-organism regulating its environment through biological interactions that influence the atmosphere, lithosphere, and hydrosphere.

Low A logic state corresponding to a binary "0". Satellite imagery is displayed on a computer monitor by a combination of highs and lows.

Low clouds A term used to signify clouds with bases below 6,000 feet and are of a stratiform or a cumuliform variety. Stratiform clouds include stratus and stratocumulus. Cumuliform clouds include cumulus and cumulonimbus. This altitude applies to the temperate zone. In the polar regions, these clouds may be found at lower altitudes. In the tropics, the defining altitudes for cloud types are generally higher.

Low latitudes The latitude belt between 30 and 0 degrees North and South of the equator. Also referred to as the tropical or torrid region.

Low Level Jet (LLJ) Strong winds that are concentrated in relatively narrow bands in the lower part of the atmosphere. It is often amplified at night. The southerly wind over the US Plains states during spring and summer is a notable example.

Low or low-pressure system A horizontal area where the atmospheric pressure is less than it is in adjacent areas. Since air always moves from areas of high pressure to areas of low pressure, air from these adjacent areas of higher pressure will move toward the low pressure area to equalize the pressure. This inflow of air toward the low will be affected by the Earth's rotation and will cause the air to spiral inward in a counterclockwise direction in the northern hemisphere. The air eventually rises near the centre of the low, causing cloudiness and precipitation.

The air in a low rotates in a counter-clockwise direction in the Northern Hemisphere, and in a clockwise direction in the Southern Hemisphere. Low-pressure cells are called cyclones.

Low pressure An area of atmospheric pressure within the Earth's atmosphere that is below average. If this system is on the Earth's surface and contains circular wind flow and enclosed isobars it is called a cyclone.

Low pressure system An area of a relative pressure minimum that has converging winds and rotates in the same direction as the earth. This is counterclockwise in the Northern Hemisphere and clockwise in the Southern Hemisphere. Also known as an cyclone, it is the opposite of an area of high pressure, or a anticyclone.

Low water Minimum height reached by a falling tide. The height may be due solely to the periodic tidal forces or it may have superimposed upon it the effects of meteorological conditions.

Low water line Intersection of the land with the water surface at an elevation of low water. Not to be confused with mean low water line.

Lower mantle Layer of the Earth's interior extending from 670 to 2,900 kilometers below the surface crust. Composed of ultramafic rock. This layer is hot and plastic and part of the mantle layer.

Lunar eclipse An eclipse of the moon occurs when the earth is in a direct line between the sun and the moon. The moon does not have any light of its own, instead, it reflects the sun's light. During a lunar eclipse, the moon is in the earth's shadow. It will often look dim and sometimes copper or orange in colour.

Luvisol soil Soil order (type) of the Canadian System of Soil Classification. This soil type is associated with forest vegetation. The most identifying traits of this soil is the presence of calcareous parent material which results in a high pH and strong eluviation of clay from the A horizon.

Lysimeter Meteorological instrument used to measure potential and actual evapotranspiration.

M

Mackerel sky The name given to cirrocumulus clouds with small vertical extent and composed of ice crystals. The rippled effect gives the appearance of fish scales.

Macroburst A large downburst with an outflow diameter of 2.5 miles (4 kilometers) or larger and damaging winds.

Macronutrient Nutritional element required by an organism in relatively large quantities.

Macros Are collections of frequently used commands which are grouped together in a file. Instead of entering commands individually, users can call the macro and the commands stored therein will be executed automatically. Macros may be stored in libraries .

Macroscale The meteorological scale for obtaining weather information covering an area ranging from the size of a continent to the entire globe.

Mafic magma Magma that is relative poor in silica but rich in calcium, magnesium, and iron content. This type of magma solidifies to form rocks relatively rich in calcium, magnesium, and iron but poor in silica.

Magma Molten rock originating from the Earth's interior.

Map A map is a graphic representation of a place. There are many different types of maps that have different uses. Since a map is 2-dimensional representation of a 3-dimensional world, compromises must be made. Different maps differ in the relative accuracy of the depiction of the area, the shapes of objects, actual distances, and compass direction. Maps that accurately reflect area are often called equal-area maps (an example is the Albers equal-area conic map). Maps that maintain the shape of objects are called conformal. Maps that correctly show the distance between areas are often called equi-distant maps (note that the shortest distance between two points on a map is generally not a straight line. but a curve). Navigational maps need accurate compass directions maintained on the map (like the Mercator map).

Mantle Layer of the Earth's interior composed of mostly solid rock that extends from the base of crust to a depth of about 2,900 kilometres.

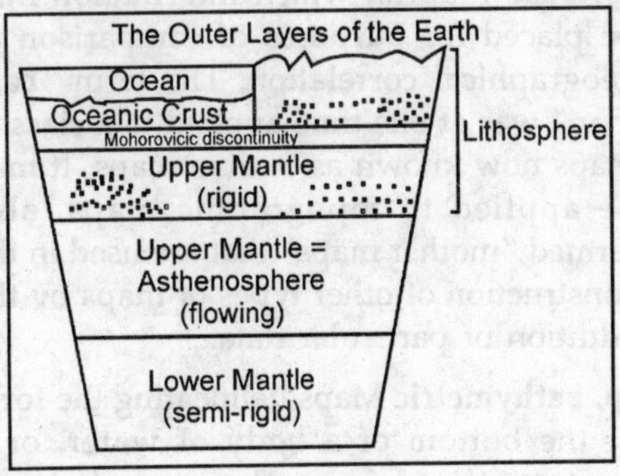

Fig. Mantle

Mangrove Treed wetlands located on the coastlines in warm tropical climates.

Mammatocumulus An obsolete term for cumulonimbus mammatus, it is a portion of a cumulonimbus cloud that appears as a pouch or udder on the under surface of the cloud. Although they do not cause severe weather, they often accompany storms.

Mammal Group of warm blooded vertebrate animals. Common characteristics found in these organisms include: hair, milk secretion, diaphragm for respiration, lower jaw composed of a single pair of bones, middle ear containing three bones, and presence of only a left systemic arch.

Magnitude 1. The quantifiable size of a natural event.

2. A quantitative measure of the size of an earthquake using the Richter scale.

Magnetosphere Region surrounding a celestial body where its magnetic field controls the motions of charged particles. The Earth's magnetic field is dipolar in nature. That is, it behaves as if produced by a giant bar magnet located near the centre of the planet with its north pole tilted several degrees from Earth's geographic north pole.

The Earth's magnetic field presents an obstacle to the solar wind, as a rock in a running stream of water This obstacle is called a bow shock. The bow shock slows down, heats, and compresses the solar wind, which then flows around the rest of Earth's magnetic field.

Magnetic reversal A change in the polarity of the Earth's magnetic field. In the past 4 million years there have been nine reversals.

Magnetic poles Either of the two points on the earth's surface where the magnetic meridians converge. They are not aligned with the geographical poles, but shift and do not lie exactly opposite of the other.

Magnetic field The space influence by magnetic force. The Earth's magnetic field is believed to be generated by the planet's core.

Magnetic declination The horizontal angle between true north and magnetic north or true south and magnetic south.

Magma plume A rising vertical mass of magma originating from the mantle.

Map digitization Conversion of map data from graphic to digital form.

Map hypsographic Map showing relief with elevations referred to the national geodetic vertical datum of 1929.

Map projections Two-dimensional representations of the three-dimensional Earth. Because projections attempt to present the spherical Earth on a flat plane, they inevitably produce distortions. Map projections are numerous and complex (e.g., there are a variety of cylindrical, conic, or azimuthal projections). Each projection has advantages and serves different purposes, and each produces different types of distortions in direction, distance, shape, and relative size of areas. One of the most famous projections is the Mercator, created by Geradus Mercator in 1569. It is a rectangular-shaped map in which all longitude and latitude lines are parallel and intersect at right angles (on a globe, meridians are not parallel, but grow narrower, eventually converging at the poles). Near the equator, the scale of the Mercator is accurate, but the farther one moves toward the poles, the greater the distortion—Antarctica in the far south and Greenland in the far north, for example, appear gigantic. The Mercator projection was used well into the 20th century, but has now been superseded by others, including the widely used Robinson projection. The Robinson projection is an elliptical-shaped map with a flat top and bottom. Developed in 1963 by Arthur H. Robinson, it is an orthophanic ("right appearing") projection, which attempts to reflect the spherical appearance of the Earth. The meridians, for example, are curved arcs, which gives the flat map a three-dimensional appearance. But to convey the likeness of a curved, three-dimensional globe, the Robinson projection must in fact distort shape, area, scale, and distance. The Albers, Lambert, Mollweide, and Winkel Tripel are some of the other commonly used map projections.

Map scale Is the relationship between distance on a map and the corresponding distance on the earth's surface. Map scale is often recorded as a representative fraction such as 1:1,000,000 (1 unit on the map represents a million units on the earth's surface) or 1:24,000 (1 unit on the map represents 24,000 units on the earth's surface). The terms "large" and "small" refer to the relative magnitude of the representative fraction. Since 1/1,000,000 is a smaller fraction than 1/24,000, the former is said to be a smaller scale. Small scales are often used to map large areas because each map unit covers a larger earth distance. Large-scale maps are employed for detailed maps of smaller areas.

Map series Family of maps conforming generally to the same specifications and designed to cover an area or a country in systematic pattern.

Map, base Map on which information may be placed for purposes of comparison or geographical correlation. The term "base map" was at one time applied to a class of maps now known as outline maps. It may be applied to topographic maps, also termed "mother maps" that are used in the construction of other types of maps by the addition of particular data.

Map, bathymetric Maps delineating the form of the bottom of a body of water, or a portion thereof, by the use of depth contours (isobaths).

Map, cadastral Map showing the boundaries of subdivisions of land, often with the bearings and lengths thereof and the areas of individual tracts, for purposes of describing and recording ownership. It may also show culture, drainage, and other features relating to land use and value.

Map, choropleth Thematic map in which areas are coloured, shaded, dotted, or hatched to create darker or lighter areas in proportion to the density of distribution of the theme subject.

Map, engineering Map showing information that is essential for planning an engineering project or development and for estimating its cost. It usually is a large-scale map of a small area or of a route. It may be entirely the product of an engineering survey, or reliable information may be collected from various sources for the purpose, and assembled on a base map.

Map, flood control Map designed for studying and planning control projects in areas subject to flooding.

Map, forestry Map prepared principally to show the size, density, kind, and value of trees in a designated area.

Map, geologic Map showing the structure and composition of geologic features.

Map, hypsometric Map showing relief by any convention, such as contours, hachures, shading, or tinting.

Map, isopleth Map consisting of lines connecting places of equal value of distribution for a given theme such as rainfall or temperature.

Map, land use Map showing by means of a coding system the various purposes for which parcels of land are being used by man.

Map, line Map composed of lines as distinguished from photographic imagery.

Map, planimetric Map that presents only the horizontal positions for features represented. distinguished from a topographic map by the omission of relief in measurable form. The features usually shown on a planimetric map include rivers, lakes, and seas; mountains, valleys, and plains; forests, and prairies; cities, farms transportation routes, and public utility facilities; and political and private boundary lines. A planimetric map intended for special use may present only those features essential to the purpose to be served.

Map, slope (clinometric map) Map showing the degree of steepness of the Earth's surface by the use of various colours or shading for critical ranges of slope.

Map, soil Map that shows the constitution, structure, and texture of the soil and identifies ongoing erosion.

Map, storm evacuation Map designed to identify coastal areas subject to flooding, to indicate recommended areas of refuge, and to emphasize available evacuation routes.

Map, thematic Map designed to provide information on a single topic, such as geology, rainfall, population.

Map, topographic Map that present the horizontal and vertical positions of the features represented; distinguished from a planimetric map by the addition of relief in measurable form.

Marble Metamorphic rock created by the recrystallization of calcite and/or dolomite.

Mare's tail The name given to thin, wispy cirrus clouds composed of ice crystals that appear as veil patches or strands, often resembling a horse's tail.

Marginal Visual Flight Rules (MVFR) Refers to the general weather conditions pilots can expect at the surface. VFR stands for visual flight rules and MVFR means minimum or marginal visual flight rules. MVFR criteria means a ceiling between 1,000 and 3,000 feet and/or 3 to 5 miles visibility. Contrast with IFR.

Marine With reference to ocean environments and processes.

Maritime air mass An air mass influenced by the sea. It is a secondary characteristic of an air mass classification, signified by the small "m" before the primary characteristic, which is based on source region. For example, mP is an air mass that is maritime polar in nature. Also known as a marine air mass.

Maritime climate A climate strongly influenced by an oceanic environment, found on islands and the windward shores of continents. It is characterized by small daily and yearly temperature ranges and high relative humidity.

Maritime effect The effect that large ocean bodies have on the climate of locations or regions. This effect results in a lower range in surface air temperature at both daily and annual scales.

Maritime Polar Air Mass (MP) Air mass that forms over extensive ocean areas of the middle to high latitudes. Around North America, these system form over the Gulf of Mexico and the eastern tropical Pacific. Maritime Polar air masses are mild and humid in summer and cool and humid in winter. In the Northern Hemisphere, maritime polar air masses are normally unstable during the winter. In the summer, atmospheric stability depends on the position of the air mass relative to a continent. Around North America, Maritime Polar air masses found over the Atlantic are stable in summer, while Pacific systems tend to be unstable.

Maritime Tropical Air Mass (MT) Air mass that forms over extensive ocean areas of the low latitudes. Around North America, these system form over the Gulf of Mexico and the eastern tropical Pacific. Maritime Tropical air masses are warm and humid in both winter and summer. In the Northern Hemisphere, maritime tropical air masses can normally stable during the whole year if they have form just west of a continent. If they form just east of a continent, these air masses will be unstable in both winter and summer.

Marsh A marsh is a type of freshwater, brackish water or saltwater wetland that is found along rivers, pond, lakes and coasts. Marsh plants grow up out of the water.

Marsh, coastal Area of salt-tolerant vegetation in brackish and (or) saline-water habitants subject to tidal inundation.

Marsh, freshwater Tract of low wet ground, usually miry and covered with rank vegetation.

Mass Refers to the amount of material found in an object (usually of unit volume).

Mass balance The relative balance between the input and output of material within a system.

Mass extinction A catastrophic, widespread perturbation where major groups of species become extinct in a relatively short time compared to normal background extinctions.

Mass movement General term that describes the downslope movement of sediment, soil, and rock material.

Mass number Total number of neutrons and protons in the nucleus of an atom. Approximate measure of the mass of an atom.

Mass storage Refers to computer equipment designed to record, store, and read "masses" of data. A variety of magnetic and optical media are currently available for mass storage, including magnetic disks, magnetic tapes, magnetic cassettes, and optical disks. The selection of a suitable mass storage device for a GIS database usually involves establishing a compromise among the factors of size, speed, and cost.

Mass wasting General term that describes the downslope movement of sediment, soil, and rock material.

Matric force Force that holds soil water from 0.0002 to 0.06 millimetres from the surface of soil particles. This force is due to two processes: soil particle surface molecular attraction (adhesion and absorption) to water and the cohesion that water molecules have to each other. This force declines in strength with distance from the soil particle. The force becomes nonexistent past 0.06 millimetres.

Matter Is the material (atoms and molecules) that constructs things on the Earth and in the Universe.

Maunder minimum Period from 1645 to 1715 during which the sun had very little sunspot activity.

Maximum The greatest value attained by a function, for example, temperature, pressure, or wind speed. The opposite of minimum.

MDC See more developed country.

Mean Statistical measure of central tendency in a set of data. The mean is calculated by adding all of the data values and dividing this quantity by the total number of data values.

Mean Anomaly (MO or MA or phase) Specifies the mean location (true anomaly specifies the exact location) of a satellite on an orbit ellipse at a particular time, assuming a constant mean motion throughout the orbit. Epoch specifies the particular time at which the satellites position is defined, while mean anomaly specifies the location of the satellite at epoch. Mean anomaly is measured from 0o to 360o during one revolution. It is defined as 0o at perigee, and hence is l80o at apogee.

Mean high water Tidal datum that is the arithmetic mean of the high water heights observed over a specific 19-year Metonic cycle (national tidal datum epoch). For stations with shorter series, simultaneous observations are made with a primary control tide station to derive the equivalent of a 19-year value. Use of "mean high tide" is discouraged.

Mean high water line Intersection of the land with the water surface at the elevation of high water.

Mean low water Tidal datum that is the arithmetic mean of the low water heights observed over a specific 19-year Metonic cycle (national tidal datum epoch). For stations with shorter series, simultaneous observations are made with a primary control tide station to derive the equivalent of a 19-year value. Use of "mean low tide" is discouraged.

Mean low water line Intersection of the land with the water surface at the elevation of low water.

Mean motion Averaged speed of a satellite in a non-circular orbit (i.e., eccentricity > 0).

Satellites in circular orbits travel at a constant speed. Satellites in non-circular orbits move faster when closer to the Earth, and slower when farther away Common practice is to compute the mean motion (average the speed), which is measured in revolutions per day.

Mean sea level The average height of the sea surface water level. For the United States, it is computed by averaging the levels of all tide stages over a nineteen year period, determined from hourly height readings measured from a fix, predetermined reference level. It is used as a basis for determining elevations, as the reference for all altitudes in upper air measurements, and as the level above which altitude is measured by a pressure altimeter for aviation. Often referred to as MSL.

Mean solar day Time it takes to complete one Earth rotation relative to the position of the sun (for example, from midnight to midnight). This measurement takes 24 hours and is longer than a sidereal day because it includes the effect of the Earth's movement (Earth revolution) around the sun.

Mean temperature The average of temperature readings taken over a specified amount of time. Often the average of the maximum and minimum temperatures.

Meander Sinuous shaped stream channel. Usually found in streams flowing over a very shallow elevation grade.

Meander line Metes-and-bounds traverse approximately along the mean high water line of a permanent body of water. By following the sinuosities of the bank or shoreline, the meander line provides data for computing the area of land remaining after the water area has been segregated. A meander line differs from other metes and bounds surveys in that it does not ordinarily determine or fix boundaries.

Meanderable Capable of being depicted by reference to a meander line.

Measured ceiling A ceiling classification applied when the ceiling value has been determined by an instrument, such as a ceilometer or ceiling light, or by the known heights of unobscured portions of objects, other than natural landmarks, near the runway.

Measurement system integrity The tracking and documentation over the long term of all causes of error or uncertainty in a final data-analysis product. These include instrument calibration, adequacy of measurement validation, data coverage and sampling density, availability and quality of ancillary data, procedures for data analysis and reduction, the results of checks against independent measurement, and quantitative error analysis.

Measurement validation The establishment of confidence in the numerical relationship between the calibrated sensor output and the actual variable being measured.

Medial moraine Deposit of material found down the centre of a glacier. Created when two glacier and their lateral moraines merge.

Median Statistical measure of central tendency in a set of data. The median is the value halfway through a data set where the values have been ordered from lowest to highest. In an even data set, the median is the average of the two halfway values.

Mediterranean climate A climate characterized by moist, mild winters and hot, dry summers.

Melting The physical process of a solid becoming a liquid. For water, this process requires approximately 80 calories of heat energy for each gram converted.

Melting level The altitude at which ice crystals and snow flakes melt as they descend through the atmosphere. Refer to a bright band.

Melting point The temperature at which a solid substance undergoes fusion, changing from a solid to a liquid state. Contrast with freezing point.

Meltwater Water produced from the melting of snow and/or glacial ice.

Mercalli scale A scale for rating the power of an earthquake.

Mercator projection A method of making maps in which the Earth's surface is shown as a rectangle with the meridians as parallel straight lines spaced at equal intervals and the parallels of latitude as parallel straight lines intersecting the meridians at right angles. Areas away from the equator appear larger than they are, with the greatest distortion near the poles.

Mercurial barometer An instrument used for measuring the change in atmospheric pressure. It uses a long glass tube, open at one end and closed at the other. After first filling the open end with mercury, it is then temporarily sealed and placed into a cistern of mercury. A nearly perfect vacuum is established at the closed end after the mercury descends. The height of the column of mercury in the tube is a measurement of air pressure. As atmospheric pressure increases, the mercury is forced from the cistern up the tube; when the atmospheric pressure decreases, the mercury flows back into the cistern. Measurement is taken in inches of mercury. Although mercurial barometers are very accurate, practicality has led observers to use aneroid barometers. First used by Evangelista Torricelli (1608-1647), an Italian physicist and mathematician, to explain the fundamental principles of hydromechanics.

Meridian A meridian is a circular arc (a great circle) of longitude that meets at the north and south poles and connects all places of the same longitude. The prime meridian (0 degrees longitude) passes through Greenwich, England.

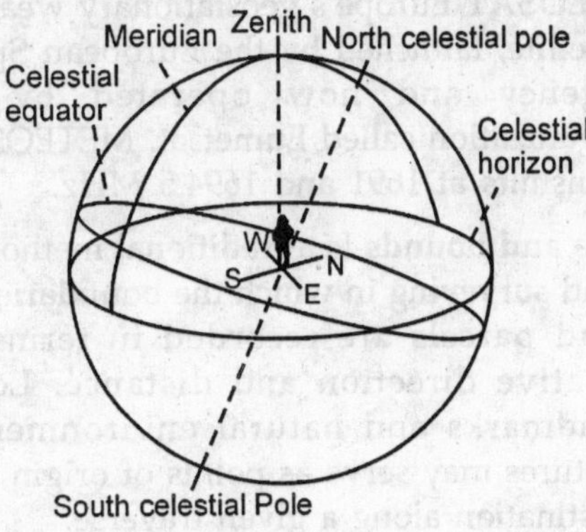

Fig. Meridian

Meridional Movement of wind or ocean waters in a direction that is roughly perpendicular to the lines of latitude.

Meridional flow Atmospheric circulation in which the north and south, or meridional, component of motion is unusually pronounced. This weakens the zonal flow.

Meridional transport Transport of atmospheric and oceanic energy from the equator to the poles.

Mesa A flat topped hill that rises sharply above the surrounding landscape. The top of this hill is usually capped by a rock formation that is more resistant to weathering and erosion.

Mesocyclone A area of rotation of storm size that may be found on the southwest part of a supercell. Its circulation is much larger than the tornado that may develop within it. Originally a radar term for a rotation signature that met certain criteria, it is best seen on Doppler radar.

Mesohigh A small, concentrated area of high pressure that may be created by the cold outflow and rain-cooled air from thunderstorms. It often forms a pseudo cold front or squall line on its leading edges. Also known as a bubble high.

Mesolow A small low pressure centre the size of an individual thunderstorm. Its presence may increase severe weather potential.

Mesopause Thin boundary layer found between the mesosphere and the thermosphere. It is usually found at an average altitude of 80 kilometres. Coldest temperatures in the atmosphere are found at the mesopause.

Mesophyte Plants that have moderate water requirements.

Mesoscale The scale of meteorological phenomena that includes MCCs, MCSs, and squall lines. These weather systems may cover fifty to several hundreds of miles. Smaller phenomena are classified as storms, while larger are classified as synoptic-scale.

Mesoscale Convective Complex (MCC) A large mesoscale convective system (MCS) which is about the size of the state of Ohio or Iowa and lasts at least 6 hours. Forming during the afternoon and evening, this is the peak time for associated severe weather. The complex normally reaches its peak intensity at night when heavy rainfall and flooding become the primary threat.

Mesoscale Convective System (MCS) A large organized convective weather system comprised of a number of individual thunderstorms. It normally persists for several hours and may be rounded or linear in shape. This term is often used to describe a cluster of thunderstorms that does not meet the criteria of a mesoscale convective complex (MCC).

Mesosphere Atmospheric layer found between the stratosphere and the thermosphere. Usually located at an average altitude of 50 to 80 kilometers above the Earth's surface. Air temperature within the mesosphere decreases with increasing altitude.

Mesotrophic lake Lake with a moderate nutrient supply.

Mesozoic Geologic era that occurred from 245 to 65 million years ago.

Mesquite A spiny deep-rooted leguminous tree or shrub that forms extensive thickets in the southwestern United States.

Metabolism Describes all of the enzymatic reactions performed by the cells of an organism.

Metadata Information describing the content or utility of a data set. For example, the dates on which data were procured are metadata.

Metamorphic rock A rock that forms from the recrystallization of igneous, sedimentary or other metamorphic rocks through pressure increase, temperature rise, or chemical alteration.

Metamorphism Process that creates metamorphic rocks.

Metasomatic metamorphism Form of metamorphism that causes the chemical replacement of elements in rock minerals when gases and liquids permeate into bedrock.

Meteor A body of matter that enters the Earth's atmosphere from space. While traveling through the atmosphere, these objects begin to burn because of friction and are sometimes seen as luminous streaks in the sky by ground observers. Many of these objects burn up completely and never reach the Earth's surface.

Meteor The former Soviet Union's series of polar-orbiting weather satellites. The Meteor satellites transmit images in a system compatible with the NOAA polar-orbiting satellites.

Meteorology The scientific study of the atmosphere and its associated phenomena.

Meteorology/meteorologist The science and study of the atmosphere and atmospheric phenomena. Various areas of meteorology include agricultural, applied, astrometeterology, aviation, dynamic, hydrometeorology, operational, and synoptic, to name a few. A scientist who studies the atmosphere and atmospheric phenomena.

METEOSAT Europe's geostationary weather satellite, launched by the European Space Agency and now operated by an organization called Eumetsat. METEOSAT transmits at 1691 and 1694.5 MHz.

Metes and bounds Is a traditional method of land surveying in which the boundaries of land parcels are recorded in terms of relative direction and distance. Local landmarks and natural environmental features may serve as points of origin and destination along a given traverse.

Methane Methane is very strong greenhouse gas found in the atmosphere. Methane concentrations in the atmosphere have increased by more than 140 % since 1750. The primary sources for the additional methane added to the atmosphere (in order of importance) are: rice cultivation, domestic grazing animals, termites, landfills, coal mining, and oil and gas extraction. Chemical formula for methane is CH_4.

Metonic cycle Period of 235 lunations or about 19 years. devised by Meton, an Athenian astronomer (5th century B.C.) for the purpose of obtaining a period at the end of which the phases of the moon recur in the same order and on the same days as in the preceding cycle.

Metric system Decimal system of weights and measures based on the meter as a unit length and the kilogram as a unit mass.

Metropolitan coalescence The merging of the urbanized areas of separate metropolitan regions; Megalopolis is an example of this process.

Metsat Generic term for meteorological (weather) satellites.

MHz (megahertz) 10^6 hertz.

Mica Silicate mineral that exhibits a platy crystal structure and perfect cleavage. Common forms of mica are biotite and muscovite.

Microbarograph A instrument designed to continuously record a barometer's reading of very small changes in atmospheric pressure.

Microburst A severe localized wind blasting down from a thunderstorm. It covers an area less than 2.5 miles (4 kilometers) in diameter and is of short duration, usually less than 5 minutes.

Micrometer One millionth of a meter; used to measure wavelengths in the electromagnetic spectrum.

Micronutrient Nutritional element required by an organism in relatively very small quantities.

Microorganism Extremely small organism that can only be seen using a microscope.

Microprocessor Controlling unit of a microcomputer; laid out on a tiny silicon chip and containing the logical elements for handling data, performing calculations, carrying out stored instructions, etc.

Microscale The smallest scale of meteorological phenomena, such as turbulence, with life spans of less than a few minutes that affect very small areas and are strongly influenced by local conditions of temperature and terrain.

Microwave Electromagnetic radiation with wavelengths between about 1000 micrometers and one meter.

Microwave radiation Form of electromagnetic radiation with a wavelength between 0.1 to 100 centimetres.

Mid latitude cyclone Cyclonic storm that forms primarily in the middle latitudes. Its formation is triggered by the development of troughs in the polar jet stream. These storms also contain warm, cold and occluded fronts. Atmospheric pressure in their centre can get as low as 970 millibars. Also called wave cyclones or frontal cyclones.

Mid oceanic ridge Chain of submarine mountains where oceanic crust is created from rising magma plumes and volcanic activity. Also associated with this feature is plate divergence which creates a rift zone.

Middle clouds A term used to signify clouds with bases between 6,000 and 18,000 feet. At the higher altitudes, they may also have some ice crystals, but they are composed mainly of water droplets. Altocumulus, altostratus, and nimbostratus are the main types of middle clouds. This altitude applies to the temperate zone. In the polar regions, these clouds may be found at lower altitudes. In the tropics, the defining altitudes for cloud types are generally higher.

Middle infrared Electromagnetic radiation between the near infrared and the thermal infrared, about 2-5 micrometers.

Middle latitudes The latitude belt roughly between 35 and 65 degrees North and South. Also referred to as the temperate region.

Migration Movement of organisms in an intentional way between two points in space. Many migrations are seasonal.

Milankovitch theory Theory proposed by Milutin Milankovitch that suggests that changes in the Earth's climate are cause by variations in solar radiation received at the Earth's surface. These variations are due to cyclical changes in the geometric relationship between the Earth and the sun.

Milky way galaxy Aggregation of about 400 billion stars in a flattened, disk-shaped structure in space. Our solar system is found in this structure.

Miller cylindrical projection Map projection that mathematically projects the Earth's surface onto a cylinder that is tangent at the equator. Directions and distances are only true at the equator. Distance, area, and shape distortion increases as one moves towards the poles. Very popular projection used in world maps.

Millibar (MB) The standard unit of measurement for atmospheric pressure used by the National Weather Service. One millibar is equivalent to 100 newtons per square meter. Standard surface pressure is 1,013.2 millibars.

Mineral Component of rocks. A naturally occurring inorganic solid with a crystalline structure and a specific chemical composition. Over 2,000 types of minerals have been classified.

Mineralization Decomposition of organic matter into its inorganic elemental components.

Minimum The least value attained by a function, for example, temperature, pressure, or wind speed. The opposite of maximum.

Mission to Planet Earth (MTPE) International research program to understand our planet's environment as a system. A major challenge of MTPE is to observe, understand, model, assess, and eventually predict global change. Meeting this challenge will help to evaluate the impact that human activity (e.g., clearing forests and burning fossil fuels) has on our environment, and to distinguish human-induced changes from the effects of natural events (e.g. volcanic eruptions, erosion).

NASA's MTPE uses space-, aircraft-, and ground~based measurements to provide the scientific basis for understanding global change. The program will produce longterm global maps of clouds, land and ocean vegetation, atmospheric ozone, sea-surface temperature, and other global processes necessary to understand the state of the Earth and to detect any patterns of change. This information will be available to scientist5 and policy makers through the earth observing system data and information system (EOSDIS).

The centrepiece of NASAs MTPE will be the Earth Observing System (EOS), a series of satellites planned for launch beginning in 1998. Measurements from EOS will be complemented by the Earth Probes, a series of discipline-specific satellites and instruments designed to observe Earth processes where smaller platforms and/or different orbits from EOS are required. Planned Earth Probes will measure tropical rainfall, ocean productivity, ozone, and ocean surface winds. In addition, MTPE includes current NASA Earth science missions collecting important data on the global environment, such as the Upper Atmosphere Research Satellite (UARS) and the Ocean Topography Experiment (TOPEX/POSEIDON), Space Shuttle experiments such as ATLAS, and aircraft campaigns.

Mississippian Geologic period that occurred roughly 320 to 360 million years ago. During this period, insects undergo major speciation and ferns first appear. Trees become a dominant plant form on continents.

Mist A collection of microscopic water droplets suspended in the atmosphere. It does not reduce visibility as much as fog and is often confused with drizzle.

Mistral Term used to describe a katabatic wind in southern France.

Mitochondria Organelle in a cell that oxidizes organic energy for use in cellular metabolism.

Mixed layer The layer of the water that is mixed through wave action or thermohaline convection.

Mixed precipitation Any of the following combinations of freezing and frozen precipitation: snow and sleet, snow and freezing rain, or sleet alone. Rain may also be present.

Mixed tide Tides that have a higher high water and lower high water as well as higher low water and lower low water per tidal period.

Mixing ratio The ratio between the weight (mass) of water vapour (or some other gas) held in the atmosphere compared to the weight of the dry air in a given volume of air. Usually measured in grams water vapour (or gas) per kilogram of dry air.

Mode Statistical measure of central tendency in a set of data. The mode is the most frequently occurring value in a data set. Data sets can contain two or more mode values that occur with the same frequency.

Model 1. Generalization of reality.

2. System describing how a phenomenon functions.

3. Mathematical representation of a system from which predictions or inferences can be made.

Model (noun) A mathematical representation of a process, system, or object developed to understand its behavior or to make predictions. The representation always involves certain simplifications and assumptions.

Modem (modulator/demodulator) Device that allows two computers (which use binary data in the form of bits) to communicate using a telephone line (which uses tones). When the computer is transmitting data, the modem is needed to modulate binary data into tones. When receiving data, the device is needed to demodulate the tones to obtain the binary data required by the computer. Since the computer must be both a transmitter and receiver of data, the modem must be able to modulate and demodulate data.

Modulation Variation in the frequency of a radio wave in accordance with some other impulse. Modulation is essential to communication systems in which a number of different signals must all share the same medium. One way this sharing can be accomplished is to place each signal in its own band of frequencies in the medium. Amplitude modulation and frequency modulation are two ways in which signals can be moved within the frequency domain to accomplish placement and sharing.

The combining of a number of signals to share a communication medium by dividing it into different frequency bands for each signal is called frequency-division multiplexing.

Amplitude modulation (AM) is technologically quite simple, and the bandwidth of the amplitude-modulated carrier is at most twice the bandwidth of the modulating signal. However; an amplitude-modulated carrier is very prone to the effects of additive noise.

Frequency modulation (FM) is more complicated than amplitude modulation, and the bandwidth of the frequency-modulated carrier can be many times that of the modulating signal. However; the process of demodulating a frequency-modulated carrier eliminates much of the deleterious effects of additive noise. This trade-off between bandwidth and noise reduction characterizes most communication situations.

Moho discontinuity The lower boundary of the crust. At this boundary seismic wave velocities show an increase in speed as they enter the upper mantle.

Moist adiabat The lifting of saturated air, or air that contains water vapour. As a parcel lifts, it cools to its saturation temperature and the relative humidity is then 100 percent. Further cooling results in condensation. When air is moist and is lifted adiabatically, then the temperature decreases at a rate of 0.55 degrees Celsius per 100 meters (2-3 degrees Fahrenheit per 1,000 feet). Contrast with a dry adiabat.

Moisture Refers to the water vapour content in the atmosphere, or the total water, liquid, solid or vapour, in a given volume of air.

Molecule Minute particle that consists of connected atoms of one or many elements.

Mollisols Soil order (type) of the United States Natural Resources Conservation Service Soil Classification System. A mollisol soil is typically found in semiarid grassland environments. They are rich in organic matter and have an abundance of calcium carbonate nodules within the profile.

Mollweide projection A Mollweide projection is a type of sinusoidal projection map in which the entire surface of the Earth is shown within an ellipse. Lines of latitude are parallel to the equator, but lines of longitude are curved in such a way that area distortion is minimal. The distortion is greatest at the edges of the ellipse. This type of projection was created by Carl B. Mollweide in 1805.

Monadnock An isolated hill or mountain of resistant rock rising above an eroded lowland.

Monera Unicellular organisms that have a prokaryotic cell type.

Monocline A fold in layered rock that creates a slight bend.

Monoscopic Pertaining to the observation of a single photograph or other view.

Monsoon The seasonal shift of winds created by the great annual temperature variation that occurs over large land areas in contrast with associated ocean surfaces. The monsoon is associated primarily with the moisture and copious rains that arrive with the southwest flow across southern India. The name is derived from the word mausim, Arabic for season. This pattern is most evident on the southern and eastern sides of Asia, although it does occur elsewhere, such as in the southwestern United States.

Fig. Monsoon

Montmorillonite A type of clay that has a large capacity to shrink and expand with wetting and drying.

Montreal protocol An international agreement to drastically reduce CFC production, the Protocol was adopted in Montreal in 1987. It was significantly strengthened at a subsequent meeting in London in 1990 that called for a complete elimination of CFCs by the year 2000. The agreement was again amended by a Meeting of the Parties in Copenhagen in November 1992. Consumption of controlled substances—such as CFCs and halons—was greatly reduced or eliminated, and many accountability dates were moved forward, often from January 2000 to 1 January 1996.

Monument (surveying) Permanent physical structure marking the location of a survey point. Common types of monuments are inscribed metal tablets set in concrete posts; and metal rods driven in the ground.

Monumentation In surveying refers to the practice of marking known horizontal and vertical control points with permanent structures such as concrete pedestals and metal plaques. Once surveyed and marked, these monuments can be used for further surveying and for the alignment of land-parcel boundaries and infrastructure.

Moraine The rocks and soil carried and deposited by a glacier. An "end moraine," either a ridge or low hill running perpendicular to the direction of ice movement, forms at the end of a glacier when the ice is melting.

More Developed Country (MDC) A highly industrialized country characterized by significant technological development, high per capita income, and low population growth rates. Examples of such countries include the United States, Canada, Japan, and many countries in Europe. Also see less developed country.

Morphological system This is a system where we understand the relationships between elements and their attributes in a vague sense based only on measured features or correlations. In other words, we understand the form or morphology a system has based on the connections between its elements. We do not understand exactly how the processes work to transfer energy and/or matter through the connections between the elements.

Morphometry The measurement of shape. Measurements are then manipulated statistically or mathematically to discover inherent properties.

Mosaic A composite picture built up from a number of image segments. An example of a mosaic is the WEFAX transmission, which includes both polar and mercator mosaics derived from TIROS-N/NOAA polar orbit image data.

Mosaic, aerial Assembly of aerial photographs whose edges usually have been torn or cut selectively and matched to the imagery on adjoining photographs to form a continuous representation of a portion of the Earth's surface.

Moss About 9,500 species of plants that belong to the division bryophyta. These low growing plants are common in moist habitats.

Mountain A mountain is a very tall high, natural place on Earth - higher than a hill. The tallest mountain on Earth is Mt. Everest.

Mountain breeze A katabatic wind, it is formed at night by the radiational cooling along mountainsides. As the slopes become colder than the surrounding atmosphere, the lower levels of air cool and drain to the lowest point of the terrain. It may reach several hundred feet in depth, and extreme cases, attain speeds of 50 knots or greater. It blows in the opposite direction of a valley breeze.

Mountain wave A wave in the atmosphere caused by a barrier, such as a mountain. Sometimes it is marked by lenticular clouds to the lee side of mountain barriers. Also called a standing wave or a lee wave.

Mouth End of a stream. Point at which a stream enters a lake, sea, or ocean.

Movement A term used in geography that deals with the migration, transport, communication, and interaction of natural and human-made phenomena across the spatial dimension.

Mud slide Fast moving soil, rocks and water that flow down mountain slopes and canyons during a heavy a downpour of rain.

Mudflow Form of mass movement where fine textured sediments and soil mix with water to create a liquid flow.

Mudstone Fine grained sedimentary rock composed of lithified silt and clay particles.

Muggy A subjective term for warm and excessively humid weather.

Multicell storm A thunderstorm made up of two or more single-cell storms.

Multiple vortex tornado A tornado which has two or more condensation funnels or debris clouds, often rotating around a common centre.

Multiplex Stereoplotter of the double-projection type characterized by its use of reduced- scale diapositives and stationary lamphouses with condensing lenses.

Multiplexer A device that combines several separate communications signals into one and outputs them on a single line.

Multispectral Scanner (MSS) A line-scanning instrument flown on Landsat satellites that continually scans the Earth in a 185 km (100 nautical miles) swath. On Landsats 1, 2, 4, and 5, the MSS had four spectral bands in the visible and near-infrared with an IFOV of 80 meters. Landsat-3 had a fifth band in the thermal infrared with an IFOV of 240 meters.

Municipal waste Unwanted by-products of modern life generated by people living in an urban area.

Muscovite Rock forming mineral of the mica group.

Muskeg Poorly drained marshes or swamps found overlying permafrost.

Mutation Change in the structure of a gene or chromosome.

Mutualism Interspecific interaction where both species experience and increase in their fitness after interacting with the other species. Mutualistic interactions between species can be of two types: symbiotic or nonsymbiotic.

Mycorrhizae Mutualistic association of a fungus with the root of higher plant. In this relationship, the fungus helps the plant in extracting certain nutrients from the soil. In exchange, the fungus is provided with a habitat and nutrition in the form of carbohydrates.

N

Nadir Point on Earth directly beneath a satellite, the opposite of zenith. Compare with subsatellite point.

Nanometer (NM) One billionth of a meter. Nanometers are used to measure wavelengths in the electromagnetic spectrum.

NASA See national aeronautics and space administration.

NASA centres The ten major NASA Centres are: Ames research centre (ARC) Located at Moffett Field, California. ARC is active in aeronautical research, life sciences, space science, and technology research. The Centre houses the world's largest wind tunnel and the world's most powerful supercomputer system.

The dryden flight research centre, Edwards Air Force Base, California, formerly part of ARC, became a separate entity March 1994. Since the 1940s, this Mojave desert site has been a testing ground for high-performance aircraft and is one of two prime landing sites for the Space Shuttle.

Goddard space flight centre (GSFC) Goddard was NASAs first major scientific laboratory devoted entirely to the exploration of space. Located in Greenbelt, Maryland, GSFC's responsibilities include design and construction of new scientific and applications satellites, as well as tracking and communication with existing satellites in orbit. GSFC is the lead centre for the Earth Observing System, a key element of Mission to Planet Earth. GSFC also directs operations at the Wallops Flight Facility on Wallops Island, Virginia, which each year launches some 50 scientific missions to suborbital altitudes on small sounding rockets.

Jet propulsion laboratory (JPL) Located in Pasadena, California, JPL is operated under contract to NASA by the California Institute of Technology. Its primary focus is the scientific study of the solar system, including exploration of the planets with automated probes. Most of the lunar and planetary spacecraft of the 1960s and 1970s were developed at JPL. JPL also is the control centre for the worldwide Deep Space Network, which tracks all planetary spacecraft.

Lyndon B. Johnson space centre (JSC) Johnson Space Centre, located between Houston and Galveston, Texas, is the lead centre for NASAs manned space flight program. JSC has been Mission Control for all piloted space flights since 1965, and now manages the Space Shuttle program. JSC's responsibilities include selecting and training astronauts; designing and testing vehicles and other systems for piloted space flight; and planning and executing space flight missions. The centre has a major role in developing the Space Station. In addition, JSC directs operations at the White Sands Test Facility in New Mexico, which conducts Shuttle-related tests. The nearby White Sands Missile Range also serves as a backup landing site for the Space Shuttle.

Kennedy space centre (KSC) Located near Cape Canaveral, Florida, KSC is NASAs primary launch site. The Centre handles the preparation, integration, checkout, and launch of space vehicles and their payloads. All piloted space missions since the Mercury program have been launched from here, including Gemini, Apollo, Skylab, and Space Shuttle flights. KSC is the Shuttle's home port, where orbiters are serviced and outfitted between missions, and then assembled into a complete Shuttle "stack" before launch. The Centre also manages the testing and launch of unpiloted space vehicles from an array of launch complexes, and conducts research programs in areas of life sciences related to human spaceflight.

Langley research centre (LaRC) Oldest of NASA's field centres, LaRC is located in Hampton, Virginia, and focuses primarily on aeronautical research. Established in 1917 by the National Advisory Committee for Aeronautics, the Centre currently devotes two-thirds of its programs to aeronautics, and the rest to space. LaRC researchers use more than 40 wind tunnels to study improved aircraft and spacecraft safety, performance, and efficiency

Lewis research centre (LeRC) Lewis Research Centre; located outside Cleveland, Ohio, conducts a varied program of research in aeronautics and space technology. Aeronautical research includes work on advanced materials and structures for aircraft. Space-related research focuses primarily on power and propulsion. Another significant area of research is in energy and power sources for spacecraft, including the Space Station, for which LeRC is developing the largest space power system ever designed.

George C. Marshall space flight centre (MSFC) The MSFC, located in Huntsville, Alabama, is responsible for developing spacecraft hardware and systems, and is perhaps best known for its role in building the Saturn rockets that sent astronauts to the Moon during the Apollo program. It is NASAs primary centre for space propulsion systems and plays a key role in the development of payloads to be flown on the shuttle (such as Spacelab). MSFC also manages two other NASA sites: the Michoud Assembly Facility in New Orleans where the Shuttle's external tanks are manufactured, and the Slidell Computer Complex in Slidell, Louisiana, which provides computer support to Michoud and to NASAs John C. Stennis Space Centre.

John C. Stennis space centre (SSC) This Centre, located on Mississippi's Gulf Coast, is NASAs prime test facility for large liquid propellant rocket engines and propulsion systems. The main mission of the Centre is to support testing, on a regular basis, of the Space Shuttle's main propulsion system. SSC is responsible for a variety of research programs in the environmental sciences and the remotesensing of Earth resources, weather; and oceans, and is the lead NASA Centre for the commercialization of space remote sensing.

National Aeronautics and Space Administration (NASA) U.S. Civilian Space Agency created by Congress. Founded in 1958, NASA belongs to the executive branch of the Federal Government.

NASAs mission to plan, direct, and conduct aeronautical and space activities is implemented by NASA Headquarters in Washington, D.C., and by ten major centres spread throughout the United States.

Dozens of smaller facilities, from tracking antennas to Space Shuttle landing strips to telescopes are located around the world. The agency administers and maintains these facilities, builds and operates launch pads, trains astronauts, designs aircraft and spacecraft, and sends satellites into Earth orbit and beyond, and processes, analyzes, and distributes the resulting data and information.

NASA shares responsibility for aviation and space activities with other federal agencies, including the Departments of Commerce, Transportation, and Defence. Much of the work on major projects such as the Space Shuttle and the Space Station is done in the private sector by aerospace companies under government contract.

From its inception, NASA has been directed to pursue the expansion of human knowledge of phenomena in the atmosphere and space. NASA's programs of basic and applied research extend from microscopic sub-atomic particles to galactic astronomy. In addition to enhancing scientific knowledge, thousands of the technologies developed for aerospace have resulted in commercial applications.

Science offices at NASA Headquarters carry out a wide range of research activities to fulfil NASAs science goals. Science offices within NASA are: Office of mission to planet earth (MTPE) Focuses on the "home planet" as a dynamic system of land, ocean, atmosphere, and life that can be investigated on a global scale from space using remote-sensing tools.

Office of life and microgravity sciences and applications explores the basic physics of how solids, liquids, and gases behave in space; seeks an understanding of the basic mechanisms that underlie space adaptation-developing more effective countermeasures to mitigate the physiological effects of space flight; and studies the role of gravity on life.

Office of space science includes the Space Physics and Astrophysics Division which studies the entire universe of stars and galaxies, including the sun. The Solar System Exploration division has launched spacecraft to all the known planets except Pluto in its quest to study the solar system.

Native species Species that normally exists and reproduces in a specific region of the Earth. Compare with immigrant species, indicator species, and keystone species.

Natural gas Hydrocarbon based gas, mainly composed of methane, commonly found in the pores of sedimentary rocks of marine origin.

Natural hazards 1. Natural phenomena that produce negative effects on life.

2. The study of the hazards of natural phenomena.

Natural ionizing radiation Ionizing radiation that comes from natural sources in the environment.

Natural selection Environment's influence on the reproductive success of individuals in a population. It results in the exclusion of maladapted genetic traits found within individuals in a population.

Nautical mile A unit of length used in marine navigation that is equal to a minute of arc of a great circle on a sphere. One international nautical mile is equivalent to 1,852 meters or 1.151 statue miles.

Nautical twilight The time after civil twilight, when the brighter stars used for celestial navigation have appeared and the horizon may still be seen. It ends when the centre of the sun is 12 degrees below the horizon, and it is too difficult to perceive the horizon, preventing accurate sighting of stars.

Navigable waters Water usable, with or without improvements, as routes for commerce in the customary means of travel on water.

Neap tide Tide that occurs every 14 to 15 days and coincides with the first and last quarter of the moon. This tide has a small tidal range because the gravitational forces of the moon and sun are perpendicular to each other. Contrasts with spring tide.

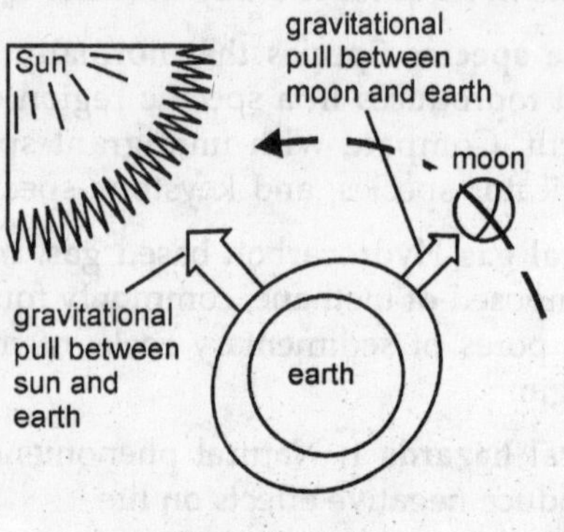

Fig. Neap tide

Near infrared Electromagnetic radiation with wavelengths from just longer than the visible (about 0.7 micrometers) to about two micrometers.

Neatline Line separating the body of a map from the map margin. On a standard quadrangle map, the neatlines are the meridians and parallels delimiting the quadrangle.

Needle ice A form of periglacial ground ice that consists of groups ice slivers at or immediately below the ground surface. Needle ice is about a few centimetres long.

Negative feedback Change in the state of a system that counteracts the measured effect of the initial alteration.

Negative vorticity advection Occurs when the rotation of the atmosphere advects lower values of vorticity into an area. This is created by divergence aloft. Contrast with positive vorticity advection.

Nephanalysis A type of analysis using satellite cloud pictures to study the relationship between cloud forms and storm systems. In classical mythology Nephele was a woman Zeus formed from a cloud.

Nepheloccygia Clouds that resemble recognizable shapes.

Net longwave radiation (balance) Balance between incoming and outgoing longwave radiation. Mathematically expressed as: L* = (LD - LU)

where L* is net longwave radiation at the surface,

LD is atmospheric counter-radiation (greenhouse effect) directed to the Earth's surface, and LU is longwave radiation lost from the Earth's surface.

Net primary productivity Total amount of chemical energy fixed by the processes of photosynthesis minus the chemical energy lost through respiration.

Net radiation (balance) Balance between incoming and outgoing shortwave and longwave radiations. Mathematically expressed as: Q* = (K + k)(1 - a) - LU + LD where Q* is surface net radiation (global annual values of Q* = 0, because input equals output, local values can be positive or negative),

K is surface direct shortwave radiation,

k is diffused shortwave radiation (scattered insolation) at the surface, a is the albedo of surface,

LD is atmospheric counter-radiation (greenhouse effect) directed to the Earth's surface, and LU is longwave radiation lost from the Earth's surface.

Net shortwave radiation (balance) Balance between incoming and outgoing shortwave radiations. Mathematically expressed as:

K* = (K + k)(1 - a)

where K* is surface net shortwave radiation,

K is surface direct shortwave radiation,

k is diffused shortwave radiation (scattered insolation) at the surface,

and a is the albedo of surface.

Network Is a system of spatially or functionally interconnected geographic features such as a system of roads, wires, pipes, or cables. With respect to computers, networks describe communication connections among machines.

Network analysis Is a range of techniques employed by engineers and planners to study the properties of networks including connectivity, capacity, and rates of flow. Network analysis can be used to estimate the capacity of a road network or to plan for additions to an electrical distribution system.

Neutral Any substance with a pH around 7.

Neutral atmosphere Condition in the atmosphere where isolated air parcels do not have a tendency to rise or sink. The parcels of air tend to be same temperature as the air that surrounds them.

Neutral solution Any water solution that is neutral (pH approximately 7) or has an equal quantity of hydrogen ions (H^+) than hydroxide ions (OH^-).

Neutralism Interspecific interaction where the species do not directly influence each other fitness.

Neutron Atomic sub-particle found in the nucleus of an atom. This particle is similar in mass to a proton but does not have an electromagnetic charge.

Névé Partially melted and compacted snow that has a density of at least 500 kilograms per cubic meter.

Newhall winds The local name for winds blowing downward from desert uplands through the Newhall Pass southward into the San Fernando Valley, north of Los Angeles.

Newton The unit of force giving a mass of about one kilogram (2.205 pounds) an acceleration of about one meter (1 yard) per second per second.

Newton's law of universal gravitation All bodies attract each other with what is called gravitational attraction. This applies to the largest stars as well as the smallest particles of matter The force of attraction between two small bodies (or between two spherical bodies of any size) is proportional to the product of their masses and inversely proportional to the square of the distance between their centres. In other words, the closer two bodies are to each other; the greater their mutual attraction. As a result, to stay in orbit, a satellite needs more speed in a low than a high orbit.

Kepler's three laws of planetary motion, which had been derived empirically by Johannes Kepler; were obtained with mathematical rigor as a consequence of Newton's law of universal gravitation in conjunction with his three laws of motion.

Newton's laws of motion Newton's three laws of motion are: Every body continues in a state of uniform motion in a straight line unless acted upon by some external force.

The time rate of change of momentum (mass×velocity) is proportional to the impressed force. In the usual case where the mass does not change, this law can be expressed in the familiar form: force = mass×acceleration or F = ma.

To every force or action, there is always an equal and opposite reaction.

Kepler's three laws of planetary motion, which had been derived empirically by Johannes Kepler; were obtained with mathematical rigor as a consequence of Newton's law of universal gravitation in conjunction with his three laws of motion.

Nibble Four bits of data.

Niche Adaptive role that a species has in a habitat. This includes its behaviour and interactions with other species.

Niche specialization Process where evolution, through natural selection, adapts a species to a particular set of abiotic and biotic characteristics within a habitat.

Nickpoint (knickpoint british spelling) Point on the long profile of a stream where the gradient is broken sudden drop in elevation. Nickpoints are the locations of rapids and waterfalls.

Night The period of the day between dusk and dawn.

Nimbostratus clouds Dark, grey low altitude cloud that produces continuous precipitation in the form of rain or snow. Found in an altitude range from the surface to 3,000 meters.

Nimbus satellite program A NASA program to develop observation systems meeting the research and development requirements of atmospheric and Earth scientists. The Nimbus satellites, first launched in 1964, carried a number of instruments: microwave radiometers, atmospheric sounders, ozone mappers, the Coastal Zone Color Scanner (CZCS), infrared radiometers, etc. Nimbus-7, the last in the series, provided significant global data on sea-ice coverage, atmospheric temperature, atmospheric chemistry (i.e. ozone distribution), the Earth's radiation budget, and sea-surface temperature.

Nitrate Form of nitrogen commonly found in the soil and used by plants for building amino acids, DNA and proteins. It is commonly produced by the chemical modification of nitrite by specialized bacteria.

Nitric acid Acid with the chemical formula: HNO_3.

Nitric oxide A gas produced by bacterial action in the soil and by high temperature combustion. Nitric oxide is a component in the production of photochemical smog. This colourless gas has the chemical formula is NO.

Nitrification The biochemical oxidation of ammonium to nitrite and nitrite to nitrate. This process is carried out by specialized bacteria.

Nitrite Form of nitrogen commonly found in the soil. It is commonly produced by the chemical modification of ammonium by specialized bacteria. This form is toxic to plants and animals at high concentrations.

Nitrogen (N_2) A colourless, tasteless, odourless gas that is the most abundant constituent of dry air. It comprises 78.09%.

Nitrogen cycle Model that describes the movement of nitrogen in its many forms between the hydrosphere, lithosphere, atmosphere and biosphere.

Nitrogen dioxide A gas produced by bacterial action in the soil and by high temperature combustion. Nitrogen dioxide is a component in the production of photochemical smog. This reddish brown gas has the chemical formula NO_2.

Nitrogen fixation Biological or chemical process where gaseous nitrogen is converted into solid forms of nitrogen. Biological fixation of nitrogen is done by specialized organisms like microorganisms like bacteria, actinomycetes, and cyano-bacteria. Chemical fixation occurs at high temperatures. One natural process that can produce enough heat to fix atmos-pheric nitrogen is lightning.

Nitrogen oxides Consists of two gases nitric oxide (NO) and nitrogen dioxide (NO_2). These gases are produced by bacterial action in the soil and by the high temperature combustion. Both gases are components in the production of photochemical smog.

Nitrogen saturation Over abundance of nitrogen in natural ecosystems because of human induced inputs related to agriculture and fossil fuel combustion.

Nitrous oxide Gas found in the atmosphere that contributes to the greenhouse effect.

Sources for nitrous oxide include: land-use conversion; fossil fuel combustion; biomass burning; and soil fertilization. Chemical formula for nitrous oxide is N_2O.

Nivation Process where snow patches initiate erosion through physical weathering, meltwater flow, and gelifluction.

Nivation hollow Ground depression found in periglacial areas that is created by nivation.

Noctilucent clouds Rarely seen clouds of tiny ice particles that form approximately 75 to 90 kilometres above the earth's surface. They have been seen only during twilight (dusk and dawn) during the summer months in the higher latitudes. They may appear bright against a dark night sky, with a blue-silver color or orange-red.

Nocturnal thunderstorms Thunderstorms which develop after sunset. They are often associated with the strengthening of the low level jet and are most common over the Plains states. They also occur over warm water and may be associated with the seaward extent of the overnight land breeze.

Nodal region A region characterized by a set of places connected to another place by lines of communication or movement.

Non clastic sedimentary rock Sedimentary rocks that are created either from chemical precipitation and crystallization, or by the lithification once living organic matter.

Non ionizing radiation A form of electromagnetic radiation that does not have enough energy to cause ionization of atoms in living tissue. Examples of this type of radiation include radio waves, microwaves, infrared light, and ordinary light.

Non parametric statistical test Statistical tests that do not assume the sample data is normally distributed.

Non renewable resource Resource that is finite in quantity and is being used faster than its ability to regenerate itself.

Nonsymbiotic mutualism Mutualistic interaction where the mutualists live independent lives yet cannot survive without each other. For example, pollinating insects like bees and some flowering plants.

Nor'easter A cyclonic storm occurring off the east coast of North America. These winter weather events are notorious for producing heavy snow, rain, and tremendous waves that crash onto Atlantic beaches, often causing beach erosion and structural damage. Wind gusts associated with these storms can exceed hurricane force in intensity. A nor'easter gets its name from the continuously strong northeasterly winds blowing in from the ocean ahead of the storm and over the coastal areas.

Normal The recognized standard value of a meteorological element as it has been averaged in a given location over a fixed number of years. Normals are concerned with the distribution of data within limits of common occurrence. The parameters may include temperatures (high, low, and deviation), pressure, precipitation (rain, snow, etc.), winds (speed and direction), thunderstorms, amount of clouds, percent relative humidity, etc.

Normal distribution A common probability distribution displayed by population data. If the values of the distribution are plotted on a graph's horizontal axis and their frequency on the vertical axis the pattern displayed is symmetric and bell-shaped. The central value or mean represents the peak or the most frequently occurring value.

Normal fault Vertical fault where one slab of the rock is displaced up and the other slab down. It is created by tensional forces acting in opposite directions.

Normal lapse rate Average rate of air temperature change with altitude in the troposphere. This value is approximately a decrease of 6.5° Celsius per 1000 meters rise in elevation.

North magnetic pole The North Magnetic Pole is the point on the Northern Hemisphere of the Earth toward which a compass' needle always points; at the North Magnetic Pole, a compass' needle will stand vertically. It is now located near in northern Canada (and its location changes over time).

North pacific high A semi-permanent, subtropical area of high pressure in the North Pacific Ocean. It is strongest in the Northern Hemispheric summer and is displaced towards the equator during the winter when the Aleutian Low becomes more dominate. Comparable systems are the Azores High and the Bermuda High.

North pole Surface location defined by the intersection of the polar axis with Earth's surface in the Northern Hemisphere. This location has a latitude of 90° North.

Northeast trade winds See trade winds.

Northern hemisphere The Northern Hemisphere is the half of the Earth that is north of the equator.

Northing Second measurement of a grid reference used to specific the location of a point on a rectangular coordinate system. The distance measured northward from the origin of a rectangular coordinate system.

Nowcast A short-term weather forecast for expected conditions in the next few hours.

Nuclear energy Energy released when the nucleus of an atom experiences a nuclear reaction like the spontaneous emission of radioactivity, nuclear fission, or nuclear fusion.

Nucleic acid Is an organic compound composed primarily of different combinations of carbon, hydrogen, nitrogen, oxygen, and phosphorus. They are very complex compounds being created by the atomic linking of thousands of individual atoms. DNA or deoxyribonucleic acid, the genetic blueprint of life, is an example of a nucleic acid.

Nucleus 1. Dense central portion of an atom that is composed of neutrons and protons.

2. Structure found in eukaryotic cells that contains the chromosomes.

Nuee ardente A glowing cloud of dense hot volcanic gas and ash that moves downslope at high speeds, incinerating the landscape.

Null hypothesis (H0) Is a hypothesis that has been suggested because it is believed to be true or because it is to be used as a starting point for scientific argument. Used in statistical testing to organize arguments.

Numerical forecasting The use of numerical models, such as the fundamental equations of hydrodynamics subjected to observed initial conditions, to forecast the weather. These models are run on high-speed computers at the National Meteorological Centre.

Nutrient Any food, chemical element or compound an organism requires to live, grow, or reproduce.

O

O horizon Topmost layer of most soils. It is composed mainly of plant litter and humus.

Oasis A spot in a desert made fertile by water; which normally originates as groundwater.

Oblique aerial photograph Photograph taken from a non-perpendicular angle from a platform in the atmosphere.

Obliquity Tilt of the Earth's polar axis as measured from the perpendicular to the plane of the Earth's orbit around the sun. The angle of this tilt varies from 22.5 to 24.5° over a 41,000 year period. Current obliquity is 23.5°.

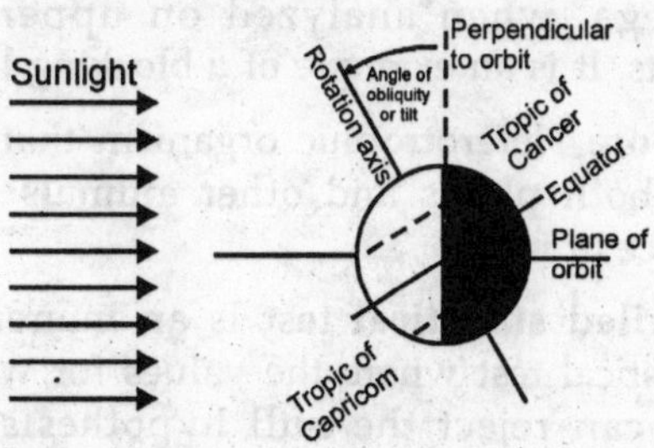

Fig. Obliquity

Obscuration Any phenomena in the atmosphere, excluding precipitation, that reduces horizontal visibility. According to the National Weather Service, some of the obstructions to visibility include blowing and widespread dust, fog (including freezing fog and patchy fog) haze, mist, sand and blowing sand smoke, blowing spray, and volcanic ash. It is reported as "X" in an observation and on the METAR.

Observation In meteorology, the evaluation of one or more meteorological elements, such as temperature, pressure, or wind, that describe the state of the atmosphere, either at the earth's surface or aloft. An observer is one who records the evaluations of the meteorological elements.

Obsidian Glassy dark coloured volcanic rock. Usually composed of rhyolite.

Occluded front (occlusion) A composite of two fronts formed as a cold front overtakes a warm front. A cold occlusion results when the coldest air is behind the cold front. The cold front undercuts the warm front and, at the Earth's surface, coldest air replaces less-cold air.

A warm occlusion occurs when the coldest air lies ahead of the warm front. Because the cold front can not lift the colder air mass, it rides piggyback up on the warm front over the coldest air.

Ocean A body of saline water found occupying all or part of the Earth's ocean basins. There are five recognized oceans: the Atlantic, the Southern Ocean, the Pacific

Ocean, the Indian Ocean, and the Arctic Ocean.

Ocean basin Part of the Earth's outer surface that is comprised of the ocean floor, mid-oceanic ridges, continental rise, and continental slope. The ocean basins are filled with saline water that makes up the oceans.

Ocean current Large scale horizontal flow of ocean water that is persistent and driven by atmospheric circulation.

Ocean floor Flat plain found at the bottom of the ocean. The ocean floor represents the surface of the oceanic crust. The ocean floor lies between the mid-oceanic ridges and the trenches, usually 5,000 to 7,000 meters below the ocean surface. Also called the abyssal plain.

Ocean trench Deep depression found at the edge of the ocean floor. Represents area of tectonic plate subduction.

Oceanic crust Basaltic portion of the Earth's crust that makes up the ocean basins. Approximately 5 to 10 kilometres thick.

Oceanic plate A rigid, independent segment of the lithosphere composed of mainly basalt that floats on the viscous plastic asthenosphere and moves over the surface of the Earth. The Earth's oceanic plates are an average 75 kilometres thick and were formed less than several hundred million years ago at one of the Earth's mid-oceanic ridges.

Oceanic survey Survey or examination of condition in the ocean or any part of it, with reference to animal or plant life, chemical elements present, temperature gradients, etc.

Oceanography The study of the ocean, embracing and integrating all knowledge pertaining to the ocean's physical boundaries, the chemistry and physics of sea water, and marine biology.

Offshore Comparatively flat zone of variable width that extends from the outer margin of the rather steeply sloping shoreface to the edge of the continental shelf.

Ohm The unit of electrical resistance, equal to the resistance of a circuit in which an electromotive force of one volt maintains a current of one ampere. Named for German physicist Georg S. Ohm (1787-1854).

Oil Hydrocarbon based liquid commonly found in the pores of sedimentary rocks of marine origin.

Old growth forest Climax forests dominated by late successional species of trees that are hundreds to thousands of years old. Examples include virgin uncut forests of Douglas fir, western hemlock, giant sequoia, and coastal redwoods located in western North America.

Oligotrophic lake Lake with a low supply of nutrients in its waters.

Olivine Common silicate mineral found in rocks formed from mafic magma. Its chemical composition varies between Mg_2SiO_4 and Fe_2SiO_4.

Omega block A warm high aloft which has become displaced and is on the polarward side of the jet stream. It frequently occurs in the late winter and early spring in the Northern Hemisphere. The name comes from its resemblance to the Greek letter, Omega, when analyzed on upper air charts. It is an example of a blocking high.

Omnivore Heterotrophic organism that can use both plants and other animals as a source of food.

One tailed statistical test Is an inferential statistical test where the values for which one can reject the null hypothesis are located entirely in one side of the centre of the probability distribution.

Online Indicates when equipment such as plotters, printers, and digitizers is turned on and is actively communicating with a computer, or is controlled by that computer. Computers in a network can also be said to be online when the network connection operative.

Onshore offshore transport The up and down movement of sediment roughly perpendicular to a shoreline because of wave action.

Opaque A condition where a material, such as a cloud, blocks the passage of radiant energy, especially light. Opaque sky cover refers to the amount of sky cover that completely hides all that might be above it.

Open range A cattle- or sheep-ranching area characterized by a general absence of fences.

Open sea That part of the ocean that extends from the continental shelf. Compare with coastal zone.

Open system Is a system that transfers both matter and energy can cross its boundary to the surrounding environment. Most ecosystems are an example of an open system.

Open talik Is a form of localized unfrozen ground (talik) in an area of permafrost. It is open to the ground surface but enclosed to permafrost below and at its sides.

Operating systems Are the software systems which manage the interaction between users and hardware. Operating systems are employed by users to tell a computer what to do with the resources at its disposal in order to run programs, store data, or send files to peripherals such as plotters and printers. Popular operating systems include DOS, VM and MVS (or IBM mainframes), VMS (for DEC machines), UNIX (including proprietary variations such as AIX, Ultrix, and UNICOS), OS/2, and Apple OS.

Opisometer Mechanical device for measuring non-linear distances on maps.

Optimum The level of an abiotic factor or condition in the environment within the tolerance range at which a species or population can function most efficiently or with the greatest positive effect to its physiological or reproductive fitness.

Orbit The path described by a heavenly body in its periodic revolution. Earth satellite orbits with inclinations near 0o are called equatorial orbits because the satellite stays nearly over the equator. Orbits with inclinations near 90o are called polar orbits because the satellite crosses over (or nearly over) the north and south poles.

Orbital plane An imaginary gigantic flat plate containing an Earth satellite's orbit. The orbital plane passes through the centre of the Earth.

Ordovician Geologic period that occurred roughly 438 to 505 million years ago. During this period, the first fish and fungi species appear.

Organ Group of cells and tissues that have a particular function for an organism.

Organelle Is a specialized structure found in cells that carry out distinct cellular functions.

Organic 1. Relating to an organism.

2. Derived from an organism.

Organic matter Mass of matter that contains living organisms or non-living material derived from organisms. Sometime refers to the organic constituents of soil.

Organic soil Soil order (type) of the Canadian System of Soil Classification. This soil type is common in fens and bogs. This soil is mainly composed of organic matter in various stages of decomposition.

Organism Any form of life.

Orientation Establishing correct relationship in direction with reference to points of the compass; the state of being in correct relationship in direction with reference to the points of the compass.

Origin The arbitrary starting point on a graph or grid coordinate system. Defined by the intersection of the x and y-axes.

Origin of coordinates Point in a system of coordinates that serves as a zero point in computing the system's elements or in prescribing its use.

Orogenesis The process of mountain building through tectonic forces of compression and volcanism.

Orogenic belt A major range of mountains on the continents.

Orographic lifting Where the flow of air is forced up and over barriers such as highlands or mountains. Moist air being forced aloft begins to cool, consequently condensation forms, and rain or snow begins to fall. By the time the air reaches the leeward side of the barrier, it sinks and warms, resulting in decreasing relative humidity, cessation of precipitation, and the dissipation of clouds. Also called an orographic uplift.

Orographic precipitation Is precipitation that forms when air is forced to rise because of the physical presence of elevated land. As the parcel rises it cools as a result of adiabatic expansion at a rate of approximately 10° Celsius per 1,000 meters until saturation. The large amounts of precipitation along the west coast of Canada are due mainly to this process.

Orographic rainfall Precipitation that results when moist air is lifted over a topographic barrier such as a mountain range.

Orographic uplift Uplift of an air mass because of a topographic obstruction. Uplift also causes the cooling of the air mass. If enough cooling occurs condensation can occur and form into orographic precipitation.

Orthophanic projection The Orthophanic (meaning 'right appearing') projection, also called the Robinson projection, is a widely-used type of map in which the Earth is shown in a flattened ellipse. In this pseudocylindrical projection, lines of latitude are parallel to the equator, but lines of longitude are elliptical arcs. In a Robinson projection, area is represented accurately, but the distances and compass directions are distorted (for example, compass lines are curved). This type of projection was first made in 1963 by Arthur H. Robinson.

Orthophotograph Photograph having the properties of an orthographic projection. It is derived from a conventional perspective photograph by simple or differential rectification so that image displacements caused by camera tilt and terrain relief are removed.

Orthophotographic map Map produced by assembling orthophotographs at a specified uniform scale in a map format.

Orthophotomap Orthophotographic map with contours and cartographic treatment, presented in a standard format, and related to standard reference systems.

Orthophotoquad Monocolour orthophotgraphic map presented in a standard quadrangle format and related to standard reference systems. It has no contours and little or cartographic treatment.

Orthophotoscope Photomechanical device used in conjunction with a double-projection stereoplotter for producing orthopho-tograph.

Outcrop Area of exposed bedrock at the Earth's surface with no overlying deposits of soil or regolith.

Outer core Outer region of the Earth's core. It is believed to be liquid nickel and iron and has a density of about 11 grams per cubic centimetre. It surrounds the inner core and has an average thickness of about 2,250 kilometres.

Outflow Also referred to as an outflow boundary, it is the outward flow of air from a system, such as a thunderstorm. It is the result of cold downdrafts and its passage includes a wind shift and temperature drop.

Outgassing The release of gas from cooling molten rock or the interior of the Earth.

Much of the atmosphere's gaseous constituents, like water vapour, nitrogen, and argon, came from outgassing.

Output Movement of matter, energy, or information out of a system.

Outwash Glaciofluvial sediments deposited by meltwater streams at the edge of a glacier.

Outwash plain A flat or gentle sloping surface of glaciofluvial sediments deposited by meltwater streams at the edge of a glacier. Usually found in close spatial association with moraines.

Overbank flow Movement of flood waters outside a stream channel during period of high discharge.

Overburden Material covering a mineral seam or bed that must be removed before the mineral can be removed in strip mining.

Overcast The amount of sky cover for a cloud layer that is 8/8ths, based on the summation layer amount for that layer.

Overedge Any portion of a map lying outside the nominal map border (neatline).

Overland flow The topographic movement of a thin film of water from precipitation to lower elevations. With time, this water will begin to organizing its flow into small channels called rills. The rills converge to form progressively larger channels until stream channels are formed. Occurs when the infiltration capacity of an area's soil has been exceeded. Also called sheet flow or runoff.

Overlay Printing or drawing on a transparent or translucent medium intended to be placed in register on a map or other graphic and which shows details not appearing or requiring special emphasis on the base material.

Overprint New material printed on a map or chart to show data of importance or special use, in addition to those data originally printed.

Overrunning This occurs when a relatively warm air mass is forced above a cooler air mass of greater density that is at the surface. Weather generally associated with this event includes cloudiness, cool temperatures, and steady rain.

Overthrust fault Fault produced by the fracturing of rock in a fold because of intense compression.

Overturned fold A fold in rock layers where one limb is pushed past the perpendicular. This results in both limbs having dips in the same direction.

Oxbow lake Is portion of abandoned stream channel filled with stagnant water and cut off from the rest of the stream. Oxbow lakes are created when meanders are cut off from the rest of the channel because of lateral stream erosion.

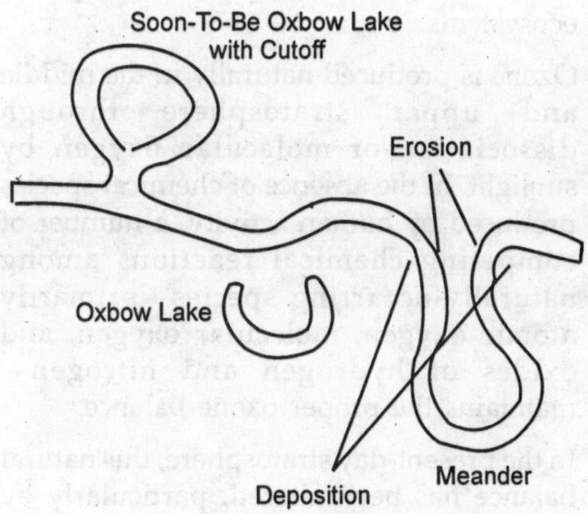

Fig. Oxbow lake

Oxidation 1. Chemical attachment of free oxygen to other elements and compounds. One of the types of chemical weathering.

2. Loss of an electron during a chemical reaction from one atom to another.

Oxisol Soil order (type) of the United States Natural Resources Conservation Service Soil Classification System. These soils are found in moist tropical environments. They are extremely weathered and very old.

Layers within the profile have accumulations of mineral oxides and have a lack of base cations available for exchange.

Oxygen (O_2) A colourless, tasteless, odourless gas that is the second most abundant constituent of dry air. It comprises 20.946%.

Ozone An almost colourless, gaseous form of oxygen with an odour similar to weak chlorine. A relatively unstable compound of three atoms of oxygen, ozone constitutes—on the average—less than one part per million (ppm) of the gases in the atmosphere (peak ozone concentration in the stratosphere can get as high as 10 ppm). yet ozone in the stratosphere absorbs nearly all of the biologically damaging solar ultraviolet radiation before it reaches the Earth's surface where it can cause skin cancer, cataracts, and immune deficiencies, and can harm crops and aquatic ecosystems.

Ozone is produced naturally in the middle and upper stratosphere through dissociation of molecular oxygen by sunlight. In the absence of chemical species produced by human activity, a number of competing chemical reactions among naturally-occurring species—primarily atomic oxygen, molecular oxygen, and oxides of hydrogen and nitrogen—maintains the proper ozone balance.

In the present-day stratosphere, this natural balance has been altered, particularly by the introduction of man-made chlorofluorocarbons. If the ozone decreases, the ultraviolet radiation at the Earth's surface will increase.

Tropospheric ozone is a by-product of the photochemical (light-induced) processes associated with air pollution. Ozone in the troposphere can damage plants and humans.

Ozone hole A large area of intense stratospheric ozone depletion over the Antarctic continent that typically occurs annually between late August and early October; and generally ends in mid November. This severe ozone thinning has increased conspicuously since the late seventies and early eighties. This phenomenon is the result of chemical mechanisms initiated by man-made chlorofluorocarbons. Continued buildup of CFCs is expected to lead to additional ozone loss worldwide.

The thinning is focused in the Antarctic because of particular meteorological conditions there. During Austral spring (September and October in the Southern Hemisphere) a belt of stratospheric winds encircles Antarctica essentially isolating the cold stratospheric air there from the warmer air of the middle latitudes. The frigid air permits the formation of ice clouds that facilitate chemical interactions among nitrogen, hydrogen, and chlorine (elevated from CFCs) atoms, the end product of which is the destruction of ozone.

Ozone layer The layer of ozone that begins approximately 15 km above Earth and thins to an almost negligible amount at about 50 km, shields the Earth from harmful ultraviolet radiation from the sun. The highest natural concentration of ozone (approximately 10 parts per million by volume) occurs in the stratosphere at approximately 25 km above Earth. The stratospheric ozone concentration changes throughout the year as stratospheric circulation changes with the seasons. Natural events such as volcanoes and solar flares can produce changes in ozone concentration, but man-made changes are of the greatest concern.

Ozone mini-hole(s) Rapid, transient, polar-ozone depletion. These depletions. which take place over a 50-kilometer squared area, are caused by weather patterns in the upper troposphere. The decrease in ozone during a mini-hole event is caused by transport, with no chemical depletion of ozone. However; the cold stratospheric temperatures associated with weather systems can cause clouds to form that can lead to the conversion of chlorine compound from inert to reactive forms. These chlorine compounds can then produce longer-term ozone reductions after the mini-hole has passed.

Ozone measuring satellite instruments Satellite-based ozone-measuring instruments can measure ozone by looking at the amount of ultraviolet absorption reflected from the Earth's surface and clouds. Some instruments provide data within the different levels of the atmosphere. The Total Ozone Mapping Spectrometer (TOMS) maps the total amount of ozone between ground and the top of the atmosphere.

The amount and distribution of ozone molecules in the stratosphere varies greatly over the globe, changing in response to natural cycles such as seasons, sun cycles. and winds. Utilizing satellites has enabled scientists to assess ozone levels simultaneously over the entire Earth, and has led them to conclude that global ozone levels are being depleted.

Ozonosphere Another name for the ozone layer.

P

P wave A seismic wave that moves material in push-pull fashion in the direction of its travel. This type of seismic wave can travel through solids, liquids, and gases. Also called a primary wave.

Pacific high High pressure system that develops over the central Pacific Ocean near the Hawaiian Islands. Also called the Hawaiian High.

Paleoclimate Climatic conditions in the geological past reconstructed from a direct or indirect data source.

Paleoclimatology Scientific study of the Earth's climate during the past.

Paleogeography The study of ancient or prehistoric geography.

Paleolake An ancient lake that existed in the past when hydrological conditions were different.

Paleosol A soil exhibiting features that are the result of some past conditions and processes.

Paleozoic Geologic era that occurred from 570 to 245 million years ago.

Palisades A line of bold cliffs.

Palmer drought index A long-term meteorological drought severity index produced by the NOAA/USDA (Department of Agriculture) Joint Agricultural Weather Facility. The index depicts prolonged times, as in months or years, of abnormal dryness or wetness. It responds slowly, changing little from week to week, and reflects long-term moisture runoff, recharge, and deep percolation, as well as evapotranspiration.

Palouser A strong, dangerous, katabatic wind that descends from the mountains into the Palouse River valley in northern Idaho and eastern Washington. Also call a Cow-Killer.

Palsa A mound of peat that develops as the result of the formation of a number ice lenses beneath the ground surface. Typical size is 1 to 7 meters high, 10 to 30 meters wide, and 15 to 150 meters long. Found in the high latitudes. Similar to a pingo.

Pan or pan 1. Collection of chemicals found in photochemical smog

2. Compact soil horizon that has a high clay content.

3. Large natural basin or depression found in desert climates.

Panchromatic Sensitive to all or most of the visible spectrum.

Panelling Is a means of subdividing the display of a GIS database or digital map. Panelling allows users to view selectively the different sections of a digital map.

Pangaea Hypothetical super continent that existed in the geological past. Its break-up created the current configuration of landmasses found on the Earth.

Panhandle A narrow projection of a larger territory (as a state).

Parallel A parallel (of latitude) is a line on a map that represents an imaginary east-west circle drawn on the Earth in a plane parallel to the plane that contains the equator.

Parallel of latitude A circle, or approximation of a circle, on the surface of the Earth, parallel tot he Equator, and connecting points of equal latitude; a circle of the celestial sphere parallel to the ecliptic, and connecting points of equal celestial latitude.

Parametric tests Statistical tests that assume the sample data is normally distributed.

Parasite Consumer organism that feeds on a host for an extended period of time. Feeding causes the host to be less fit and may eventually cause premature death.

Parasitism Biological interaction between species where a parasite species feeds on a host species.

Parcel A volume of air small enough to contain uniform distribution of its meteorological properties and large enough to remain relatively self-contained and respond to all meteorological processes.

Parent material The mineral material from which a soil forms.

Parhelion The scientific name for sun dogs. Either of two colored luminous spots that appear at roughly 22 degrees on both sides of the sun at the same elevation. They are caused by the refraction of sunlight passing through ice crystals. They are most commonly seen during winter in the middle latitudes and are exclusively associated with cirriform clouds. They are also known as mock suns.

Parity The addition of one or more redundant bits to information to verify its accuracy.

Partial obscuration Denotes that 1/8th or more of the sky, but not all of the sky, is hidden by any surface-based phenomena in the atmosphere, excluding precipitation. It often reduces horizontal visibility but not the vertical. It is reported as "-X" in an observation and on the METAR.

Particulate matter Particles of dust, soot, salt, sulphate compounds, pollen, or other particles suspended in the atmosphere.

Partly cloudy The state of the weather when the clouds are conspicuously present, but do not completely dull the sky or the day at any moment. The National Weather Service does not have an amount of sky cover for this condition. Refer to clear, few, scattered, broken, and overcast.

Parts Per Billion (PPB) Number of parts of a substance found in one billion parts of a particular gas, liquid, or solid.

Parts Per Million (PPM) Number of parts of a substance found in one million parts of a particular gas, liquid, or solid.

Parts per thousand (PPT) Number of parts of a substance found in one thousand parts of a particular gas, liquid, or solid.

Pascal (Pa) Unit of atmospheric pressure named in honour of Blaise Pascal (I 632-1 662), whose experiments greatly increased knowledge of the atmosphere. A Pascal is the force of one Newton acting on a surface area of one square meter. It is the unit of pressure designated by the International System.

Pascal's law When an external pressure is applied to any confined fluid at rest, the pressure is increased at every point in the fluid by the amount of external pressure applied. It means that the pressure of the atmosphere is exerted not only downward on the surface of an object, but also in all directions against a surface which is exposed to the atmosphere. Formulated by Blaise Pascal (1623-1662), a French mathematician, theologian, and physicist.

Passive remote sensing Form of remote sensing where the sensor passively captures electromagnetic radiation reflected or emitted by an object.

Passive system A system sensing only radiation emitted by the object being viewed or reflected by the object from a source other than the system.

Paternoster lakes A linear series of mountain valley lakes that are formed from glacial erosion. They form behind glacial moraines or in glacially carved out rock basins. The name of this feature is related to the series of lakes looking like a string of beads.

Pathogen Microscopic parasite organism that causes disease in a host. Disease causes the host to be less fit and may eventually cause premature death.

Patterned ground Term used to describe a number of surface features found in periglacial environments. These features can resemble circles, polygons, nets, steps, and stripes. The development of some of these shapes is thought to be the result of freeze-thaw action.

Payload The instruments that are accommodated on a spacecraft.

Peak annual flow The largest discharge produced by a stream during a one year period.

Peak gust The highest instantaneous wind speed observed or recorded.

Peat Partially decomposed remains of plants that once flourished in a waterlogged environment.

Pebbles A rounded piece of rock that is larger than gravel.

Pediment A gradually sloping bedrock surface located at the base of fluvial-eroded mountain range. Found in arid locations and normally covered by fluvial deposits.

Pediplain An arid landscape of little relief that is occasionally interrupted by the presence of scattered inselbergs. Formed by the coalescence of several pediments.

Pedogenesis Process of soil formation.

Pedogenic regime The particular soil forming process that operates in a certain climate. Some of the main processes are: laterization, salinization, podzolization, calcification, and gleization.

Pedology The scientific study of soils.

Pedon A basic soil sampling unit. Often viewed as a soil profile that goes from the surface to a depth where soil meets bedrock.

Peninsula A peninsula is a body of land that is surrounded by water on three sides.

Pennsylvanian Geologic period that occurred roughly 286 to 320 million years ago. During this period, the first reptiles and winged insects appear.

Perched water table Water table that is positioned above the normal water table for an area because of the presence of a impermeable rock layer.

Percolation Vertical movement or infiltration of water from the Earth's surface to its

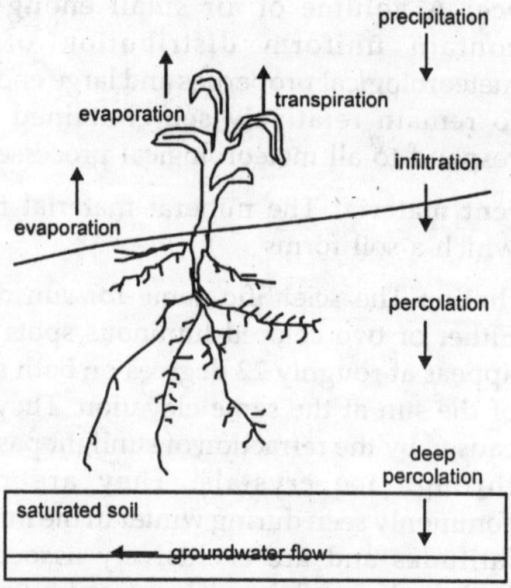

Fig. Percolation

subsurface. Movement usually stops when the flowing water reaches the water table.

Perennial plant Plant species that lives for more than two years.

Peridotite Coarse grained ultramafic igneous rock composed mainly of olivine and pyroxene. The mantle is though to be composed primarily of this rock type.

Perigee The point nearest the earth on the moon's orbit. This term can be applied to any other body orbiting the earth, such as satellites. It is the opposite of apogee.

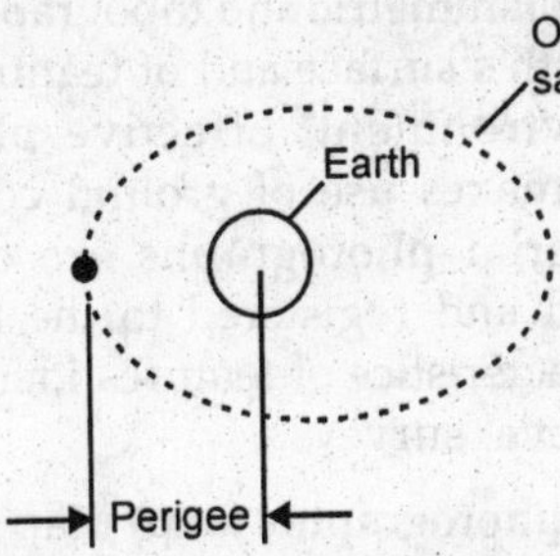

Fig. Perigee

Periglacial Landforms created by processes associated with intense freeze-thaw action in an area high latitude areas or near an alpine or continental glacier.

Perihelion The point of the earth's orbit that is nearest to the sun. Although the position is part of a 21,000 year cycle, currently it occurs around January, when the earth is about 3 million miles closer to the sun than at aphelion. This term can be applied to any other celestial body in orbit around the sun. It is the opposite of aphelion.

Period decay (aka decay) The tendency of a satellite to lose orbital velocity due to the influence of atmospheric drag and gravitational forces. A decaying object eventually impacts the surface of the Earth or burns up in the atmosphere. This parameter directly affects the satellite's mean motion.

Period Geologic time unit that is shorter than an era but longer than a epoch.

Periodic table Table that describes some of the chemical properties of the known elements.

Peripherals Are items of computer equipment such as plotters, printers, scanners, digitizers, and graphics displays. Although essential to a complete GIS system, these pieces of equipment are considered peripheral to central computer resources such as the processor, memory, keyboard, and display screen.

Permafrost Zone of permanently frozen water found in high latitude soils and sediments. Five types of permafrost have been recognized: continuous permafrost, discontinuous permafrost, sporadic permafrost, alpine permafrost, and subsea permafrost.

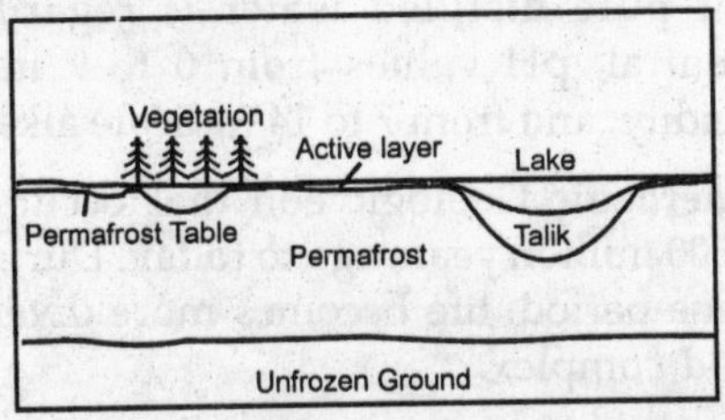

Fig. Permafrost

Permeability A measure of the ability of soil, sediments, and rock to transport water horizontally and vertically. Permeability is dependent on the porosity of the medium the water is flowing through. Some rocks like granite have very poor permeability, while rocks like shale are actually quite pervious. As for soils, sand is the most pervious, while clay has the lowest permeability. Silt usually is somewhere in the middle.

Permian Last geologic period in the Paleozoic era. Occurred from 286 to 245 million years ago. This period saw the mass extinction of many corals, brachiopods, and trilobites. It also saw the diversification and growing dominance of the reptiles.

Peroxyacetyl Nitrate (PAN) Chemical found in photochemical smog. Formed from photochemical reactions involving nitric oxide (NO) and volatile organic compounds (VOCs). Quite damaging to plants.

Perturbations Minor corrections to the Keplerian model of a satellite orbit as an ellipse of constant shape and orientation. Since satellite orbits are affected by Earth's gravity and drag caused by the Earth's atmosphere (causing satellites to spiral downward), minor adjustments must be made to the orbit.

PH A symbol for the degree of acidity or alkalinity of a solution. Expressed as a negative logarithm of the hydrogen ion concentration in a solution, pH = -log10(H^+). If the hydrogen ion concentration of a solution increases, the pH will decrease, and vice versa. The value for pure distilled water is regarded as neutral, pH values from 0 to 7 indicate acidity, and from 7 to 14 indicate alkalinity.

Phanerozoic Geologic eon that occurs from 2500 million years ago to today. During this time period, life becomes more diversified and complex.

Phase change Reorganization of a substance at the atomic or molecular level resulting in a change of the physical state of matter. For example, a change from solid to liquid to a gas.

Phase interval In direct readout, the time between the end of a satellite image start tone and the start of the actual frame data. The phase interval repre sents white level video, interrupted by a black level pulse marking the start of each line and is used to set up phasing prior to image display.

Phenology Subdiscipline of agriculture, a science that treats relations between climate and periodic biological phenomena that are related to or caused by climatic conditions, such as the budding of trees and the migration of birds.

Phloem Food conducting tissue in vascular plants.

Photochemical smog A type of smog that forms in large cities when chemical reactions take place in the presence of sunlight, its principal component is ozone. Ozone and other oxidants are not emitted into the air directly but form from reactions involving nitrogen oxides and hydrocarbons. Because of its smog-making ability, ozone in the lower atmosphere (troposphere) is often referred to as "bad" ozone.

Photodissociation The splitting of a molecule by photon normally from the sun.

Photogrammetry Uses aerial photographs to produce planimetric and topographic maps of the earth's surface and of features of the built environment. Effective photogrammetry makes use of ground control by which aerial photographs are carefully compared and registered to the locations and characteristics of features identified in ground-level surveys.

Photomap (photographic map) Map made by adding marginal information, descriptive data, and a reference system to a photograph or assembly of photographs.

Photometer Any of a number of atmospheric phenomena which appear as luminous patterns in the sky. They do not directly cause adverse weather. They include halos, coronas, rainbows, and fogbows.

Photon A quantum (smallest unit in which waves may be emitted or absorbed) of light.

Photoperiod The duration of the daylight period.

Photoperiodism 1. Mechanism possessed by some organisms to use photoperiod to sense seasonal time.

2. Response by organisms to changes in the duration of day and night.

Photosphere The intensely bright portion of the sun visible to the unaided eye; the "surface" of the sun. Reaching temperatures estimated at about 11,000 degrees Fahrenheit, it is the portion of the sun's atmosphere which emits continuous electromagnetic radiation.

Photosynthesis Is the chemical process where plants and some bacteria can capture and organically fix the energy of the sun. This chemical reaction can be described by the following simple equation: $6CO_2 + 6H_2O +$ light energy >>> $C_6H_{12}O_6 + 6O_2$

The main product of photosynthesis is a carbohydrate, such as the sugar glucose, and oxygen which is released to the atmosphere. All of the sugar produced in the photosynthetic cells of plants and other organisms is derived from the initial chemical combining of carbon dioxide and water with sunlight. This chemical reaction is catalyzed by chlorophyll acting in concert with other pigment, lipid, sugars, protein, and nucleic acid molecules. Sugars created in photosynthesis can be later converted by the plant to starch for storage, or it can be combined with other sugar molecules to form specialized carbohydrates such as cellulose, or it can be combined with other nutrients such as nitrogen, phosphorus, and sulphur, to build complex molecules such as proteins and nucleic acids.

Photosynthetic autotroph An organism that produces food molecules inorganically by using light and the chemical process of photosynthesis. Plants are the dominant photosynthetic autotrophs on the Earth. This organism does not require outside sources of organic food energy for survival.

Photosynthetically active radiation Electromagnetic radiation in the part of the spectrum used by plants for photosynthesis.

Phylogenic classification Classification of organisms based on genetic connections between other species.

Phylum A group or category used in the taxonomic and/or phylogenic classification of organisms. A phylum is composed of one or more classes. In the classification of plants the category division is often used synonymously.

Physical climate system The system of processes that regulate climate, including atmospheric and ocean circulation, evaporation, and precipitation.

Physical geography Field of knowledge that studies natural features and phenomena on the Earth from a spatial perspective. Subdiscipline of Geography.

Physical map A physical map is a map that shows an areas natural physical features, like mountains, lakes, and rivers.

Physical weathering Breakdown of rock and minerals into small sized particles through mechanical stress.

Physiographic region A portion of the Earth's surface with a basically common topography and common morphology.

Physiography Physical geography.

Phytoplankton Small photosynthetic organisms, mostly algae and bacteria, found inhabiting aquatic ecosystems.

Piedmont Lying or formed at the base of mountains; in the United States, an area in the southern states at the base of the Blue Ridge Mountains.

Piedmont glacier A large glacier formed from the merger of several alpine glaciers.

Pigment Organic substance found in plant and animal cells that creates colouring.

Pilot balloon A small balloon whose ascent is used to determine the direction and speed of low level atmospheric winds. Also known as a pibal.

Pilot report A report of in-flight weather by an aircraft pilot or crew member. Often referred to as a pirep.

Pingo A large conical mound that contains an ice core. This feature can be up to 60 to 70 meters in height. Form in regions of permafrost. Common in the Mackenzie Delta region of Canada.

Pioneer community Community dominated by pioneer species of plants.

Pioneer species Plant species that dominate a community in the early stages of succession.

Pitted topography Landscape characterized by numerous kettle holes on a glacial outwash plain.

Pixel Smallest part (addressable element) of an electronically-coded image, such as a computer display. Pixel is a contraction of "picture element.".

Place A term used in geography that describes the factors that make the location of natural and human-made phenomena unique.

Plagioclase feldspar A type of feldspar that is rich in sodium and calcium. Common rock forming mineral.

Plain Region of uniform general slope, comparatively level, of considerable extent, and not broken by marked elevations and depressions (it may be an extensive valley floor or a plateau summit); an extent of level or nearly level land; a flat, gently sloping, or nearly level region of the sea floor.

Plan position indicator Also known as a PPI Scope, it is a radar indicator scope displaying range and azimuth of targets in polar coordinates.

Planar projection A planar projection is a type of map in which the details of the globe are projected onto a plane (a flat surface) yielding a rectangular-shaped map. Cylindrical maps have a lot of distortion towards the edges.

Plane of the ecliptic Hypothetical two-dimensional surface in which the Earth's orbit around the sun occurs.

Planet 1. Any one of the nine primary celestial bodies that orbit the sun: Mercury, Venus, Earth, Mars, Jupiter, Saturn, Uranus, Neptune, and Pluto.

2. A similar body orbiting another star.

Planetable Instrument consisting essentially of a drawing board on a tripod and some type of sighting device (alidade) with attached straightedge, used for plotting the lines of survey directly from observation in the field.

Planetary albedo The fraction of incident solar radiation that is reflected by a planet and returned to space. The planetary albedo of the Earth-atmosphere system is approximately 30 percent, most of which is due to backscatter from clouds in the atmosphere.

Planimetric map A planimetric map is a map that represents only the horizontal positions of features (and not the vertical positions, like heights, which a topographic map shows).

Planimetry Plan details of a map - those having no indication of relief or contour.

Plankton Minute plant (phytoplankton) and animal organisms (zooplankton) that are found in aquatic ecosystems.

Plant Organisms belonging to the kingdom Plantae. These organisms have the following general characteristics: lack of locomotion, lack of a nervous system, and cellulose cell walls. Most plants can photosynthesize.

Plasma A fourth state of matter (in addition to solid, liquid, and gas) that exists in space. In this state, atoms are positively charged and share space with free negatively-charged electrons. Plasma can conduct electricity and interact strongly with electric and magnetic fields. The solar wind is actually hot plasma blowing from the sun.

Plastic deformation Irreversible change in the shape of a material without fracture as the result of the force of compression or expansion.

Plat Diagram drawn to scale showing all essential data pertaining to the boundaries and subdivisions of a tract of land, as determined by survey or protraction. As used by the Bureau of Land Management, the drawing which represents the particular area included in a survey, such as township, private land claim, or mineral claim, and the lines surveyed, established, or retraced, showing the direction and length of each such line; The relation to the adjoining official surveys; the boundaries, descriptions, and area of each parcel of land subdivided; and, as nearly as may be practicable, a representation of the relief and improvements within the limits of the survey.

Plate tectonics Theory suggesting that the Earth's surface is composed of a number of oceanic and continental plates. Driven by convection currents in the mantle, these plates have the ability to slowly move across the Earth's plastic asthenosphere. This theory is very important to geology and geomorphology because it helps to explain the occurrence and formation of mountains, folds, faults, volcanoes, earthquakes, ocean trenches, and the mid-oceanic ridges.

Plateau A plateau is a large, flat area of land that is higher than the surrounding land.

Plateau basalt An accumulation of horizontal flows of basaltic lava. Also called flood basalts.

Platform Is a another term for computer hardware, including microcomputers, workstations, and mainframe computers. When discussing software, platform independence implies the software can be run on any computer.

Platted land Land that has been divided into surveyed lots.

Playa A dry lake bed found in a desert.

Pleistocene epoch (ice age) Period of time from about 2 million years ago to 10,000 years ago. During this period areas of land at higher and middle latitudes where covered with glacial ice.

Plow/plough wind The spreading downdraft and strong straight-line winds preceding a thunderstorm. So called in the American Midwest because of its ability to flatten tall grasses as it passes.

Plucking Erosive process of particle detachment by moving glacial ice. In this process, basal ice freezes in rock surface cracks. As the main body of the glacial ice moves material around the ice in the cracks is pulled and plucked out. Also called quarrying.

Plural society A situation in which two or more culture groups occupy the same territory but maintain their separate cultural identities.

Pluton Any mass of intrusive igneous rock.

Podzol soil 1. Soil order (type) of the Canadian System of Soil Classification. This soil type is often found under coniferous forests. Its main identifying traits are a poorly decomposed organic layer, an eluviated A horizon, and a B horizon with illuviated organic matter, aluminium, and iron.

2. Soil commonly found under coniferous forests.

Podzolization Soil forming process that produces a strongly leached soil with a distinctive iron hardpan layer in the B horizon. Common in cool, moist forest environments.

Point bar Stream bar deposit that is normally located on the inside of a channel bend.

Polar air mass An air mass that forms over a high latitude region. Continental polar air (cP) is formed over cold surface regions and is typically very stable with low moisture. Maritime polar air (mP), produced over warmer waters, is less stable with high moisture.

Polar axis Is a line drawn through the Earth around the planet rotates. The point at which the polar axis intercepts the Earth's surface in the Northern Hemisphere is called the North Pole. Likewise, the point at which the polar axis intercepts the Earth's surface in the Southern Hemisphere is called the South Pole.

Polar cell Three-dimensional atmospheric circulation cell located at roughly 60 to 90° North and South of the equator. Vertical air flow in the Polar cell consists of rising air at the polar font and descending air at the polar vortex.

Polar easterlies Winds that originate at the polar highs and blow to the subpolar lows in a east to west direction.

Polar front Weather front located typically in the mid-latitudes that separates arctic and polar air masses from tropical air masses. Along the polar front we get the development of the mid-latitude cyclone. Above the polar front exists the polar jet stream.

Polar high Surface area of atmospheric high pressure located at about 90° North and South latitude. These high pressure systems produced by vertically descending air currents from the polar vortex.

Polar jet Marked by a concentration of isotherms and strong vertical shear, this jet is the boundary between the polar air and the subtropical air. It often divides into two branches, the north and the south, and marks the high speed core of the prevailing westerlies. It is associated with the location and motion of the high and low pressure areas of the middle latitudes, and therefore, is variable in position, elevation, and wind speed. Its position tends to migrate south in the Northern Hemispheric winter and north in the summer, and its core winds increase during the winter and become less strong in the summer.

Polar jet stream Relatively fast uniform winds concentrated within the upper atmosphere in a narrow band. The polar jet stream exists in the mid-latitudes at an altitude of approximately 10 kilometres. This jet stream flows from west to east at speeds between 110 to 185 kilometres per hour.

Polar orbit An orbit with an orbital inclination of near 90o where the satellite ground track will cross both polar regions once during each orbit. The term is used to describe the near-polar orbits of spacecraft such as the USA's NOAA/TIROS and Landsat satellites.

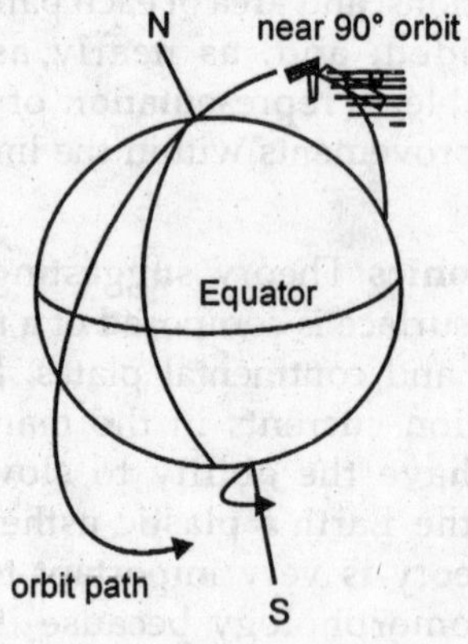

Fig. Polar orbit

Polar stratospheric clouds High altitude clouds found in the stratosphere where the temperature is less than -85° Celsius. Commonly found over Antarctica. Have a role in the creation of the ozone hole over Antarctica.

Polar vortex High pressure system located in the upper atmosphere at the polar regions. In this system, air in the upper troposphere moves into the vortex centre and then descends to the Earth's surface to create the polar highs.

Polar-orbiting satellite A satellite whose orbit passes over both of the earth's between poles. Compare with a geostationary satellite.

Poles/polar The poles are the geographic point at 90 degrees latitude North and South on the earth's surface. They are equal distance from the equator. The polar region is considered to be that area between 60 and 90 degrees latitude, both North and South.

Political map A political map is a map that shows cultural features, like the political boundaries of countries, states, provinces, and cities.

Pollutant Particles, gases, or liquid aerosols in the atmosphere which have an undesirable effect on man or his surroundings. Something unfavorable to health and life that has been added to the environment.

Pollution Physical, chemical, or biological change in the characteristics of some component of the atmosphere, hydrosphere, lithosphere, or biosphere that adversely influences the health, survival, or activities of humans or other living organisms.

Polycyclic landform Landform that shows the repeated influence of one or more major geomorphic processes over geological time. Major geomorphic processes are: weathering, erosion, deposition, and massive Earth movements caused by plate tectonics.

Polygenetic landform Landform that shows the influence of two or more major geomorphic processes. Major geomorphic processes are: weathering, erosion, deposition, and massive earth movements caused by plate tectonics.

Polynodal Many-centred.

Polypedon An identifiable soil with distinct characteristics found in a location or region. Composed of numerous pedons.

Pond A pond is a small body of water surrounded by land. A pond is smaller than a lake.

Pool Scoured depression found on the bed of streams. Associated with riffles.

Population 1. Refers to all the individuals of a given species in a specific area or region at a certain time. Its significance is more than that of a number of individuals because not all individuals are identical. Populations contain genetic variation within themselves and between other populations.

2. A statistical population is the entire collection of people, animals, plants or things from which we may collect data from.

Population crash Sudden decline in the number of individuals found in a population because of a scarcity of environmental resources that are required for survival, growth, and reproduction.

Population density Number of individuals of a particular species found in a specified area.

Population parameter A value used to represent a certain quantifiable characteristic of a population. For example, the population mean is a parameter that is often used to indicate the central value of quantity.

Pore ice A form of periglacial ground ice that is found in the spaces that exist between particles of soil.

Porosity The void spaces found in rock, sediment, or soil. Commonly measured as the percentage of void space in a volume of substance.

Positive feedback Change in the state of a system that enhances the measured effect of the initial alteration.

Positive vorticity advection The rotation of the atmosphere advects higher values of vorticity into an area. This is created by cyclonic turning and is often associated with upward motion of air. When found in advance of a short wave aloft, it often enhances thunderstorm potential. Contrast with negative vorticity advection.

Post-industrial An economy that gains its basic character from economic activities developed primarily after manufacturing grew to predominance. Most notable would be quaternary economic patterns.

Potential energy Is the energy that a body possesses by virtue of its position and that is potentially transformable into another form of energy.

Potential evapotranspiration Is a measure of the ability of the atmosphere to remove water from the surface through the processes of evaporation and transpiration assuming no limitation on water supply.

Pounds per square inch (PSI) A unit for measuring pressure. One PSI equals the pressure resulting from a force of one pound force acting over an area of one square inch.

PPB See parts per billion.

PPM See parts per million.

PPT See parts per thousand.

Prairie A prairie is a wide, relatively flat area of land that has grasses and only a few trees.

Precambrian Span of geologic time that dates from 4.6 billion to 570 million years ago. Made up of three geologic eras: Hadean, Archean, and Proterozoic.

Precambrian rock The oldest rocks, generally more than 600 million years old.

Precambrian shield Another term for shield.

Precession of the equinox Wobble in the Earth's polar axis. This motion influences the timing aphelion and perihelion over a cyclical period of 23,000 years.

Precession The comparatively slow torquing of the orbital planes of all satellites with respect to the Earth's axis, due to the bulge of the Earth at the equator which distorts the Earth's gravitational field. Precession is manifest by the slow rotation of the line of nodes of the orbit (westward for inclinations less than 90o and eastward for inclinations greater than 90o).

Precipitable water Amount of water potentially available in the atmosphere for precipitation. Usually measured in a vertical column that extends from the Earth's surface to the upper edge of the troposphere.

Precipitate Solidification of a previously dissolved substance from a solution.

Precipitation 1. Is any aqueous deposit, in liquid or solid form, that develops in a saturated atmosphere (relative humidity equals 100 %) and falls to the ground generally from clouds. Most clouds, however, do not produce precipitation. In many clouds, water droplets and ice crystals are too small to overcome natural updrafts found in the atmosphere. As a result, the tiny water droplets and ice crystals remain suspended in the atmosphere as clouds.

2. The state of being precipitated from a solution.

Precipitation Moisture that falls from clouds. Although clouds appear to float in the sky they are always falling, their water droplets slowly being pulled down by gravity. Because their water droplets are so small and light, it can take 21 days to fall 1,000 feet and wind currents can easily interrupt their descent. Liquid water falls as rain or drizzle. All raindrops form around particles of salt or dust. (Some of this dust comes from tiny meteorites and even the tails of comets.) Water or ice droplets stick to these particles, then the drops attract more water and continue getting bigger until they are large enough to fall out of the cloud. Drizzle drops are smaller than raindrops. In many clouds, raindrops actually begin as tiny ice crystals that form when part or all of a cloud is below freezing. As the ice crystals fall inside the cloud, they may collide with water droplets that freeze onto them. The ice crystals continue to grow larger; until large enough to fall from the cloud. They pass through warm air; melt, and fall as raindrops.

When ice crystals move within a very cold cloud (10ºF and -40ºF) and enough water droplets freeze onto the ice crystals, snow will fall from the cloud. If the surface temperature is colder than 32ºF the flakes will land as snow.

Precipitation Weights: one raindrop .000008 lbs

One snowflake .0000003 lbs

One cumulus cloud 10,000,000 lbs

One thunderstorm 10,000,000,000 lbs

One hurricane 10,000,000,000,000 lbs

Precision Refers to the level of measurement and exactness of description in a GIS database. Precise locational data may measure position to a fraction of a unit. Precise attribute information may specify the characteristics of features in great detail. It is important to realize, however, that precise data—no matter how carefully measured—may be inaccurate. Surveyors may make mistakes or data may be entered into the database incorrectly. Therefore, a distinction is made between precision and accuracy .

Predation Biological interaction between species where a predator species consumes a prey species.

Predator Consumer organism who feeds on prey. Consumption usually causes the death of the prey.

Prediction Forecast or extrapolation of the future state of a system from current or past states.

Pre-frontal squall line A line of thunderstorms that precedes an advancing cold front. Refer to a squall line.

Pre-frontal trough An elongated area of relatively low pressure preceding a cold front that is usually associated with a shift in wind direction. Refer to a trough.

Pressure The force per unit area exerted by the weight of the atmosphere above a point on or above the earth's surface. Also known as atmospheric pressure or barometric pressure.

Pressure altimeter An aneroid barometer calibrated to indicate altitude in feet instead of unit of pressure. It read accurately only in a standard atmosphere and when the correct altimeter setting is used.

Pressure altitude The altitude in standard atmosphere at which a given pressure will be observed. It is the indicated altitude of a pressure altimeter at an altitude setting of 29.92 inches of mercury, and is therefore the indicated altitude above the 29.92 constant pressure surface.

Pressure change The net difference between the barometric pressure at the beginning and ending of a specified interval of time, usually the three hour period preceding an observation.

Pressure characteristic The pattern of the pressure change during the specified period of time, usually the three hour period preceding an observation. This is recorded in three categories: falling, rising, or steady.

Pressure gradient The amount of pressure change that occurs over a fixed distance at a fixed altitude.

Pressure gradient force Force due to spatial differences in atmospheric pressure. Usually expressed in millibars or kilopascals per unit distance (meters or kilometres). This force is primarily responsible for the formation of wind.

Pressure jump A sudden increase in the observed atmospheric pressure or station pressure.

Pressure melting point Temperature at which minerals deep within the Earth and ice below the surface of a glacier are caused to melt because of the introduction of pressure.

Pressure tendency The pressure characteristic and amount of pressure change during a specified time period, usually the three hour period preceding the observation.

Prevailing visibility It is considered representative of visibility conditions at the observation station. It is the greatest distance that can be seen throughout at least half the horizon circle, but not necessarily continuous.

Prevailing westerlies Winds in the middle latitudes (approximately 30o to 60o) that generally blow from west to east. The subtropical high pressure regions at the horse latitudes (30o) forces surface air poleward, and the rotation of the Earth causes these winds to bear to the right (east) in the Northern Hemisphere and to the left (east) in the Southern Hemisphere. This is, to some extent, an idealized picture of the atmospheric circulation. The actual circulation on individual days includes modifications and variations due to the migratory cyclones and anticyclones of middle latitudes, causing rapid and often violent weather changes, as warm semi-tropical air from the horse latitudes meets cold polar air from the high latitudes.

Prevailing wind A wind that blows from one direction more frequently than any other during a given period, such as a day, month, season, or year.

Prey Organism that is consumed by a predator.

Primary consumer Organisms that occupy the second trophic level in the grazing food chain. These organisms are herbivores.

Primary pollutant Air pollutants that enter the atmosphere directly.

Primary producer Organisms that occupy the first trophic level in the grazing food chain. These organisms are photosynthetic autotrophs.

Primary product A product that is important as a raw material in developed economies; a product consumed in its primary (i.e., unprocessed) state.

Primary sector That portion of a region's economy devoted to the extraction of basic materials (e.g., mining, lumbering, agriculture).

Primary succession Succession on soil or sediments that do not contain an active seed bank.

Prime meridian Zero degrees longitude (0°). The prime meridian runs through the Royal Greenwich Observatory in Greenwich, England (the location was established in 1884 by international agreement). The prime meridian divides the globe into the Western and Eastern hemispheres. The Earth's time zones are measured from the prime meridian. The time at 0° is called Universal Time (UT) or Greenwich Mean Time (GMT). With the Greenwich meridian as the starting point, each 15° east and west marks a new time zone. The 24 time zones extend east and west around the globe for 180° to the International Date Line. When it is noon along the prime meridian, it is midnight along the International Date Line.

Printed circuit A fibre card on which integrated circuits and other electronic components can be mounted. Connections between the components are etched in the correct circuit patterns.

Probability Statistical chance that an event will occur.

Process An association of phenomena governed by physical, chemical, or biological laws. An example of a process is the vertical mixing of ocean waters in the so-called surface-mixed layer; the state variables for this process include temperature, salinity in the water on a vertical scale of tens of meters, and heat flow and wind stress at the sea surface. Other examples include the volcanic deposition of dust and gases into the atmosphere, eddy formation in the atmosphere and oceans, and soil development.

Process study An organized, systematic investigation of a particular process designed to identify all of the state variables involved and to establish the relationships among them. Process studies yield numerical algorithms that connect the state variables and determine their rates of change; such algorithms are essential ingredients of Earth system models.

Process response system This is a system that integrates the characteristics of both morphological and cascading systems. In a process-response system, we can model the processes involved in the movement, storage, and transformation of energy and/or matter between system elements and we fully understand how the form of the system in terms of between measured features.

Producer An organism that can synthesize the organic nutrients in requires for growth through processes like photosynthesis.

Productivity Rate of energy fixation or storage of biomass by plants. Usually expressed per unit area and time.

Profiler A type of Doppler radar that measures both wind speed and direction from 1,500 feet up through 55,000 feet in the atmosphere.

Prognostic chart A chart of forecast predictions that may include pressure, fronts, precipitation, temperature, and other meteorological elements. Also known as a prog.

Progradation The natural extension of a shoreline seaward.

Prograde orbit Orbits of the Earth in the same direction as the rotation of the Earth (west-to-east).

Progressive succession Succession where the developing plant community becomes complex and contains more species and biomass over time.

Projection A projection is a representation of one thing onto another, such as a curved 3-dimensional surface (like the Earth) onto a flat 2-dimensional map. A projection is a representation of one thing onto another, such as a curved 3-dimensional surface (like the Earth) onto a flat 2-dimensional map. There are 3 major types of projections, cylindrical, conic, and planar.

Prokaryote Organisms whose cells have their genetic material in the form of loose strands of DNA found in the cytoplasm. They also do not have a membrane-bound nucleus and have few specialized structures located within their cell boundary.

Propagule Structure that develops into a plant.

Proportional Cause and effect relationship between two variables where a positive or negative change in the quantity of one causes a predictable similar quantity change in the other.

Protein Organic substances primarily composed of carbon, hydrogen, nitrogen, and some other minor elements which are arranged in about 20 different compounds known as amino acids. The various amino acids found in a protein are linked together by peptide bonds.

Proterozoic Geologic eon that occurred from 570 to 2500 million years ago. During this time period, the first single-celled and multi-celled eukaryotic organisms evolved and developed.

Protista Organisms that are mainly unicellular and have a eukaryotic cell type. A few multicellular members exist.

Protocol Is a set of rules or standards which govern communication between computers and between computers and peripherals.

Proton A sub-particle of an atom that contains a positive charge.

Protoplasm Substances making up a cell including its exterior membrane.

Protozoa Heterotrophic eukaryotic unicellular organisms that belong to the kingdom protista.

Proxy data Data that measures the cause and effect relationship between two variables indirectly.

Psychrometer Instrument used to measure atmospheric humidity. It consists of two thermometers (wet-bulb and a dry-bulb) one of which has its bulb covered by a moistened wick. Humidity is determined by the difference in readings between the two thermometers after air has passed over both of them for a specific time period.

Psychrometric table Table of values that allows for the determination of relative humidity and dew point from dry-bulb and wet-bulb temperatures recorded on a psychrometer.

Pueblo A type of Indian village constructed by some tribes in the southwestern United States. A large community dwelling, divided into many rooms, up to five stories high, and usually made of adobe. Also, a Spanish word for town or village.

Pulse A very short duration of time. In regard to a radar, it is a brief burst of a electromagnetic radiation emitted by the radar.

Pyramid of biomass Graphic model describing the distribution of biomass in an ecosystem or community at the trophic level.

Pyramid of numbers Graphical model describing the number of organisms that exist at each trophic level in a community or an ecosystem.

Pyroclastic material Pieces of volcanic rock thrown out in a volcanic explosion.

Pyroxene A group of single chained inosilicate minerals whose basic chemical unit is the silica tetrahedron (SiO_4). They are common rock forming minerals and are found in most igneous and metamorphic rocks. They form at high temperatures with very little water in the crystallization environment.

Q

Quad centreed photograph Middle exposure of a phototriplet (three consecutive aerial photographs) take so that the middle photograph is exposed directly above the centre of the quadrangle and the preceding and following photographs are exposed directly above the boundaries of the quadrangle. The flying height is set such that the quad-centred photograph covers the entire quadrangle.

Quadrangle Four-sided area, bounded by parallels of latitude and meridians of longitude used as an area unit in mapping (dimensions are not necessarily the same in both directions). Also, a geometric figure of significance in geodetic surveying.

Quantitative Precipitiation Forecast (QPF) A forecast of rainfall, snowfall or liquid equivalent of snowfall.

Quantitative revolution Time in the history (after 1950) of physical geography when measurement became the central focus of research. This measurement was used primarily for hypothesis testing. With measurement came mapping, models, statistics, and mathematics. Researchers began investigating process rather than mere description of the environment.

Quartz Mineral with the chemical formula SiO_2. Quartz is common in continental crust but rare in oceanic crust.

Quartzite Metamorphic rock rich in quartz created by the recrystallization of sandstone.

Quasi stationary front A front which is nearly stationary or moves very little since the last synoptic position. Also known as a stationary front.

Quaternary Geologic period that occurred roughly 1.6 million years ago to today. During much of this period continental glaciers in the Northern Hemisphere covered large regions of land surface in the high and mid-latitudes. Homo sapiens appear about 200,000 years BP (before present) and become the first species to alter the Earth's environment on a large-scale.

Quaternary sector That portion of a region's economy devoted to informational and idea-generating activities (e.g., basic research, universities and colleges, and news media).

R

R horizon Soil horizon found beneath the C horizon. Consists of consolidated rock showing little sign of weathering or pedogenesis.

R selected species (Malthusian strategy) A species that shows the following characteristics: short life span; early reproduction; low biomass; and the potential to produce large numbers of usually small offspring in a short period of time.

R&D Research and development.

RADAR Acronym for Radio detection and ranging. An electronic instrument used to detect distant objects and measure their range by how they scatter or reflect radio energy. Precipitation and clouds are detected by measuring the strength of the electromagnetic signal reflected back. Doppler radar and nexrad are examples.

Radarsat Satellite program established by the Canadian Space Agency for the purpose of remotely sensing the Earth's resources. Radarsat uses an active sensing system that transmits microwaves.

Radarsonde observation An upper air observation used to determine winds, and other meteorological data, by tracking the range, elevation, and azimuth of a radar target carried aloft . A type of rawinsonde.

Radial velocity A type of velocity that expresses motion toward or away from a given location. In Doppler radar, it is the component of motion that is parallel to the radar beam.

Radial-line plotting Determination of the location of points by the successive intersection and resection of direction lines radiating from the radial centres of overlapping aerial photographs.

Radiant 1. In optics, the point or object from which light proceeds. 2. In geometry a straight line proceeding from a given point, or fixed pole, about which it is conceived to revolve. 3. In astronomy the point in the heavens from which a shower of meteors seems to proceed.

Radiant energy Energy in the form of electromagnetic waves. In some cases it refers to the radiation emitted from the sun.

Radiation Energy transfer in the form of electromagnetic waves or particles that release energy when absorbed by an object.

Radiation budget A measure of all the inputs and outputs of radiative energy relative to a system, such as Earth.

Radiation fog Fog that is created when radiational cooling at the earth's surface lowers the temperature of the air near the ground to or below its initial dew point. Formation is best when there is a shallow surface layer of relatively moist air beneath a drier layer, clear skies, and light surface winds. This primarily occurs during the night or early morning. It may also be called ground fog.

Radiative cooling Cooling process of the Earth's surface and adjacent air; which occurs when infrared (heat) energy radiates from the surface of the Earth upward through the atmosphere into space. Air near the surface transfers its thermal energy to the nearby ground through conduction, so that radiative-cooling lowers the temperature of both the surface and the lowest part of the atmosphere.

Radiative transfer Theory dealing with the propagation of electromagnetic radiation through a medium.

Radio Frequency (RF) A frequency that is useful for radio transmission, usually between 10 kHz and 300,000 MHz.

Radio spectrum The complete range of frequencies or wave lengths of electromagnetic waves, specifically those used in radio and television.

Radio wave An electrical impulse sent through the atmosphere at radio frequency.

Radioactive Giving off or capable of giving off radiant energy in the form of particles or rays, as in alpha, beta, and gamma rays.

Radioactive decay Natural decay of the nucleus of an atom where alpha or beta particle and/or gamma rays are released at a fixed rate.

Radioisotope or radioactive isotope A unstable isotope of an element. This material decays spontaneously and releases subatomic particles and electromagnetic energy.

Radiometer An instrument that quantitatively measures electromagnetic radiation. Weather satellites carry radiometers to measure radiation from snow, ice, clouds, bodies of water; the Earth's surface, and the sun.

Radiosonde A balloon-borne instrument that measures meteorological parameters from the Earth's surface up to 20 miles in the atmosphere. The radiosonde measures temperature, pressure, and humidity and transmits or "radios" these data back to Earth. Upper air winds also are determined through tracking of the balloon ascent. Radiosonde observations generally are taken twice a day (0000 and 1200 UTC) around the globe. NOAA's National Weather Service (NWS) operates a network of about 90 radiosonde observing sites in the U.S and its territories. When the balloons burst radiosondes return to Earth on a parachute.

Approximately 25 percent are recovered and returned to NWS for reconditioning and reuse.

Rail gauge The distance between the two rails of a railroad.

Rain Precipitation in the form of liquid water droplets greater than 0.5 mm. If widely scattered, the drop size may be smaller. It is reported as "R" in an observation and on the METAR. The intensity of rain is based on rate of fall. "Very light" (R−) means that the scattered drops do not completely wet a surface. "Light" (R-) means it is greater than a trace and up to 0.10 inch an hour. "Moderate" (R) means the rate of fall is between 0.11 to 0.30 inch per hour. "Heavy" (R+) means over 0.30 inch per hour.

Rain forest An evergreen woodland of the tropics distinguished by a continuous leaf canopy and an average rainfall of about 100 inches per year. Rain forests play an important role in the global environment. The Earth sustains life because of critical balances and interactions among many factors. Were there not processes at work that limit the effects of other essential processes, Earth would become uninhabitable. Destruction of tropical rain forests reduces the amount of leaf area in the tropics, and consequently the amount of carbon dioxide absorbed, causing increases in levels of carbon dioxide and other atmospheric gases. It is estimated that cutting and burning of tropical forests contributes about 20 percent of the carbon dioxide added to the atmosphere each year. The World Resources Institute and the International Institute for Environment and Development have reported that the world's tropical forests are being destroyed at the rate of fifty-four acres per minute, or twenty-eight million acres lost annually. Rain forest destruction also means the loss of wide spectrumand biological life, erosion of soil, and possible desertification.

Rain gauge An instrument used to measure the amount of rain that has fallen. Measurement is done in hundredths of inches (0.01").

Rain shadow Also referred to as a precipitation shadow, it is the region on the lee side of a mountain or similar barrier where the precipitation is less than on the windward side. For example, the relatively dry Washoe Valley of western Nevada is in the rain shadow of the Sierra Nevada.

Rainbow A luminous arc featuring all colours of the visible light spectrum (red, orange, yellow, green, blue, indigo, and violet). It is created by refraction, total reflection, and the dispersion of light. It is visible when the sun is shining through air containing water spray or raindrops, which occurs during or immediately after a rain shower. The bow is always observed in the opposite side of the sky from the sun.

Raindrop impact Force exerted by a falling raindrop on a rock, sediment, or soil surface.

Rainfall The amount of precipitation of any type, primarily liquid. It is usually the amount that is measured by a rain gauge. Refer to rain for rates of intensity and the quantitative precipitation for forecasting.

Rainshadow An area of diminished precipitation on the lee (downwind) side of a mountain or mountain range.

Rainshadow effect Reduction of precipitation commonly found on the leeward side of a mountain. The reduction in precipitation is the result of compression warming of descending air.

Rainsplash Soil erosion caused from the impact of raindrops.

Rainwash The erosion of soil by overland flow. Normally occurs in concert with rainsplash.

Random Process or event that occurs by chance.

Range A statistical measure of the dispersion of observation values in a data set. Determined by taking the difference between the largest and the smallest observed value.

Range resolution The ability of radar to distinguish between targets on the same azimuth but at different ranges.

Rangeland Land-use type that supplies vegetation for consumption by grazing and browsing animals. This land-use type is normally not intensively managed.

Raster Displays and databases build all geographic features from grid cells in a matrix. A raster display builds an image from pixels, pels, or elements of coarse or fine resolution. A raster database maintains a similar "picture" of reality in which each cell records some sort of information averaged over the cell's area. The size of the cell may again be coarse or fine, ranging from centimetres to kilometres. Many satellites, like Landsat and SPOT, transmit raster images of the earth's surface. Reflectance at a certain wavelength is measured for each cell in an image. The cells may cover areas on the earth's surface several hundreds of meters square, the area covered being a function of a particular satellite's resolution.

Rawinsonde An upper air observation that evaluates the winds, temperature, relative humidity, and pressure aloft by means of a balloon-attached radiosonde that is tracked by a radar or radio direction-finder. It is a radiosonde observation combined with a winds-aloft observation, called a rawin.

Re entrants A prominent indentation in an escarpment, ridge or shoreline.

Reach An expanse of a stream channel.

Real time As it happens.

Realized niche Describes the part of the fundamental niche that a species actually occupies.

Receiver sensitivity The ability of a receiver to detect weak signals through the noise level of the receiving system, which includes the antenna and internal thermal noise of the receiver.

Recessional moraine Moraine that is created during a pause in the retreat of a glacier. Also called a stadial moraine.

Recharge area The area on the Earth's surface that receives water for storage into a particular aquifer.

Reconnaissance (recco) code An aircraft weather reconnaissance code that has come to refer primarily to in-flight tropical weather observations, but actually signifies any detailed weather observation or investigation from an aircraft in flight.

Rectangular coordinate system System that measures the location of points on the Earth on a two-dimensional coordinate plane.

Rectification, differential The process of scanning and reprojecting a photograph onto a horizontal plane in differential elements to remove displacements caused by tilt and relief. The process may be accomplished by any one of a number of instruments developed specifically for the purpose.

Rectification, simple Projection of an aerial photograph (mathematically, graphically, or photographically) from its plane onto a horizontal plane by translation, rotation, and (or) scale change to remove displacement due to tilt of the camera.

Recumbent fold A fold in which the axial plane is almost horizontal.

Recurrence interval The average time period that separates natural events of a specific magnitude. For example, floods of a specific stream discharge level.

Red tide A phenomena that occurs near the surface of the sea and is caused by an abnormal growth of dinoflagellates, single-celled plankton. The toxin secreted by the dinoflagellates is capable of killing fish through a cyclic effect. As decomposition continues, it creates more nutrients, which feed the plankton bloom, and the greater quantities can color the water red.

Reduction 1. Chemical process that involves the removal of oxygen from a compound.

2. A form of chemical weathering.

Reef A ridge of rocks found in the tidal zone along a coastline. One common type of reef is the coral reef.

Reference map Map that shows natural and human-made objects from the geographical environment with an emphasis on location. Compare with thematic map.

Reflected infrared radiation Form of electromagnetic radiation with a wavelength between 0.7 to 3.0 micrometers (μm).

Reflected wave A water wave that reflects off the shore or another obstacle and is redirected towards the sea or lake.

Reflection (atmospheric) Process where insolation is redirect by 180° after striking a particle. This redirection causes 100 %

loss. Most of the reflection in the Earth's atmosphere occurs in clouds because of light's interception with particles of liquid and frozen water. The reflectivity of a cloud can range from 40-90 %.

Reflectivity A measure of the process by which a surface can turn back a portion of incident radiation back into the medium through which the radiation approached. Reflectivity of a given surface is know as albedo. It is also the degree by which a precipitation type is able to reflect a radar beam.

Refraction The bending of light or radar beam as it passes through a zone of contrasting properties, such as atmospheric density, water vapour, or temperature.

Reg A rocky desert landscape.

Region An area having some characteristic or characteristics that distinguish it from other areas. A territory of interest to people and for which one or more distinctive traits are used as the basis for its identity.

Regional metamorphism Large scale metamorphic modification of existing rock through the heat and pressure of plutons created at tectonic zones of subduction.

Regolith Loose layer of rocky material overlying bedrock.

Regosol soil Soil order (type) of the Canadian System of Soil Classification. This type is any young underdeveloped soil that lacks identifying soil horizons.

Relational database Stores data in such a way that it can be added to, and used independently of, all other data stored in the database. Users can query a relational database without knowing how the information has been organized. Although relational databases have the advantages of ease-of-use and analytical flexibility, their weakness can be slower retrieval speed, SQL is one example of a relational database.

Relative humidity The ratio between the actual amount of water vapour held in the atmosphere compared to the amount required for saturation. Relative humidity is influenced by temperature and atmospheric pressure.

Relative vorticity The sum of the rotation of an air parcel about the axis of the pressure system and the rotation of the parcel about its own axis. Refer to vorticity.

Relief Elevations and depressions of the land or sea bottom.

Relief map A relief map is a topographic map that uses different colours or shades to indicate elevations.

Relief shading Technique for making hypsography on a map appear three dimensional by the use of graded shadow effects. Generally, the features are shaded as though illuminated from the northwest.

Remapping Flattening the Earth into a standard map projection. When the spherical Earth is photographed by satellites, areas lying near the outer edge of the picture are distorted. Remapping rectifies the distortion.

Remote sensing The technology of acquiring data and information about an object or phenomena by a device that is not in physical contact with it. In other words, remote sensing refers to gathering information about the Earth and its environment from a distance, a critical capability of the Earth Observing System

For example, spacecraft in low-Earth orbit pass through the outer thermosphere, enabling direct sampling of chemical species there. These samples have been used extensively to develop an understanding of thermospheric properties. Explorer-17, launched in 1963, was the first satellite to return quantitative measurements of gaseous stratification in the thermosphere. However, the mesosphere and lower layers cannot be probed directly in this way—global observations from space require remote sensing from a spacecraft at an altitude well above the mospause. The formidable techological challenges of atmospheric remote sensing, many of which are now being overcome, have delayed detailed study of the stratosphere and mososphere by comparison with thermospheric research advances.

Some remote-sensing systems encountered in everyday life include the human eye and brain, and photographic and video cameras.

Remote sensor Mechanical devices used to remotely sense an object or phenomenon.

Representational fraction Scale of a map or chart expressed as a fraction or ratio that relates unit distance on the map to distance measured in the same unit on the ground.

Reproduction Summation of all processes involved in printing copies from an original drawing. A printed copy of an original drawing made by the processes of reproduction.

Reptile Group of terrestrial vertebrate animals that includes turtles, tortoises, snakes, lizards, crocodiles, and alligators.

Reservoir A reservoir is a man-made lake that stores water for future use.

Resolution A measure of the ability to separate observable quantities. In the case of imagery, it describes the area represented by each pixel of an image. The smaller the area represented by a pixel, the more accurate and detailed the image. APT has a resolution of 4 km, i.e., each pixel represents a square, 4 km on each side. HRPT has a resolution of 1.1 km at nadir (4 km at edge of scan), and WEFAX of 8 km.

Resolution cell The smallest unit of area in an image of discrete elements. The area represented by a pixel.

Resource Anything that is both naturally occurring and of use to humans.

Resource partitioning The evolutionary process of species living in the same ecosystem dividing up resources so that each species develops dissimilar resource requirements to avoid competition.

Respiration Is the typical process where mitochondria of cells of organisms release chemical energy from sugar and other organic molecules through chemical oxidation. This process occurs in both plants and animals. In most organisms, respiration releases the energy required for all metabolic processes. This chemical reaction can be described by the following simple equation: $C_6H_{12}O_6 + 6O_2 >>> 6CO_2 + 6H_2O$ + released energy.

Retrogression In meteorology, it is the movement of a weather system in a direction opposite to the direction of the basic flow in which it is embedded. Often used in reference to a long wave trough or other macroscale feature. For example, a long wave trough that may move slightly westward when the "normal" movement and flow is eastward.

Retrogressive succession Succession where the plant community becomes simplistic and contains fewer species and less biomass over time.

Reverse fault This vertical fault develops when compressional force causes the displacement of one block of rock over another.

Revolution Process of the Earth circling the sun in its orbit. Revolution determines the seasons, and the length of the year. In addition, differences in seasons occur because of Earth's inclination (tilt on its axis) of about 23.5 degrees as it revolves around the sun. Compare with rotation.

Rhumb line A line of constant compass direction or bearing which crosses the meridians at the same angle. A part of a great circle.

Rhyolite A fine grained extrusive igneous rock that is rich in quartz and potassium feldspar. Derived from felsic magma.

Ria coast An extensively carved out coast with conspicuous headlands and deep re-entrants.

Ribbon falls Spectacular narrow waterfalls that occur at the edge of a hanging valley.

Ribonucleic Acid (RNA) Form of nucleic acid. Ribonucleic acid is used by most organisms to read the genetic information found in DNA and to produce specific organic molecules used in the development and functioning of cells.

Richter scale A logarithmic measurement scale of earthquake magnitude. This scale measures the energy released by the largest seismic wave associated with the earthquake.

Ridge An elongated area of high atmospheric pressure that is associated with an area of maximum anticyclonic circulation. The opposite of a trough.

Riffle Bar deposit found on the bed of streams. Associated with these deposits are pools.

Rift Zone between two diverging tectonic plates. The mid-oceanic ridge is an area where such plate divergence is occurring.

Rift valley Steep sided valley found on the Earth's surface created by tectonic rifting.

Right ascension of ascending node One of six Keplerian elements, it indicates the rotation of the orbit plane from some reference point. Two numbers orient an orbital plane in space; inclinations is the first, this is the second. After specifying inclination, an infinite number of orbital planes are possible. The intersection of the equatorial plane and the orbital plane must be specified by a location on the equator that fully defines the orbital plane. The line of nodes occurs in two places. However, by convention, only the ascending node (where the satellite crosses the equator going from south to north) is specified. The descending node (where the satellite crosses the equator going from north to south) is not. Because the Earth spins, conventional latitude and longitude points are not used to separate where the lines of node occur. Instead, an astronomical coordinate system is used, known as the right-ascension/declination coordinate system, which does not spin with the Earth. Right ascension of ascending node is an angle, measured at the centre of the Earth, from the vernal equinox to the ascending node. For example, draw a line from the centre of the Earth to the point where the satellite crossed the equator (going from south to north). If this line points directly at the vernal equinox, then RAAN = 0o.

Rill A very small steep sided channel carrying water. This landscape feature is intermittent and forms for only a short period of time after a rainfall.

Rime Deposit of ice crystals that occurs when fog or super cooled water droplets comes in contact with an object with a temperature below freezing (0° Celsius). This deposit develops outward on the windward side of the object.

Rip current It is formed by a strong surface water movement, or current, of a short duration that flows seaward from the shore. The return flow is piled up onshore by the incoming waves and wind. It is localized, of narrow width, and its position relative to the beach can change as the wave condition changes. Therefore, the higher the waves, the stronger the current.

Riparian rights The rights of water use possessed by a person owning land containing or bordering a water course or lake.

Ripple Stream bed deposit found streams. Ripples are only a few centimetres in height and spacing and are found in slow moving streams with fine textured beds.

River A long narrow channel of water that flows as a function of gravity and elevation across the Earth's surface. Many rivers empty into lakes, seas, or oceans.

Riverine Located on or inhabiting the banks or the area near a river or lake.

Road map A road map shows major and minor highways, plus cities and towns. This type of map is used by road travellers, and often shows other information useful for travellers, including parks and campgrounds.

Robinson projection The Robinson projection is a widely-used type of map in which the Earth is shown within an ellipse with a flat top and bottom. In this pseudocylindrical. projection, lines of latitude are parallel to the equator, but lines of latitude are elliptical arcs. In a Robinson projection, area is represented accurately, but the distances and compass directions are distorted (for example, compass lines are curved). This type of projection was first made in 1963 by Arthur H. Robinson; it is also called the Orthophanic projection (meaning 'right appearing').

Roche moutonnee A feature of glacial erosion that resembles an asymmetrical rock mound. It is smooth and gently sloping on the side of ice advance. The lee-side of this feature is steep and jagged.

Rock A compact and consolidated mass of mineral matter. Three types of rock are recognized: igneous, sedimentary, and metamorphic.

Rock cycle General model describing the geomorphic and geologic processes involved in the creation, modification and recycling of rocks.

Rock flour Very finely ground rock fragments that form between the base of a glacier and the underlying bedrock surface.

Rock slide Large scale mass movement of rock materials downslope.

Rocketsonde A type of radiosonde that is shot into the atmosphere by a rocket, allowing it to collect data during its parachute descent from a higher position in the atmosphere than a balloon could reach.

Rockfall Type of mass movement that involves the detachment and movement of a small block of rock from a cliff face to its base. Normally occurs when the rock has well defined bedding planes that are exaggerated by freeze-thaw action or thermal expansion and contraction.

Roll cloud A relatively rare, low-level, horizontal, tube-shaped cloud. Although they are associated with a thunderstorm, they are completely detached from the base of the cumulonimbus cloud.

Rossby waves The movement of ridges and troughs in the upper wind patterns, primarily the jet stream, circling the earth. Named for Carl-Gustaf Rossby, a U.S. Weather Bureau (NWS) employee, who first theorized about the existence of the jet stream in 1939.

Rotation Process of the Earth turning on its axis. Rotation determines day and night, and the length of the day. Compare with revolution.

Rotation The spinning of a body, such as the earth, about its axis.

Rotational slip Form of mass movement where material moves suddenly along a curvilinear plane. Also called a slump.

Rotor cloud An altocumulus cloud formation that can be found in the lee of a mountain or similar barrier. The air rotates around a horizontal axis, creating turbulence. Altocumulus lenticularis is an example.

Rubber sheeting Is a technique for edge matching and is another name for warping.

Runoff The topographic flow of water from precipitation to stream channels located at lower elevations. Occurs when the infiltration capacity of an area's soil has been exceeded. It also refers to the water leaving an area of drainage. Also called overland flow.

Runway Visual Range (RVR) It is the maximum distance at which the runway, or the specified lights or markers delineating it, can be seen from a position above a specified point on its centre line. This value is normally determined by visibility sensors located alongside and higher than the centre line of the runway. RVR is calculated from visibility, ambient light level, and runway light intensity.

S

S band One of the segments or bands into which the radio frequency spectrum above 1000 MHz is divided, designated by letters. Signals from GOES and other geostationary spacecraft transmitting on or near 1691 MHz are transmitting on S-Band.

S wave A seismic wave that moves material it encounters perpendicular to its direction of travel. This type of seismic wave causes shear stress in the material it moves through. Also called a secondary wave or a shear wave.

Saffir-Simpson scale damage-potential scale Developed in the early 1970s by Herbert Saffir, a consulting engineer, and Robert Simpson, then Director of the National Hurricane Centre, it is a measure of hurricane intensity on a scale of 1 to 5. The scale categorizes potential damage based on barometric pressure wind speeds, and storm surge.

Salinity Concentration of dissolved salts found in a sample of water. Measured as the total amount of dissolved salts in parts per thousand. Seawater has an average salinity of about 34 parts per thousand (ppt).

Salinization Pedogenic process that concentrates salts at or near the soil surface because evapotranspiration greatly exceeds water inputs from precipitation.

Salt 1. The mineral sodium chloride.

2. Compounds that are produced as the result of a metal atom replacing a hydrogen atom in an acid.

Salt marsh Coastal wetland ecosystem that is inundated for some period of time by seawater. Plants that exist in this community have special adaptation to survive in the presence of high salinities in their immediate environment. Generally, found poleward of 30° North and South latitude.

Salt water The water of the ocean, distinguished from fresh water by its appreciable salinity.

Saltation Transport of sediment initiated by moving air or water where particles move from a resting surface to the transport medium in quick continuous repeated cycles.

Saltwater intrusion The invasion of saltwater into freshwater aquifers in coastal and inland areas. This condition can be cause when groundwater, which charges the aquifer, is withdrawn faster than it is recharged by precipitation.

Sample A sample is a subset group of data selected from a larger population group. Most samples are drawn at random to guaranty equal representation in the data.

Sampling The process of obtaining a sequence of discrete digital values from a continuous sequence of analog data.

Sand Loose particles of hard, broken rock or minerals. In observing, sand is reported when particles of sand are raised to sufficient height that reduces visibility. It is reported as "SA" in an observation and on the METAR.

Sand dune A hill or ridge of aeolian sand deposits with a minimum height of less than one meter and a maximum height of about 50 meters. Found in hot deserts and along sandy coastlines.

Sand ripples Another term used for wind ripples.

Sand sea A large region of sand and sand dunes in a desert. Common to erg deserts.

Sand sheet Deposit of sometimes stratified less well sorted sand that almost resemble dunes. Common in northern Europe. Believed to form when windblown materials settle on areas of patchy snow.

Sand wedge A form of ice wedge that contains accumulations of wind blown sand in long vertical layers. A form of periglacial ground ice.

Sandstone A type of sedimentary rock that contains a large quantity of weathered quartz grains.

Sandstorm A strong wind carrying sand particles through the air. They are low level occurances, usually only ten feet in height to not more than fifty feet above the surface. Due to the frequent winds created by surface heating, they are most predominate during the day and die out in the night. Visibility is reduced to between 5/8ths and 6/16ths statute mile, and if less than 5/16ths, then the storm is considered a heavy sandstorm. It is reported as "SS" in an observation and on the METAR.

Santa Ana winds The hot, dry winds, generally from the east, that funnel through the Santa Ana river valley south of the San Gabriel and San Bernadino Mountains in southern California, including the Los Angeles basin. Classified as katabatic, it occurs most often during the winter and it is an example of a foehn wind.

SAR SAT Search and rescue tracking system carried on NOAA polar-orbiting satellites that receives emergency signals from persons in distress. The satellites transmit these signals to ground receiving stations in the U.S. and overseas. Signals are forwarded to the nearest rescue coordination centre which computes the location from which the emergency signals came and provides the coordinates of the emergency site to a rescue team.

Sargasso sea An area of the North Atlantic Ocean between Bermuda and the Azores. It is in the middle of the North Atlantic oceanic gyre, with converging surface waters. Consequently, it has less biological features than any other region of the ocean because the lack of not mixing with more nutrient-rich waters.

Satellite A free-flying object that orbits the Earth, another planet, or the sun.

Satellite A manufactured object that orbits a celestial body, either in a geostationary or a polar manner. Some of the information that is gathered by weather satellites, such as GOES9, includes upper air temperatures and humidity, recording the temperatures of cloud tops, land, and ocean, monitoring the movement of clouds to determine upper level wind speeds, tracing the movement of water vapour, monitoring the sun and solar activity, and relaying data from weather instruments around the world.

Satellite dish (parabolic reflector) Bowl shaped antennas that collect and focus the signals that a satellite beams down to Earth. The dish reflects the incoming radio frequency energy to a focal point where it can be picked up by a feedhorn antenna to transfer the RF energy to a transmission line. The bigger the dish, the greater will be the intercepted RE energy and hence, the gain. For example, a satellite dish is used to receive goes wefax imagery.

Satellite orbital elements See keplerian elements.

Satellite pictures Pictures taken by a weather satellite that reveal information, such as the flow of water vapour, the movement of frontal systems, and the development of a tropical system. Looping individual pictures aids meteorologists in forecasting. One way a picture can be taken is as a visible shot, that is best during times of visible light (daylight). Another way is as an IR (infrared) shot, that reveals cloud temperatures and can be used day or night.

Satellite positioning A procedure by which satellites are used to locate precise objects or particular points on Earth.

Satellite revolution The time from one perigee (the point of an elliptical orbit path where a satellite is closest to Earth) to the next.

Saturate To treat or charge something to the point where no more can be absorbed, dissolved, or retained. In meteorology, it is used when discussing the amount of water vapour in a volume of air.

Saturated Adiabatic Lapse Rate (SALR) The rate of decline in the temperature of a rising parcel of air after it has reached saturation. This rate is less than the dry adiabatic lapse rate (9.8° Celsius per 1000 meters) because of the heat energy added to the ascending air parcel from condensation and deposition processes.

Saturation Atmospheric condition where water is changing its phase to liquid or solid. At saturation, relative humidity is 100 % unless there is a shortage of deposition nuclei or condensation nuclei. Generally, this process is caused by the cooling of the atmosphere.

Saturation mixing ratio Mass of water vapour that a kilogram of dry air can hold at saturation. Measured in grams.

Saturation point The point when the water vapour in the atmosphere is at its maximum level for the existing temperature.

Savanna A tropical or sub-tropical plant community characterized by trees and shrubs scattered among a cover of grasses, herbs and forbs. The climate of a savanna is tropical with a dry season occurring in the low sun period of the year.

Scale The scale of a map is the ratio between the distance between two points found on the map as compared to the actual distance between these points in the real world.

Scanner A system that optically scans its detector(s) across a scene and records or stores the data in a two-dimensional format to form an image.

Scanning Is a process by which data and maps are converted to digital form using optical or video equipment. Scanning differs from digitizing in that entire pages of data or map sheets are captured as images all at once. Digitizing involves entering discreet points and digits one at a time. Scanning may eventually lower the cost of adding information to a database. At the moment, scanning and digitizing take about the same amount of time. This is because scanned images, once in a computer, still require coding—the operator must indicate the relationships and data to be stored and where to store the information. This operation is normally part of the digitizing process. In the future, the use of artificially intelligent coding algorithms may make scanning faster and less expensive.

Scanning radiometer An imaging system consisting of lenses, moving mirrors, and solid-state image sensors used to obtain observations of the Earth and its atmosphere. Scanning radiometers, which are the sole imaging systems on all current operational weather satellites, have far better long4erm performance than the vidicon IV camera tubes used with earlier spacecraft.

Scarification Extensive movements of soil, sediment, and rock material caused by humans.

Scarp Also "escarpment." A steep cliff or steep slope, formed either as a result of faulting or by the erosion of inclined rock strata.

Scattered The amount of sky cover for a cloud layer between 3/8ths and 4/8ths, based on the summation layer amount for that layer.

Scattering (atmospheric) Is an atmospheric process where small particles and gas molecules diffuse part of the incoming solar radiation in random directions without any alteration to the wavelength of the electromagnetic energy. Scattering does, however, reduce the amount of incoming radiation reaching the Earth's surface. A significant proportion of scattered shortwave solar radiation is redirected back to space. The amount of scattering that takes place is dependent on two factors: wavelength of the incoming radiation and the size of the scattering particle or gas molecule. In the Earth's atmosphere, the presence of a large number of particles with a size of about 0.5 µm results in shorter wavelengths being preferentially scattered. This factor also causes our sky to look blue because this colour corresponds to those wavelengths that are best diffused.

Schist A medium to coarse grained metamorphic rock with well developed bedding planes derived from the foliated recrystrallization of platy like minerals like mica.

Science Science is a way of acquiring knowledge. To do science, one must follow a specific universal methodology. The central theme in this methodology is the testing of hypotheses and the ability to make predictions. The overall goal of science is to better understand nature and our Universe.

Scientific method The approach science uses to gain knowledge. This method tries to be unbias and neutral. Involves inductive and deductive reasoning, hypothesis testing and falsification, and predictive model testing.

Sclerophyllous vegetation Term used to describe drought resistant vegetation common in Mediterranean climates. Some common adaptations present in this type of vegetation include: deep roots, reduced leaf area exposed to the atmosphere, and waxy thick leaves with closing stomata which resist water loss.

Scour 1. Refers to the erosive power of water. 2. Abrasive effects of rocks and sediments incorporated in the ice base of a glacier.

Screaming eagles Cloud pattern so named because some observers maintain they can see the head of an eagle facing west in these cloud patterns. The pattern is similar to a comma, only the pattern is disorganized and not solid. Weather associated with screaming eagles consists of rain showers and gusty surface winds up to about 25 knots. The eagles can intensify and enlarge when moving into areas east of troughs; in that case, intense thunder storms can develop. Screaming eagles are common in the Pacific Ocean between Hawaii and the equator and are uncommon in the western Atlantic.

Scree An accumulation of weathered rock fragments at the base of a steep rock slope or cliff.

Scud Low fragments of clouds, usually stratus fractus, that are unattached and below a layer of higher clouds, either nimbostratus or cumulonimbus. They are often along and behind cold fronts and gust fronts, being associated with cool moist air, such as an outflow from a thunderstorm. When observed from a distance, they are sometimes mistaken for tornadoes.

Sea A sea is a large body of salty water that is often connected to an ocean. A sea may be partly or completely surrounded by land.

Sea arch A coastal landform composed of rock that resembles an arch. These landforms are created when waves erode through a thin headland from both sides.

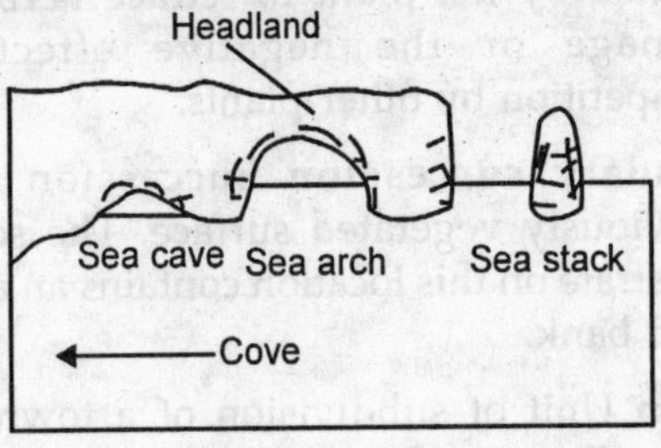

Fig. Sea arch

Sea breeze A diurnal coastal breeze that blows onshore, from the sea to the land. It is caused by the temperature difference when the surface of the land is warmer than the adjacent body of water. Predominate during the day, it reaches its maximum early to mid afternoon. It blows in the opposite direction of a land breeze.

Sea breeze front A coastal phenomena, it is restricted to large bodies of water and their immediate coast lines. This is usually the landward extent of the sea breeze. Due to the imbalance of heating between land and water, a region of maximum upward motion or convergence occurs by mid-afternoon in the summer some 10 to 15 miles inland. Several air mass thunderstorms will form, or a line of towering cumulus clouds with showers. At the beach, there are blue skies and a light breeze. This often occurs along the coast of the Gulf of Mexico and Flordia's east coast.

Sea breeze Local coastal wind that blows from the ocean to land. Sea breezes usually occur during the day because the heating differences of land and sea cause pressure differences. Cooler heavier air from the sea moves in to replace rising warm air on the coastline.

Sea-floor spreading The process of oceanic crust creation and sea-floor movement that occurs at the mid-oceanic ridge.

Sea fog A type of advection fog which forms in warm moist air cooled to saturation as the air moves across cold water.

Sea ice Ice that is formed by the freezing of sea water. It forms first as small crystals, thickens into sludge, and coagulates into sheet ice, pancake ice, or ice floes of various shapes and sizes.

Sea level The datum against which land elevation and sea depth are measured. Mean sea level is the average of high and low tides.

Sea level pressure The atmospheric pressure at mean sea level, usually determined from the observed station pressure.

Sea mile A unit of length distinguished from a nautical mile. One sea mile is equivalent to 1,000 fathoms (6,000 feet).

Sea spray Sometimes called salt spray, it is the drops of sea water (salt water) blown from the top of a wave.

Sea stack A steep pillar of rock located in the ocean a short distance from the coastline. These landforms are created when waves erode through a thin headland from both sides.

Sea Surface Temperature (SST) The temperature of the water's surface. It is measured using buoy and ship data, infrared satellite imagery, and coastal observations.

Seamount A volcanic mountain found on an ocean basin that has an origin not related to a mid-oceanic ridge or a tectonic subduction zone.

Search and rescue International satellite-aided search and rescue project. COSPASI SARSAT satellites monitor the entire surface of the Earth, and transmit distress signals to special ground receiving stations. The receiving stations compute the location of the signal, and notify the nearest rescue coordination centre. Satellite search has cut recovery time from days to hours, and has aided downed airplanes, capsized boats, and persons in other emergencies.

Season A division of the year according to some regularly recurring phenomena, usually astronomical or climatic. For example, in the Northern Hemisphere, winter is said to begin on the winter solstice and end on the vernal equinox when spring begins, covering the months of December, January, and February. In the tropics, there is the dry and the rainy season, depending on the amount of precipitation.

Seaward Positioned or located away from land but towards an ocean or sea.

Seawater The mixture of water and various dissolved salts found in the world's oceans and seas.

Second growth forest Stand of forest that is the result of secondary succession.

Second home A seasonally occupied dwelling that is not the primary residence of the owner. Such residences are usually found in areas with substantial opportunities for recreation or tourist activity.

Second law of thermodynamics This law states that heat can never pass spontaneously from a colder to a hotter body. As a result of this fact, natural processes that involve energy transfer must have one direction, and all natural processes are irreversible. This law also predicts that the entropy of an isolated system always increases with time.

Secondary consumer Organisms that occupy the third trophic level in the grazing food chain. These organisms are carnivores. Also known as a primary carnivore.

Secondary pollutant Atmospheric pollutants that are created chemically in the atmosphere when primary pollutants and other components of the air react.

Secondary sector That portion of a region's economy devoted to the processing of basic materials extracted by the primary sector.

Secondary substance Organic chemical produced by a plant that has no direct function in its metabolism. Many of these chemicals are toxic and are believed to be created by the plant to reduce herbivore damage or the negative effects of competition by other plants.

Secondary succession Succession on a previously vegetated surface. The soil or substrate on this location contains an active seed bank.

Section Unit of subdivision of a township; normally a quadrangle 1 mile square with boundaries conforming to meridians and parallels within established limits, and containing 640 acres as nearly as practicable.

Sedge Grass-like plant that is adapted to grow in moist habitats.

Sediment rating curve Numerical expression or graphical curve that describes the quantitative relationship between stream discharge and the sediment transported by a particular stream.

Sediment Solid material that has been or is being eroded, transported, and deposited. Transport can be due to fluvial, marine, glacial or aeolian agents.

Sedimentary rock Rocks formed by the deposition, alteration and/or compression, and lithification of weathered rock debris, chemical precipitates, or organic sediments.

Seed Fertilized ovule of a plant that contains an embryo and food products for germination. Once germinated, the embryo can grow into a mature individual.

Seed bank Collection of seeds available for germination in the soil.

Seed dispersal Movement of a plant seed away from the parent plant by a passive or active mechanism.

Seepage 1. The gradual movement of water into the soil layer.

2. Slow movement of sub-surface water to the surface. This flow is not great enough to call it a spring.

Seepage lake A lake that gets its water primarily from the seepage of groundwater.

Segregated ice A form of periglacial ground ice that consists of almost pure ice that often exists as an extensive horizontal layer. The ice layer grows because of the active migration of water from around the feature. These features are found just below the active layer.

Seif 1. A large sand dune that is elongated in the general direction of the dominant winds.

2. A sand dune formed by winds from multiple directions.

Seismic Shaking displacement usually caused by an earthquake.

Seismic wave Successive wave-type displacement of rock usually caused by an earthquake.

Seismograph Instrument that measures the energy contained in seismic waves from an earthquake or other type of ground displacement.

Seismology A branch of science focused on the study of earthquakes and seismic activity.

Self regulation The ability of some systems to maintain a steady state equilibrium through positive and negative feedbacks.

Semi diurnal tide Tides that have two high and two low waters per tidal period.

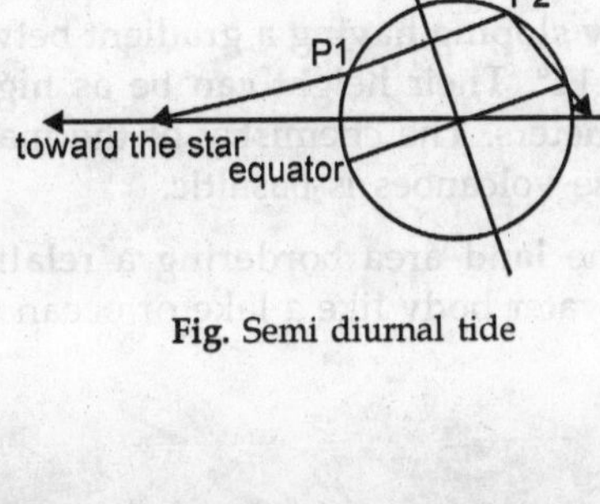

Fig. Semi diurnal tide

Semi-major axis One of the six Keplerian elements, it indicates the size of an orbit. The semi-major axis is one half of the longest diameter of an orbital ellipse, e.g., one-half of the distance between the apogee and perigee of an Earth orbit. (The semi-major axis is related to the orbital period and mean motion by Kepler's third law).

Semi permanent pressure systems A relatively stable, stationary pressure-and-wind system where the pressure is predominately high or low with the changing season. They are not of a transitory nature, like migratory lows that develop from temperature and density differences. Examples include the Icelandic Low and the Bermuda High in the North Atlantic.

Sensible heat Heat that can be measured by a thermometer and thus sensed by humans.

Sensible heat flux Process where excess heat energy is transferred into the atmosphere. The process first involves the movement of heat energy from the Earth's surface to the atmosphere by conduction and convection. The heat energy then can move horizontally advection (atmospheric circulation).

Sensitivity analysis Is a term applied to a range of techniques employed to assess how solutions derived from GIS systems are affected by the accuracy and platforms.

Sensor Device that produces an output (usually electrical) in response to stimulus such as incident radiation Sensors aboard satellites obtain information about features and objects on Earth by detecting radiation reflected or emitted in different bands of the electromagnetic spectrum. Analyzing the transmitted data provides valuable scientific information about Earth. Weather satellites commonly carry radiometers, which measure radiation from snow ice, clouds, and bodies of water Spaceborne radars are used for Earth observations, bouncing radar waves off land and ocean surfaces to study sea-surface conditions, ice thickness, and land surface features. A wind scatterometer is a special type of radar designed to measure ocean surface winds indirectly by boun & mg signals off the water and measuring them from various angles. Infrared (IR) detectors measure heat generated by Earth features in the IR band of the spectrum.

Photographic reconnaissance sensors in their simplest form are large telescope-camera systems used to view objects on Earths' surface. The bigger the lens, the smaller the object that can be detected. Camera-telescope systems now incorporate all sorts of Sophisticated electronics to produce better images, but even these systems need cloudless skies, excellent lighting, and good colour contrast between objects and their surroundings to detect objects the size of a basketball. Some of the satellites produce film images that must be returned to Earth, but a more convenient method is to record the image as a series of digital code numbers, then reconstruct the image from the electronic code using a computer at a ground station.

Sensor calibration The relationship between input and output for a given measurement.

Severe thunderstorm A thunderstorm with winds measuring 50 knots (58 mph) or greater, 3/4 inch hail or larger, or tornadoes. Severe thunderstorms may also produce torrential rain and frequent lightning.

Severe weather Generally, any destructive weather event, but usually applies to localized storms, such as blizzards, intense thunderstorms, or tornadoes.

Shale Fine grained sedimentary rock composed of lithified clay particles.

Sharecropping A form of agricultural tenancy in which the tenant pays for use of the land with a predetermined share of his crop rather than with cash rent.

Shear It is the rate of change over a short duration. In wind shear, it can refer to the frequent change in wind speed within a short distance. It can occur vertically, such as a change with height, or horizontally. Directional shear is a frequent change in direction within a short distance, which can also occur vertically or horizontally. This can also occur vertically or horizontally. When used in reference to Doppler radar, it describes the change in radial velocity over short distances horizontally.

Shear line A line of maximum horizontal wind shear. A narrow zone across which there is an abrupt change in the horizontal wind component parallel to it.

Shear stress Stress caused by forces operating parallel to each other but in opposite directions.

Shear wave A seismic wave that creates wave-like motion perpendicular to the direction of seismic energy propagation. Also called S-wave.

Sheeting A form of physical weathering of rock where surface sheets of material fracture and exfoliate because of pressure release. Also.

Sheetwash The removal of loose surface materials by overland flow. Process of erosion.

Shield A large stable area of exposed very old (more than 600 million years) igneous and metamorphic rock found on continents. This rock forms the nucleus of the continents.

Shield volcano Volcano created from alternate layers of lava flows. Shield volcanoes are slightly sloping having a gradient between 6 and 12°. Their height can be as high as 9000 meters. The chemistry of the magma of these volcanoes is basaltic.

Shore The land area bordering a relatively large water body like a lake or ocean.

Shoreline The line that separates a land surface from a water body.

Short wave A progressive wave of smaller amplitude, wave length, and duration than a long wave. It moves in the same direction as the basic current in which it is embedded, and may induce upward motion ahead of it, possibly contributing to thunderstorm development. They are more numerous than long waves and often disappear with height in the atmosphere.

Shortwave radiation Electromagnetic radiation with a wavelength between 0.1 and 0.7 micrometers (μm). Commonly used to describe the radiation emitted from the sun.

Showalter stability index A measure of the local static stability of the atmosphere. It is determined by lifting an air parcel to 500 millibars and then comparing its temperature to that of the environment. If the parcel is colder than its new environment, then the atmosphere is more stable. If the parcel is warmer than its new environment, then the atmosphere is unstable and the potential for thunderstorm development and severe weather increases.

Shower It is precipitation from a convective cloud. It occurs in the form of rain (SHRA), snow (SHSN), or ice (SHPE) ,and is characterized by its sudden beginning and ending, changes in intensity, and rapid changes in the appearance of the sky. It is reported as "SH" in an observation and on the METAR.

Shrub A woody plant species that is smaller than a tree. Shrubs usually do not have a trunk.

Sial layer The part of the crust that forms the continents and is composed of relatively light, granitic rocks.

Siberian express A fierce, cold flow of air that originates in Siberia, then moves into Alaska and northern Canada before moving southward into the United States.

Siberian high The semi-permanent high pressure area that forms over Siberia during the winter. The average central pressure exceeds 1030 millibars from late November to early March. It is one the main contributors to a monsoon in southern Asia.

Sidereal day Time it takes to complete one Earth rotation relative to the position of a fixed star. This measurement takes 23 hours, 56 minutes, and 4.09 seconds. Compare with mean solar day.

Sidereal time The measure of time as defined by the diurnal motion of the vernal equinox. A sidereal day is equivalent to one complete rotation of the earth relative to the equinox, which is 23 hours, 56 minutes, and 4.091 seconds. A sidereal year is the interval required for the earth to make one absolute revolution around the sun, which is 365 days, 6 hours, 9 minutes, and 9.5 seconds. Compare with the solar day.

Signal Electrical impulses, sound or picture elements, etc., received or transmitted. Signals can exist in many different forms and media (electrical/wires, acoustic/air light/transparent fibres, etc.), but all signals will vary with time.

The signal shape plotted as a function of time is called the waveshape or waveform. Some waveforms are repetitive or periodic, that is, a small segment of the waveform repeats itself regularly. Other waveforms, such as noise, are nonperiodic or aperiodic. All waveforms can be distilled into the combination of pure waves called sine waves. The frequency of a sine wave is the rate at which the fundamental shape repeats itself.

Most signals occupy a limited range of frequencies between a lower limit and an upper limit. This range or band of frequencies occupied by a signal is called the bandwidth of the signal.

Communication medium or channel can pass only a specific range or band of frequencies, which is called the bandwidth of the channel. The bandwidths of the channel and the signal determine the number and types of signals that can be transmitted by a particular communication channel. Signals often are too small and need to be made larger through a process called amplification. The amount of amplification is measured in decibels. However amplification is an imperfect process, and inadvertently introduces various distortions, noise, and bandwidth limitations. Often, multiple signals must share the same medium. One way the sharing can be accomplished is to place each signal in its own band of frequencies within the total band of the medium. The combining of a number of signals to share a medium by dividing it into different frequency bands for each signal is called frequency-division multiplexing.

Frequency-division multiplexing requires the ability to move signals around so that each multiplexed signal occupies its own band. This is accomplished through a process called modulation, in which a high-frequency sine wave carries the signal into the specified band. Either the amplitude or the frequency of the carrier wave can be varied, or modulated, in synchrony with the information-bearing signal. These methods are called amplitude modulation (AM) and frequency modulation (FM). FM is the more complex process of the two, and the bandwidth of the FM carrier can be many times that of the modulating signal. The process of demodulating a frequency-modulated signal eliminates much of the deleterious effects of additional noise. (The trade-off between bandwidth and noise immunity characterizes most communication systems. Both are analog modulation schemes for multiplexing signals in the frequency spectrum.)

Digitizing a signal requires a number of steps and results in a binary digital signal that takes on one of two discrete values. This process results in considerable immunity to additive noise, but requires a considerable increase in bandwidth.

Signal to Noise Ratio (SNR) In decibels (dB), the difference between the amplitude of a desired radio frequency (RF) signal and the internal or external RF noise level in a system. A negative SNR indicates the signal is below the system noise level and unusable. The greater the positive SNR, the less effect noise will have on the final quality. SNR of at least + 12dB is necessary to produce imagery with minimal noise effects.

Silage Fodder (livestock feed) prepared by storing and fermenting green forage plants in a silo.

Silcretes Sedimentary rock created by the chemical precipitation of silica.

Silica Mineral that is composed of silicon dioxide, SiO_2.

Silicate Group of minerals that have crystal structures based on a silica tetrahedron (SiO_4).

Silicate magma Magma that is felsic in composition.

Sill Horizontal planes of igneous rock that run parallel to the grain of the original rock deposits. They form when magma enters and cools in bedding planes found within the crust.

Silo Usually a tall, cylindrical structure in which fodder (animal feed) is stored; may be a pit dug for the same purpose.

Silt Mineral particle with a size between 0.004 and 0.06 millimetres in diameter.

Siltstone Fine grained sedimentary rock composed of lithified silt particles.

Silurian Geologic period that occurred roughly 408 to 438 million years ago. During this period, the first plant and insect species appear.

Sima layer The part of the crust that forms the ocean basins and lower layers in the crust and is composed of relatively heavy, basaltic rocks.

Sine wave A smoothly varying wave that repeats itself; its frequency is the rate at which the fundamental shape repeats itself. Any waveform can be distilled into a combination of pure sine waves of varying frequencies and amplitudes.

Sink The process of providing storage for a substance. For example, plants—through photosynthesis—transform carbon dioxide in the air into organic matter which either stays in the plants or is stored in the soils. The plants are a sink for carbon dioxide.

Sinkhole A pit like hole in found in areas of karst. These features are caused by the weathering of limestone or dolomite by subsurface drainage. Also called a sink or doline.

Sinusoidal equal-area projection Map projection that represents areas in their true form on a two-dimensional map. Distances are only correct along parallels and central meridian. Shapes become more distorted away from the central meridian and close to the poles.

Sinusoidal projection A sinusoidal projection is a type of map projection in which lines of latitude are parallel to the equator, and lines of longitude are curved around the prime meridian.

Site Features of a place related to the immediate environment on which the place is located (e.g., terrain, soil, subsurface, geology, ground water).

Situation Features of a place related to its location relative to other places (e.g., accessibility, hinterland quality).

Skew t-log p diagram A thermodynamic diagram, using the temperature and the logarithm of pressure as coordinates. It is used to evaluate and forecast air parcel properties. Some values that can be determined are the Convective condensation level (CCL), the Lifting condensation level (LCL), and the Level of free convection (LFC).

Sky The vault-like apparent surface against which all aerial objects are seen from the earth.

Sky cover The amount of the celestial dome that is hidden by clouds and/or obscurations.

Skylab The first U.S. space station, launched unmanned in May 1973 and soon after occupied in succession by three crews through November 1973.

Slate A fine grained metamorphic rock with well developed bedding planes derived from the slight recrystrallization of shale.

Sleet Also known as ice pellets, it is winter precipitation in the form of small bits or pellets of ice that rebound after striking the ground or any other hard surface. It is reported as "PE" in an observation and on the METAR.

Sling psychrometer Psychrometer that uses a rotating handle and a whirling motion to ventilate its wet-bulb thermometer.

Slip face The lee side of a dune where material accumulates and slides or rolls downslope.

Slope aspect Main compass direction (North, North East, East, South East, South, South West, West, and North West) that a slope faces.

Slope failure The downslope movement of soil and sediment by processes of mass movement.

Slush Snow or ice on the ground that has been reduced to a softy watery mixture by rain and/or warm temperatures.

Small circle A circle on the globe's surface that does not bisect the centre of the Earth.

Parallels of latitude are examples of small circles.

Small craft advisory An advisory issued for marine interests, especially for operators of small boats or other vessels. Conditions include wind speeds between 20 knots (23 mph) and 34 knots (39 mph).

Smog Generic term used to describe mixtures of pollutants in the atmosphere.

Smoke Small particles produced by combustion that are suspended in the air. A transition to haze may occur when the smoke particles have travelled great distance (25 to 100 miles or more), and when the larger particles have settled out. The remaining particles become widely scattered through the atmosphere. It is reported as "FU" in an observation and on the METAR.

Snout Front end of a glacier. Also called the terminus.

Snow Type of precipitation that forms in air with temperatures below freezing. Snow forms when water vapour deposits directly as a solid on a deposition nuclei, by passing the liquid state. A snowflake forms first as a very tiny crystal developing on a six-sided hexagonal deposition nuclei. The ice crystal then grows fastest at the six points as these area are more directly exposed to the atmosphere's water vapour. Snow is most common in winter just north of the centre of mid-latitude cyclones. As the warm moist air travels around the centre of lowest pressure, it overrides colder air located north of the low and is cooled to its saturation temperature, producing rainfall and snow. Snow generally occurs with east winds, since the winds at locations north of a mid-latitude cyclone are from the east.

Snow advisory A statement or advisory issued when snow is expected to create hazardous travel conditions. It warns of less severe weather conditions than a winter storm warning.

Snow banner A plume of snow blown off a mountain crest, resembling smoke blowing from a volcano.

Snow blindness Temporary blindness or impaired vision that results from bright sunlight reflected off the snow surface. The medical term is niphablepsia.

Snow cover The areal extent of ground covered by the snow. It is usually expressed as a percent of the total area of a given region.

Snow creep A continuous, extremely slow, downhill movement of a layer of snow.

Snow crust The crisp, almost icy, surface on fallen snow, usually formed by the slight melting and refreezing of the surface snow.

Snow depth The actual depth of snow on the ground at any instant during a storm, or after any single snowstorm or series of storms.

Snow devil A small, rotating wind that picks up loose snow instead of dirt (like a dust devil) or water (like a waterspout). Formed mechanically by the convergence of local air currents. Also called a snowspout.

Snow eater Any warm downslope wind, or foehn, that blows over snowy terrain and melts the snow. Generally applies to the Chinook wind in the Rocky Mountains.

Snow flurry/flurries Light showers of snow, generally very brief without any measurable accumulation. May be reported as "SHSN−" in an observation and on the METAR.

Snow garland Snow appearing as a beautiful long thick rope draped on trees, fences and other objects. Formed by the surface tension of thin films of water bonding individual snow crystals.

Snow grains Frozen precipitation in the form of very small, white, opaque grains of ice. The solid equivalent of drizzle. It is reported as "SG" in an observation and on the METAR.

Snow level The elevation in mountainous terrain where the precipitation changes from rain to snow, depending on the temperature structure of the associated air mass.

Snow line Altitudinal or latitudinal limit separating zones where snow does not melt during the summer season from areas in which it does. Similar to the concept of firm limit except that it is not limited to glaciers.

Snow melt Conversion of snow into runoff and groundwater flow with the onset of warmer temperatures.

Snow pellets Type of precipitation. Snow pellets are white, spherical grains of ice 2 to 5 millimetres in diameter. They can be distinguished from packed snowflakes since snow pellets are firm enough to bounce when they hit the ground. Snow pellets develop as super cooled droplets freeze on ice crystals. They may fall for a brief period as the precipitation changes from ice pellets to snow.

Snow roller The product of moist, cohesive snow that when initiated by wind rolls down a hill, gathering snow until it can no longer move, either due to size or lack of slope. It is shaped like a rolled sleeping bag, some reaching four feet across and seven feet in diameter.

Snow shower Frozen precipitation in the form of snow, characterized by its sudden beginning and ending. It is reported as "SHSN" in an observation and on the METAR.

Snow squall A heavy snow shower accompanied by sudden strong winds, or a squall.

Snowburn A burn of the skin, like a sunburn, but caused by the sun's rays reflected off the snow surface.

Snowfall The rate at which snow falls, usually expressed in inches of snow depth over a six hour period.

Snowfield An area of permanent snow accumulation. Usually at high altitudes or latitudes.

Snowflakes An ice crystal or an aggregate of ice crystals which fall from clouds.

Snowpack The amount of annual accumulation of snow at higher elevations.

Soil Layer of unconsolidated material found at the Earth's surface that has been influenced by the soil forming factors: climate, relief, parent material, time, and organisms. Soil normally consists of weathered mineral particles, dead and living organic matter, air space, and the soil solution.

Soil colloids Very small organic and inorganic particles found in a soil. Inorganic colloids are often clay particles. Soil colloids carry a negative electrical charge and are the primary sites for cation exchange. Soil colloids hold large quantities of elements and compounds which are used by plants for nutrition.

Soil creep Slow mass movement of soil downslope. Occurs where the stresses on the slope material are too small to create a rapid failure.

Soil erosion Transport of soil mineral particles and organic matter by wind, flowing water, or both. Human activities that disturb the soil surface or remove vegetation can enhance this natural process.

Soil fertility The ability of a soil to provide nutrients for plant growth.

Soil heat flux The rate of flow of heat energy into, from, or through the soil.

Soil horizon Layer within a soil profile that differs physically, biologically or chemically from layers above and/or below it.

Soil moisture recharge The process of water filling the pore space found in a soil (storage).

Soil organic matter Organic constituents of soil.

Soil permeability The rate at which water and air move vertically through a soil.

Soil porosity The volume of water that can be held in a soil. Also refers to the ratio of the volume of voids to the total volume of the soil.

Soil profile Vertical arrangement of layers or horizons in a soil.

Soil science The study of soils from an interdisciplinary perspective.

Soil solution Aqueous liquid found within a soil. This liquid normally contains ions released from mineral particles, organic matter or plant roots and leaves.

Soil structure General term that describes how mineral and particles organic matter of are organized and clumped together in a soil.

Soil taxonomy The classification of a soil in a hierarchical system based on its various properties. Grouping goes from general to specific.

Soil texture The relative quantities of the different types and sizes of mineral particles in a soil.

Soil water The water found occupying the pore spaces between soil particles.

Solar altitude Height of the sun above the horizon from either True North or True South.

Solar Backscatter Ultraviolet Radiometer (SBUV) Instrument that measures the vertical distribution and total ozone in the Earth's atmosphere. Data is used for the continuous monitoring of ozone distribution to estimate long-term trends. SBUV instruments are flown on NOAA polar-orbiting satellites.

Solar constant A term used to describe the average quantity of solar insolation received by a horizontal surface at the edge of the Earth's atmosphere. This value is approximately 1370 Watts per square meter.

Solar cycle Eleven-year cycle of sunspots and solar flares that affects other solar indexes such as the solar output of ultraviolet radiation and the solar wind. The Earth's magnetic field, temperature, and ozone levels are affected by this cycle.

Solar day The complete rotation of the earth in relation to the sun. Although it varies, an average has determined a mean solar day of 24 hours. It is universally used for civil purposes. Compare with the sidereal day.

Solar eclipse An eclipse of the sun occurs when the moon is in a direct line between the sun and the earth, casting some of the earth's surface in its shadow. The moon's disk shaped outline appears to cover the sun's brighter surface, or photosphere. That part of the earth that is directly in the moon's shadow will see a total eclipse of the sun, while the areas around it will see a partial eclipse.

Solar noon Point of time during the day when the sun is aligned with True North and True South.

Solar radiation Energy received from the sun is solar radiation. The energy comes in many forms, such as visible light (that which we can see with our eyes). Other forms of radiation include radio waves, heat (infrared), ultraviolet waves, and x-rays. These forms are categorized within the electromagnetic spectrum.

Solar system The collection of celestial bodies that orbit around the sun.

Solar wind A continuous plasma stream expanding into interplanetary space from the sun's corona. The solar wind is present continuously in interplanetary space. After escaping from the gravitational field of the sun, this gas flows outward at a typical speed of 400 km per second to distances known to be beyond the orbit of Pluto.

Besides affecting Earth's weather solar activity gives rise to a dramatic visual phenomena in our atmosphere. The streams of charged particles from the Sun interact the Earth's magnetic field like a generator to create current systems with electric potentials of as much as 100,000 volts. Charged electrons are energized by this process, sent along the magnetic field lines towards Earth's upper atmosphere, excite the gases present in the upper atmosphere and cause them to emit light which we call the auroras. The auroras are the northern (aurora borealis) and southern (aurora Australis) lights.

Solar year The time it takes the Earth to make one orbit around the sun. This is approximately 365.2422 days.

Solid A state of matter where molecules where the mass of the substance does not have the property of flow.

Solifluction Form of mass movement in environments that experience freeze-thaw action. It is characterized by the slow movement of soil material downslope and the formation of lobe-shaped features.

Solonetzic soil Soil order (type) of the Canadian System of Soil Classification. This soil type is associated with grassland habitats where high levels of evapotranspiration cause the accumulation of salts at or near the soil surface. These soils are common in the dry regions of the prairies where evapotranspiration greatly exceeds precipitation input.

Solstice The point at which the sun is the furthest on the ecliptic from the celestial equator. The point at which sun is at maximum distance from the equator and days and nights are most unequal in duration. The Tropic of Cancer and the Tropic of Capricorn are those parallels of latitude which lies directly beneath a solstice. In the Northern Hemisphere, the winter solstice falls on or about December 21 and the summer solstice on or about June 21.

Soluble Capable of being dissolved; in this case, the characteristic of soil minerals that leads them to be carried away in solution by water.

Solum Part of the soil that is capable of supporting life.

Solution 1. Form of chemical weathering where rocks and minerals are dissolved by water. Materials entering the mixture can alter the chemical nature of the solution and can increase the strength of this weathering agent. For example, the mixing of carbon dioxide and water can form carbonic acid.

2. The dissolving of a substance into a liquid.

Sound A sound is a wide inlet of the sea or ocean that is parallel to the coastline; it often separates a coastline from a nearby island.

Sounder A special kind of radiometer that measures changes in atmospheric temperature with height, as well as the content of various chemical species in the atmosphere at various levels. The High Resolution Infrared Radiation Sounder (HIRS), found on NOAA polar-orbiting satellites, is a passive instrument.

Sounding A plot of the atmosphere, using data from upper air or radiosonde observations. Usually confined to a vertical profile of the temperatures, dew points, and winds above a fixed location.

Source A source is the beginning of a river.

Source region Area where air masses originate and come to possess their moisture and temperature characteristics.

South magnetic pole The South Magnetic Pole is the point on the Southern Hemisphere of the Earth toward which a compass' needle always points; at the South Magnetic Pole, a compass' needle will stand vertically. It is now located just off the coast the continent of Antarctica (and its location changes over time).

South pole Surface location defined by the intersection of the polar axis with Earth's surface in the Southern Hemisphere. This location has a latitude of 90° South.

Southern hemisphere The Southern Hemisphere is the half of the Earth that is south of the equator.

Southern oscillation A periodic reversal of the pressure pattern across the tropical Pacific Ocean during El Nino events. It is represents the distribution of temperature and pressure over an oceanic area.

Space 1. A distance, area, or volume.

2. An infinite three-dimensional area in which objects have relative coordinates to each other.

3. The region beyond the outer limits of the Earth's atmosphere.

Space economy The locational pattern of economic activities and their interconnecting linkages.

Space Environment Monitor (SEM) Instrument that measures the condition of the Earth's magnetic field and the solar activity and radiation around the spacecraft, and transmits these data to a central processing facility. NOAA polar orbiting and geostationary satellites both carry SEMs.

Space physics Scientific study of magnetic and electric phenomena that occur in outer space, in the upper atmosphere of the planets, and on the sun.

Space shuttle NASA's manned, recoverable spacecraft designed to be used as a launch vehicle for Earth-orbiting experiments and as a short-term research platform.

Spacelab A manned laboratory module built by the European Space Agency (ESA) that accommodates dozens of experiments on each flight, mainly in the categories of materials science and life science.

Spatial analysis The examination of the spatial pattern of natural and human-made phenomena using numerical analysis and statistics.

Spatial complementarity The occurrence of location pairing such that items demanded by one place can be supplied by another.

Spatial interaction Movement between locationally separate places.

Spatial isolation Reproductive isolation of two or more populations of a species by distance or physical barriers. Over long periods of time geographic isolation leads to speciation through divergent evolution because of environmental heterogeneity. Also called geographic isolation.

Spatial tradition Academic tradition in modern Geography that investigates geographic phenomena from a strictly spatial perspective.

Specialist species Species that have a relatively narrow ecological niche. These species may be able to live in only one type of habitat, tolerate only specific environmental conditions, or use only a few types of food.

Speciation The process by which new species originate through mutations, natural selection, and evolution.

Species 1. The different kinds of organisms found on the Earth as defined by taxonomic and/or phylogenic classification.

2. A group of interbreeding organisms that do not ordinarily breed with members of other groups.

Species association A particular grouping of species in an area.

Specific gravity The ratio of the mass of a body to the mass of an identical volume of water at a specific temperature.

Specific heat Is the heat capacity of a unit mass of a substance or heat needed to raise the temperature of 1 gram (g) of a substance 1 degree Celsius.

Specific humidity The ratio of the density of the water vapour to the density of the air, a mix of dry air and water vapour. It is expressed in grams per gram or in grams per kilograms. The specific humidity of an air parcel remains constant unless water vapour is added to or taken from the parcel.

Spectral band A finite segment of wavelengths in the electromagnetic spectrum.

Spectrum 1. The series of coloured bands diffracted and arranged in the order of their respective wave lengths by the passage of white light through a prism or other diffracting medium and shading continuously from red (produced by the longest visible wave) to violet (produced by the shortest visible wave).

2. Any of various arrangements of coloured ands or lines, together with invisible components at both ends of the spectrum, similarly formed by light from incandescent gases or other sources of radiant energy, which can be studied by a spectrograph.

3. In radio, the range of wave lengths of radio waves, from 3 centimetres to 30,000 meters, or of frequencies of radio waves, from 10 to 10,000,000 kilocycles. Also radio spectrum.

4. The entire range of radiant energies.

Speed of light Velocity of light in a vacuum. This velocity is approximately 3×10^8 meters per second. It takes light from the sun 8 minutes and 20 seconds to reach the Earth.

Spheroid Mathematical figure closely approaching the geoid in form and size and used as a surface of reference for geodetic surveys. A reference spheroid or ellipsoid is a spheroid determined by revolving an ellipse about its shorter (polar) axis and used as a base for geodetic surveys of a large section of the Earth (such as the Clarke spheroid of 1866 which is used for geodetic surveys in the United States).

Spheroidal weathering A type of below ground chemical weathering where the corners of jointed rocks become rounded over time. Rock changes from a rectangular to more round shape.

Spit A long and narrow accumulation of sand and/or gravel that projects into a body of ocean water. These features form as the result of the deposition of sediments by longshore drift.

Spodosols Soil order (type) of the United States Natural Resources Conservation Service Soil Classification System. A spodosol soil has pronounced layers of illuviation characterized by accumulations of iron and aluminium oxides in the B horizon. These soils form in moist, cool climates.

Sporadic permafrost Form of permafrost that exists as small islands of frozen ground in otherwise unfrozen soil and sediments.

SPOT (systeme pour l'observation de la terre) French, polar-orbiting Earth observation satellite(s) with ground resolution of 10 meters. SPOT images are available commercially and are intended for such purposes as environmental research and monitoring, ecology management, and for use by the media, environmentalists, legislators, etc.

Spot elevation Point on a map or chart whose height above a specified datum is noted, usually by a dot or a small sawbuck and elevation value. Elevations are shown, on a selective basis, for road forks and intersections, grade crossings summit of hills, mountain.

Spring The season of the year which occurs as the sun approaches the summer solstice, and characterized by increasing temperatures in the mid-latitudes. Customarily, this refers to the months of March, April, and May in the North Hemisphere, and the months of September, October, and November in the Southern Hemisphere. Astronomically, this is the period between the vernal equinox and the summer solstice.

Spring tide Tide that occurs every 14 to 15 days and coincides with the new and full moon. This tide has a large tidal range because the gravitational forces of the moon and sun are complementary to each other. Contrasts with neap tide.

Squall A sudden onset of strong winds with speeds increasing to at least 16 knots (18 miles per hour) and sustained at 22 or more knots (25 miles per hour) for at least one minute. The intensity and duration is longer than that of a gust. It is reported as "SQ" in an observation and on the METAR.

Squall line A narrow band or line of active thunderstorms that is not associated with a cold front. It may form from an outflow boundary or the leading edge of a mesohigh.

St. Elmo's fire A luminous, and often audible, electric discharge that is intermediate in nature. It occurs from objects, especially pointed ones, when the electrical field strength near their surfaces attains a value near 1000 volts per centimetre. It often occurs during stormy weather and might be seen on a ship's mast or yardarm, aircraft, lightning rods, and steeples. Also known as corposant or corona discharge.

Stability The capability of a system to tolerate or recover from disturbance or an environmental stress.

Stable Occurs when a rising air parcel becomes denser than the surrounding air. It will then return to its original position. When the density of the air parcel remains the same as the surrounding air after being lifted, it is also considered stable, since it does not have the tendency to rise or sink further. Contrast with unstable air and instability.

Stable atmosphere Condition in the atmosphere where isolated air parcels have a tendency to sink. The parcels of air tend to be cooler than the air that surrounds them.

Stable equilibrium In a stable equilibrium the system displays tendencies to return to the same equilibrium after disturbance.

Stadia Technique of distance measurement wherein the observer reads the intercept subtended on a graduated rod between two marks on the reticle of the telescope.

Stage The elevation of the water surface in a stream channel.

Stagnation area An area that has a combination of stable stratification, weak horizontal wind speed, and little, if any, significant precipitation. It is usually associated with a warm area of high pressure.

Standard atmosphere A standard atmosphere has been defined by the International Civil aeronautical organization (ICAO). It assumes a mean sea level temperature of 15 degrees Celsius, a standard sea level pressure of 1,013.25 millibars or 29.92 inches of mercury, and a temperature lapse rate of 0.65 degrees Celsius per 100 meters up to 11 kilometres in the atmosphere.

Standard atmospheric pressure A pressure of 101.32 kilopascals or 1013.2 millibars.

Standard deviation A statistical measure of the dispersion of observation values in a data set. Calculated by determining the square root of the variance.

Standard surface pressure The measurement of one atmosphere of pressure under standard conditions. It is equivalent to 1,013.25 millibars, 29.92 inches of mercury, 760 millimetres of mercury, 14.7 pounds per square inch, or 1.033 grams per square centimetre.

Standing cloud Any type of isolated cloud, generally formed over peaks or ridges of mountainous areas, that appears stationary or standing over the terrain. An example is altocumulus lenticularis.

Standing wave An atmospheric wave that is stationary with respect to the medium in which it is embedded. A mountain wave is an example.

Staple product A product that becomes a major component in trade because it is in steady demand; thus, a product that is basic to the economies of one or more major consuming populations.

Star A large and very massive, self-luminous celestial body of gas that illuminates via the radiation derived from its internal source of energy.

Starch Complex carbohydrate composed of thousands of glucose units. Main compound that plants use to store their food energy.

State of matter Form of matter. Matter can exist in three different forms gas, liquid, and solid.

Static equilibrium Static equilibrium occurs where force and reaction are balanced and the properties of the system remain unchanged over time.

Station elevation The vertical distance above mean sea level that is the reference level for all current measurements of atmospheric pressure at that station.

Station pressure The atmospheric pressure with respect to the station elevation.

Stationary front A transition zone in the atmosphere where there is little movement of opposing air masses and winds blow towards the front from opposite directions.

Steady state equilibrium In this type of equilibrium the average condition of the system remains unchanged over time.

Steam fog A type of advection fog that is produced by evaporation when cool air passes over a warm wet surface and the fog rises, giving the appearance of steam. Also called sea smoke when it occurs over the ocean. An example is Arctic Sea Smoke.

Stemflow Is the process that directs precipitation down plant branches and stems. The redirection of water by this process causes the ground area around the plant's stem to receive additional moisture. The amount of stemflow is determined by leaf shape and stem and branch architecture. In general, deciduous trees have more stemflow than coniferous vegetation.

Stephan Boltzmann law This radiation law suggests the amount of radiation given off by a body is proportional to the 4th power of its temperature as measured in Kelvin units. This law can be expressed by the following simple equation: E* = sT 4

where E* is the amount of radiation emitted by the body in Watts per square meter,

s is a constant equal to 0.0000000567,

and T is the temperature of the body in Kelvins.

Steppe Russian term for mid-latitude grasslands.

Stereocompilation Production of a map or chart manuscript from aerial photographs and geodetic control data by means of photogrammetric instruments.

Stereoplotter Instrument for plotting a map by observation of stereomodels formed by pairs of photographs.

Stereoscopic Pertaining to the use of binocular vision for observation of a pair of overlapping photographs or other perspective views, giving impression of depth.

Stoma (pl. stomata) Small opening on the surface of a plant that is used for gas exchange.

Storm An individual low pressure disturbance, complete with winds, clouds, and precipitation. Examples include thunderstorms, tornadoes, or even tropical cyclones. The name is associated with destructive or unpleasant weather. Storm-scale refers to disturbances the size of individual thunderstorms.

Storm surge The increase in sea water height from the level that would normally occur were there no storm. Although the most dramatic surges are associated with hurricanes, even smaller low pressure systems can cause a slight increase in the sea level if the wind and fetch is just right. It is estimated by subtracting the normal astronomic tide from the observed storm tide.

Storm track The path taken by a storm (thunderstorm, mid-latitude cyclone or hurricane) or the average path taken by storms.

Storm winds On the Beaufort Wind Scale, a wind with speeds from 56 to 63 knots (64 to 72 miles per hour).

Stoss Side of a slope that faces the direction of flow of ice, wind, or water. Opposite of lee.

Straight line winds Any surface wind that is not associated with rotation. An example is the first gust from a thunderstorm, as opposed to tornado winds.

Strait A strait is a narrow body of water that connects two larger bodies of water.

Strata The layers or beds found in sedimentary rock.

Stratified drift A type of glacial drift that has been partially sorted by glaciofluvial meltwater.

Stratiform Clouds composed of water droplets that exhibit no or have very little vertical development. The density of the droplets often blocks sunlight, casting shadows on the earth's surface. Bases of these clouds are generally no more than 6,000 feet above the ground. They are classified as low clouds, and include all varieties of stratus and stratocumulus. The opposite in type are the vertical development of cumuliform clouds.

Stratigraphy Subdiscipline of geology that studies sequence, spacing, composition, and spatial distribution of sedimentary deposits and rocks.

Stratocumulus clouds Low altitude grey coloured cloud composed of water droplets that has a patchy appearance. Each cloud patch consists of a rounded mass. This cloud has a somewhat uniform base and normally covers the entire sky. Between the patches blue sky can be seen. Found in an altitude range from the surface to 3,000 meters.

Stratopause The stratopause is a relatively thin atmospheric transition layer found between the stratosphere and the mesosphere. The height of this layer is about 50 kilometres above the Earth's surface.

Stratosphere Region of the atmosphere between the troposphere and mesosphere, having a lower boundary of approximately 8 km at the poles to 15 km at the equator and an upper boundary of approximately 50 km. Depending upon latitude and season, the temperature in the lower stratosphere can increase, be isothermal, or even decrease with altitude, but the temperature in the upper stratosphere generally increases with height due to absorption of solar radiation by ozone.

Stratus fractus Stratus clouds that appear in irregular fragments, as if they had been shred or torn. Also appears in cumulus clouds (called cumulus fractus), but not in cirrus clouds.

Stratus One of the three basic cloud forms (the others are cirrus and cumulus). It is also one of the two low cloud types. It is a sheet like cloud that does not exhibit individual elements, and is, perhaps, the most common of all low clouds. Thick and grey, it is seen in low, uniform layers and rarely extends higher than 5,000 feet above the earth's surface. A veil of stratus may give the sky a hazy appearance. Fog may form from a stratus cloud that touches the ground. Although it can produce drizzle or snow, it rarely produces heavy precipitation. Clouds producing heavy precipitation may exist above a layer of stratus.

Stream A long narrow channel of water that flows as a function of gravity and elevation across the Earth's surface. Many streams empty into lakes, seas or oceans.

Stream bank Sides of the stream channel.

Stream bed Bottom of the stream channel.

Stream channel Long trough-like depression that is normally occupied by the water in a stream.

Stream discharge A river or stream's rate of flow over a particular period of time. Usually measured by a current meter and expressed in cubic meters per second. Stream discharge depends on the volume and velocity of the flow.

Stream flow The flow of water in a river or stream channel.

Stream gradient The change in elevation from a stream's headwaters to its mouth expressed in degrees, percentage, or as a distance ratio (rise/run).

Stream load Refers to the material or sediment carried by a stream. In normally consists of three components: bed load (pebbles and sand which move along the stream bed without being permanently suspend in the flowing water), suspended load (silts and clays in suspension) and dissolved load (material in solution).

Stream long profile Vertical and horizontal profile of the stream. Most streams have a profile that is concave shaped.

Stream order The relative position, or rank, of a stream channel segment in a drainage network.

Striations Grooves of scratches found in surface rock that are the result of glacial abrasion.

Strike One of the directional properties of a geologic structure such as a fold or a fault. Strike is the horizontal directional taken by an imaginary line drawn on the plane of the formation.

Strike slip fault Fault that primarily displays horizontal displacement.

Structural landform Is a landform created by massive Earth movements due to plate tectonics. This includes landforms with some of the following geomorphic features: fold mountains, rift valleys, and volcanoes.

Subatomic particles Extremely small particles that make up the internal structure of atoms.

Subduction (tectonic) Process of plate tectonics where one lithospheric plate is pushed below another into the asthenosphere.

Subduction zone Linear area where tectonic subduction takes place.

Sublimation The process of a solid (ice) changing directly into a gas (water vapour), or water vapour changing directly into ice, at the same temperature, without ever going through the liquid state (water). The opposite of crystallization.

Submarine canyon V-shaped canyons cut into the continental slope to a deep of up to 1200 meters. These features are normally associated with major rivers.

Subpolar glacier Glacier in which the ice found from it's surface to base has a temperature as cold as -30° Celsius throughout the year. This is well below the pressure melting point. However, melting does occur in the accumulation zone in the summer. One of the three types of glaciers: cold glacier; temperate glacier; and subpolar glacier.

Subpolar lows Surface zone of atmospheric low pressure located at about 60° North and South latitude. These low pressure systems are produced by the frontal lifting of subtropical air masses over polar air.

Subpolar The region bordering the polar region, between 50 and 70 degrees North and South latitude. This is generally an area of semi-permanent low pressure that exists and where the Aleutian and Icelandic Lows may be found. However, a dome of high pressure may form over the cold continental surfaces during the winter, for example, the North American High and the Siberian High.

Subrefraction Less than normal bending of light or a radar beam as it passes through a zone of contrasting properties, such as atmospheric density, water vapour, or temperature. Contrast with superrefraction.

Subsatellite point Point where a straight line drawn from a satellite to the centre of the Earth intersects the Earth's surface.

Subsea permafrost Form of permafrost that exists beneath the sea in ocean sediments.

Subsidence A sinking or downward motion of air, often seen in anticyclones. It is most prevalent when there is colder, denser air aloft. It is often used to imply the opposite of atmospheric convection.

Subsolar point The location on the Earth where the sun is directly overhead.

Subsystem A subunit of either the physical climate system (e.g., ocean dynamics) or the biogeochemical cycles (e.g., terrestrial ecosystems).

A subunit of a spacecraft, e.g., the telemetry subsystem, the power subsystem, the sensor subsystem, etc.

Subtropical The region between the tropical and temperate regions, an area between 35 and 40 degrees North and South latitude. This is generally an area of semi-permanent high pressure that exists and is where the Azores and North Pacific Highs may be found.

Subtropical air An air mass that forms over the subtropical region. The air is typically warm with a high moisture content due to the low evaporative process.

Subtropical high pressure zone Surface zone of atmospheric high pressure located at about 30° North and South latitude. These high pressure systems produced by vertically descending air currents from the Hadley cell.

Subtropical jet stream Relatively fast uniform winds concentrated within the upper atmosphere in a narrow band. The subtropical jet stream exists in the subtropics at an altitude of approximately 13 kilometres. This jet stream flows from west to east and has a speed that is somewhat slower that the polar jet stream.

Succession Directional cumulative change in the types plant species that occupy a given area, through time.

Succulent vegetation Group of plants that have the ability to survive in deserts and other dry climates by having no leaves. Instead their branches and stems that are photosynthetic. This adaptation reduces the surface area for evaporation thus reducing the loss of scarce water.

Sugar Type of carbohydrate chemically based on carbon, oxygen, and hydrogen.

Sulfate aerosol Type of solid compound commonly found in the atmosphere. These particles play an important role in reflecting, absorbing, and scattering incoming insolation. The source of these compounds is both natural and human-made. Most of the human-made particles come from the combustion of fossil fuels.

Sulfur dioxide A gas produced from volcanic eruptions, ocean spray, organic decomposition and the burning of fossil fuels. Sulphur dioxide is a component in the creation of acid precipitation. This colourless gas has the chemical formula SO_2.

Sulfuric acid Acid with the chemical formula H_2SO_4.

Summation layer amount The amount of sky cover for each layer is given in eighths of sky cover attributable to clouds or obscurations. The summation amount for any given layer is equal to the sum of the sky cover for the layer being evaluated plus the sky cover for all lower layers, including partial obscuration. A summation amount for a layer can not exceed 8/8ths.

Summer Astronomically, this is the period between the summer solstice and the autumnal equinox. It is characterized as having the warmest temperatures of the year, except in some tropical regions. Customarily, this refers to the months of June, July, and August in the North Hemisphere, and the months of December, January, and February in the Southern Hemisphere.

Summer solstice Date when the declination of the sun is at 23.5° North of the equator. This date is usually June 21 or 22.

Sun dog Either of two coloured luminous spots that appear at roughly 22 degrees on both sides of the sun at the same elevation. They are caused by the refraction of sunlight passing through ice crystals. They are most commonly seen during winter in the middle latitudes and are exclusively associated with cirriform clouds. The scientific name for sun dogs is parhelion and they are also known as mock suns.

Sun pillar Horizontal ice crystals in the form of plates, which occur in clouds and ice fog near the earth's surface, reflect sunlight into vertical sun pillars for a spectacular display.

Sun The closest star to Earth (149,599,000 km away on average). The sun dwarfs the other bodies in the solar system, representing approximately 99.86 percent of all the mass in the solar system. One hundred and nine Earths would be required to fit across the Sun's disk, its interior could hold over 1.3 million Earths.

The source of the Sun's energy is the nuclear reactions that occur in its core. There, at temperatures of 15 million degrees Celsius (27 million degrees Fahrenheit) hydrogen atom nuclei, called protons, are fused and become helium atom nuclei. The energy produced through fusion at the core moves outward, first in the form of electromagnetic radiation called photons. Next, energy moves upward in photon heated solar gas-this type of energy transport is called convection. Convective motions within the solar interior generate magnetic fields that emerge at the surface as sunspots and loops of hot gas called prominences. Most solar energy finally escapes from a thin layer of the Sun's atmosphere called the photosphere-the part of the Sun observable to the naked eye.

The sun appears to have been active for 4.6 billion years and has enough fuel for another 5 billion years or so. At the end of its life, the Sun will start to fuse helium into heavier elements and begin to swell up, ultimately growing so large that it will swallow Earth. After a billion years as a "red giant" it will suddenly collapse into a "white dwarf." It may take a trillion years to cool off completely.

Sunrise The daily appearance of the sun on the eastern horizon as a result of the earth's rotation. In the United States, it is considered as that instant when the upper edge of the sun appears on the sea level horizon. In Great Britain, the centre of the sun's disk is used instead. Time of sunrise is calculated for mean sea level.

Sunset The daily disappearance of the sun below the western horizon as a result of the earth's rotation. In the United States, it is considered as that instant when the upper edge of the sun just disappears below the sea level horizon. In Great Britain, the centre of the sun's disk is used instead.

Time of sunset is calculated for mean sea level.

Sunspot Dark coloured region on the sun that represents an area of cooler temperatures and extremely high magnetic fields.

Sun-synchronous Describes the orbit of a satellite that provides consistent lighting of the Earth-scan view. The satellite passes the equator and each latitude at the same time each day. For example, a satellite's sun-synchronous orbit might cross the equator twelve times a day each time at 3:00 p.m. local time. The orbital plane of a sun-synchronous orbit must also precess (rotate) approximately one degree each day, eastward, to keep pace with the Earth's revolution around the sun.

Supercell A severe thunderstorm characterized by a rotating, long-lived, intense updraft. Although not very common, they produce a relatively large amount of severe weather, in particular, extremely large hail, damaging straight-line winds, and practically all violent tornadoes.

Supercooled water Cooling of water below 0° Celsius without freezing. Common in clouds where there is a deficiency of condensation nuclei.

Supercooling The reduction of the temperature of any liquid below the melting point of that substance's solid phase. Cooling a substance beyond its nominal freezing point. Supercooled water is water that remains in a liquid state when it is at a temperature that is well below freezing. The smaller and purer the water droplets, the more likely they can become supercooled.

Superrefraction Greater than normal bending of light or radar beam as it passes through a zone of contrasting properties, such as atmospheric density, water vapour, or temperature. Contrast with subrefraction.

Super-saturation Atmospheric condition where saturation occurs at a relative humidity greater than 100 % because of a shortage of deposition or condensation nuclei.

Surface boundary layer The lowest layer of the earth's atmosphere, usually up to 3,300 feet, or one kilometer, from the earth's surface, where the wind is influenced by the friction of the earth's surface and the objects on it. May also be called the boundary layer or the friction layer.

Surface creep The sliding and rolling movement of soil particles on the Earth's surface because of wind. Eolian process of soil particle movement.

Surface heat flux Process where excess heat energy is transferred into the ground.

Surface tension Tension of a liquid's surface. Due to the forces of attraction between molecules.

Surface wave Type of seismic wave that travels across the Earth's surface. These earthquake generated waves cause the Earth's surface to roll or sway like waves on the ocean.

Surge A large, destructive ocean wave caused by very low atmospheric pressure and strong winds. Hurricanes often cause a surge of the ocean surface.

Survey Orderly process of determining data relating to any physical or chemical characteristics of the Earth. The associated data obtained in a survey. An organization engaged in making a survey.

Survey mode Refers to observational emphasis upon frequent global coverage, usually with restricted spatial and spectral resolution, aimed at developing a consistent, long-term data product for later interpretation.

Suspended load Portion of the stream load that is carried almost permanently suspended in flowing water.

Suspension Erosional movement of sediment continually held in the transport medium of air, water or ice.

Sustainable development Forms of economic growth and other human activities that meet the requirements of the present without jeopardizing the ability of future generations of individuals to meet their own needs.

Sustainable yield The amount of a naturally self-reproducing community, such as trees or fish, that can be harvested without diminishing the ability of the community to sustain itself.

Swamp A swamp is a type of freshwater wetland that has spongy, muddly land and a lot of water. Many trees and shrubs grow in swamps.

Swash A thin sheet of water that moves up the beach face after a wave of water breaks on the shore.

Swath The area observed by a satellite as it orbits the Earth.

Swell Ocean waves that have travelled out of their generating area. Swell characteristically exhibits a more regular and longer period and has flatter wave crests than waves within their fetch.

Symbiotic Mutual relationship between two organisms which is necessary for either to survive.

Symbiotic mutualism Mutualistic interaction where the species interact physically and their relationship is biologically essential for survival.

Syncline A fold in rock layers that forms a trough-like bend.

Synoptic chart Any map or chart that depicts meteorological or atmospheric conditions over a large area at any given time.

Synoptic scale The size of migratory high and low pressure systems in the lower troposphere that cover a horizontal area of several hundred miles or more. Contrast with macroscale, mesoscale, and storms.

Synoptic view The ability to see large areas at the same time.

Synthetic aperture radar (SAR) A high-resolution ground-mapping technique that effectively synthesizes a large receiving antenna by processing the phase of the reflected radar return.

System A system is a set of interrelated components working together towards some kind of process.

System attribute A system attribute is the perceived characteristic of a system element. For example, number, size, colour, volume, and temperature may be some of the perceived characteristics of clouds in the atmospheric system.

System boundary Outer edge of system. Zone between one system and another system.

System element System elements are the kinds of things or substances composing the system. They may be atoms or molecules, or larger bodies of matter-sand grains, rain drops, plants, or cows.

System relationship Is the association that exist between the elements and attributes of a system based on cause and effect.

System state Current value of a system's elements, attributes, and/or relationships.

Syzygy The points in the moon's orbit about the earth at which the moon is new or full.

T

Tacheometer (tachymeter) Surveying instrument designed for use in the rapid determination of distance, direction, and difference of elevation from a single observation, using a short base which may be an intergraph part of the instrument.

Taiga A moist subarctic coniferous forest that begins where the tundra ends and is dominated by spruces and firs.

Taku Name for a katabatic type of cold wind that occurs in Alaska.

Talik An unfrozen section of ground found above, below, or within a layer of discontinuous permafrost. These layers can also be found beneath water bodies in a layer of continuous permafrost. A number of different types of talik have been distinguished: closed talik, open talik, and through talik.

Talus An accumulation of angular rock debris from rockfalls.

Talus slope A slope that is composed of talus.

Tarn A small mountain lake that occurs inside a cirque basin.

Taxon A classification category for a group of organisms.

Taxonomic classification Classification of organisms based on structural and physiological connections between other species.

Tectonic plate An extensive layer of lithosphere that moves as a discrete unit on the surface of the Earth's asthenosphere.

Teleconnections Information used by forecasters to determine what the weather might be elsewhere when compared with past weather conditions at the same degree of longitude.

Telemetry Telecommunications transmission to a distance of measured magnitude by radio or telephony with suitably coded modulation, e.g., amplitude, frequency phase, pulse.

Transmission of data collected at a remote location over communications channels to a central station.

Surveying measurement of linear distances by use of tellurometer-a device that uses microwaves to measure distance.

Temperate climate Climates with distinct winter and summer seasons, typical of regions found between the Tropics of Cancer and Capricorn and the Arctic and Antarctic Circles. Considered the climate of the middle latitudes.

Temperate deciduous forest Forested biome found in the mid-latitudes and dominated by deciduous vegetation.

Temperate glacier Glacier in which the ice found below 10 to 20 meters from its surface is at the pressure melting point. One of the three types of glaciers: cold glacier; temperate glacier; and subpolar glacier.

Temperate rain forest An ecosystem that is dominated by large and very tall evergreen trees. This biome occurs along the Pacific Northwest coast of North America where annual precipitation is high and temperatures are mild.

Temperature inversion Situation where a layer of warmer air exists above the Earth's surface in a normal atmosphere where air temperature decreases with altitude. In the warmer layer of air, temperature increases with altitude.

Temperature Measure of the energy in a substance. The more heat energy in the substance, the higher the temperature. The Earth receives only one twobillionth of the energy the sun produces. Much of the energy that hits the Earth is reflected back into space. Most of the energy that isn't reflected is absorbed by the Earth's surface. As the surface warms, it also warms the air above it.

Tephra Fragmented rock material ejected by a volcanic explosion. Also called pyroclastic material.

Terabit A trillion (10^{12}) bits.

Terminal Doppler Weather Radar (TDWR) Doppler radar installed at major airports throughout the United States to detect microbursts.

Terminal fall velocity Velocity at which a particle being transported by wind or water falls out of the moving medium. This velocity is dependent on the size of the particle.

Terminal moraine Moraine that marks the maximum advance of a glacier.

Terminal velocity Maximum speed that can be achieve by a body falling through a fluid like water or air.

Terminus End or snout of a glacier.

Terrace An elevated surface above the existing level of a floodplain or shore that is created by stream or ocean wave erosion.

Terrestrial radiation Long wave radiation that is emitted by the earth back into the atmosphere. Most of it is absorbed by the water vapour in the atmosphere, while less than ten percent is radiated directly into space.

Territory A specific area or portion of the Earth's surface; not to be confused with region.

Tertiary consumer Organisms that occupy the fourth trophic level in the grazing food chain. These organisms are carnivores. Also known as a secondary carnivore.

Tertiary Geologic period that occurred roughly 1.6 to 65 million years ago. During this period, mammals become a dominant species on the planet.

Tertiary sector That portion of a region's economy devoted to service activities (e.g., transportation, retail and wholesale operations, insurance).

Tetrahedron Silicon atom joined by four oxygen atoms (SiO_4). The atomic properties of this molecule cause it to develop a unique three dimensional crystal lattice that is pyramid shaped.

Texas norther Local name in the south-central Great Plains for strong winter winds blowing north or northwest following a sharp cold front with dropping temperatures. Marked by a dark, blue-black sky. Also called a Blue Norther.

Texture The relative quantities of the different types and sizes of mineral particles in a deposit of sediment.

Thalweg Line of deepest water in a stream channel as seen from above. Normally associated with the zone of greatest velocity in the stream.

Thaw A warm spell of weather when ice and snow melt. To free something from the binding action of ice by warming it to a temperature above the melting point of ice.

Thematic map Map that displays the geographical distribution of one phenomenon or the spatial associations that occur between a few phenomena. Compare with reference map.

Thematic Mapper (TM) A Landsat multispectral scanner designed to acquire data to categorize the Earth's surface. Particular emphasis was placed on agricultural applications and identification of land use. The scanner continuously scans the surface of the Earth, simultaneously acquiring data in seven spectral channels. Overlaying two or more bands produces a false colour image. The ground resolution of the six visible and shortwave bands of the Thematic Mapper is 30 meters, and the resolution of the thermal infrared band is 120 meters. Thematic mappers have been flown on Landsats-4 and -5.

Theodolite An optical instrument used to track the motion of a pilot balloon, or pibal, by measuring the elevation and azimuth angles.

Theory Proposed explanation for the causal mechanisms responsible for a phenomenon or a set of facts.

Thermal circulation Atmospheric circulation caused by the heating and cooling of air.

Thermal equator Continuous area on the globe that has the highest surface temperatures because of the presence of the Intertropical Convergence Zone.

Thermal high Area of low pressure in the atmosphere caused by surface temperatures.

Thermal infrared Electromagnetic radiation with wavelengths between about 3 and 25 micrometers.

Thermal low Also known as heat low, it is an area of low pressure due to the high temperatures caused by intensive heating at the surface. It tends to remain stationary over its source area, with weak cyclonic circulation. There are no fronts associated with it. An example is the low that develops over southwestern United States and northwestern Mexico during the summer months.

Thermal metamorphism Is the metamorphic alteration of rock because of intense heat released from processes related to plate tectonics.

Thermocline Boundary in a body of water where the greatest vertical change in temperature occurs. This boundary is usually the transition zone between the layer of warm water near the surface that is mixed and the cold deep water layer.

Thermodynamic equilibrium This type of equilibrium describes a condition in a system where the distribution of mass and energy moves towards maximum entropy.

Thermodynamic laws Laws that describe the physical processes, relationships, and phenomena associated with heat.

Thermodynamics Study of the processes that involve the transformation of heat into mechanical work, of mechanical work into heat, or the flow of heat from a hotter body to a colder body.

Thermograph Essentially, a self-recording thermometer. A thermometer that continuously records the temperature on a chart.

Thermohaline In oceanography, it pertains to when both temperature and salinity act together. An example is thermohaline circulation which is vertical circulation induced by surface cooling, which causes convective overturning and consequent mixing.

Thermokarst Landscape dominated by depressions, pits, and caves that is created by the thawing of ground ice in high latitude locations. Resembles karst landscape but is not created by chemical weathering.

Thermometer An instrument used for measuring temperature. The different scales used in meteorology are Celsius, Fahrenheit, and Kelvin or Absolute.

Thermosphere Atmospheric layer above the mesosphere (above 80 kilometers) characterized by air temperatures rising rapidly with height. The thermosphere is the hottest layer in the atmosphere. In the thermosphere, gamma, X-ray, and specific wavelengths of ultraviolet radiation are absorbed by certain gases in the atmosphere. The absorbed radiation is then converted into heat energy. Temperatures in this layer can get as high as 1300-1800° Celsius.

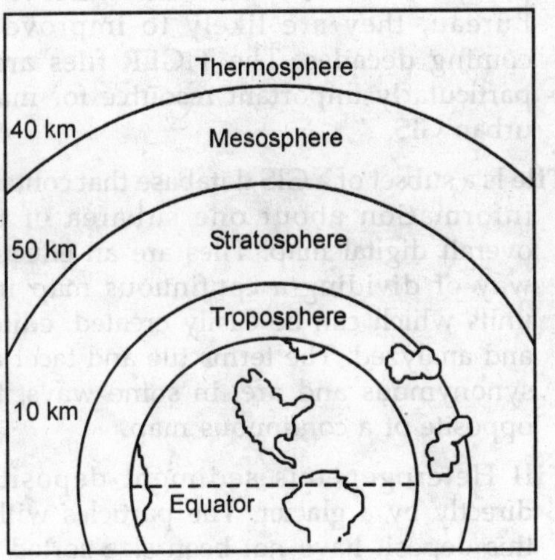

Fig. Thermosphere

Thickness The thickness of a layer in the atmosphere is proportional to the mean temperature of that whole layer. The layer most often used in meteorology is between 1000 and 500 millibars. There can be different temperature profiles in the lowest layer of the atmosphere with the same 1000-500 millibar thickness value, depending on what is happening above that lowest layer. For example, if the lower levels are warming but higher levels are cooling, the overall mean temperature, the thickness, could remain the same. Likewise, on a sunny day, the amount of incoming solar radiation, affects the temperature right at the earth's surface, without necessarily having much effect on the thickness of the whole layer.

Third law of thermodynamics This law states if all the thermal motion of molecules (kinetic energy) could be removed, a state called absolute zero would result and all energy would be randomly distributed.

Threatened species Species that is still plentiful in its natural range but is likely to become endangered because of declining population numbers.

Threshold The minimum-sized market for an economic activity. The activity will not be successful until it can reach a population larger than this threshold size.

Threshold velocity Velocity required to cause entrainment in the erosional agents of wind, water or ice. Threshold velocity is usually higher than the velocity required for transport because factors like particle cohesion.

Through talik Is a form of localized unfrozen ground (talik) in an area of permafrost. It is open to the ground surface and to an area of unfrozen ground beneath it. Permafrost encases it along the sides.

Throughfall Describes the process of precipitation passing through the plant canopy. This process is controlled by factors like: plant leaf and stem density, type of the precipitation, intensity of the precipitation and duration of the precipitation event. The amount of precipitation passing through varies greatly with vegetation type.

Thrust fault A geologic fault where the hanging wall is forced over the foot wall.

Thunder The sound emitted by rapidly expanding gases along the channel of a lightning discharge. Over three-quarters of lightning's electrical discharge is used in heating the gases in the atmosphere in and immediately around the visible channel. Temperatures can rise to over 10,000 degrees Celsius in microseconds, resulting in a violent pressure wave, composed of compression and rarefaction. The rumble of thunder is created as one's ear catches other parts of the discharge, the part of the lightning flash nearest registering first, then the parts further away.

Thunder snow A wintertime thunderstorm from which falls snow instead of rain. Violent updrafts and at or below freezing temperatures throughout the atmosphere, from surface to high aloft, discourage the melting of snow and ice into rain. Intense snowfall rates often occur during these situations.

Thunderstorm Local storm resulting from warm humid air rising in an unstable environment. Air may start moving upward because of unequal surface heating, the lifting of warm air along a frontal zone, or diverging upper-level winds (these diverging winds draw air up beneath them). The scattered thunderstorms that develop in the summer are called air mass thunderstorms because they form in warm, maritime tropical air masses away from other weather fronts. More violent severe thunderstorms form in areas with a strong vertical wind shear that forces the updraft into the mature stage, the most intense stage of the thunderstorm. Severe thunderstorms can produce large hail, forceful winds, flash floods, and tornadoes.

Tidal current Regional scale ocean current that is created the tidal rise and fall of the ocean surface.

Tidal period Time it takes for one tidal cycle.

Tidal zone Area along the coastline that is influence by the rise and fall of tides.

Tide Periodic rise and fall of the water resulting from gravitational interactions between the Sun, Moon, and Earth. The vertical component of the particulate motion of a tidal wave. Although the accompanying horizontal movement of the water is part of the same phenomenon, it is preferable to designate this motion as tidal current.

TIGER Is an acronym for topologically integrated geographic encoding and referencing file. This is a type of digital map developed by the U.S. Bureau of the Census to support the 1990 population census. Census maps in TIGER format succeed the previous DIME format. TIGER files are available for every county in the United States and for the millions of census blocks in urban areas. Although the accuracy of TIGER files varies from county to county, partly for reasons beyond the control of the Bureau, they are likely to improve in coming decades. The TIGER files are a particularly important resource for many urban GIS.

Tile Is a subset of a GIS database that contains information about one subarea of the overall digital map. Tiles are an effective way of dividing a continuous map into units which can be easily created, edited, and analyzed. The terms tile and facet are synonymous and are, in some ways, the opposite of a continuous map.

Till Heterogeneous sediment deposited directly by a glacier. The particles within this deposit have not been size sorted by the action of wind or water.

Till plain Extensive flat plain of till that forms when a sheet of ice becomes detached from the main body of the glacier and melts in place depositing the sediments it carried.

Tilt The inclination to the vertical of a significant feature of the pressure pattern or of the field of moisture or temperature. For example, midlatitide troughs tend to display a westward tilt with altitude through the troposphere.

Time Measurable period in which cause and effect occurs and systems function.

Time-distance A time measure of how far apart places are (how long does it take to travel from place A to place B?). This may be contrasted with other distance metrics such as geographic distance (how far is it?) and cost-distance (how much will it cost to get there?).

TIROS-NINOAA satellites NOAA satellites that continuously orbit the Earth from North to South Pole (hence, polar orbiting) at an altitude of approximately 470 nautical miles (870.44 km or 540.86 statute miles). These environmental satellites collect visible and infrared imagery and provide atmospheric-sounding data and meteorological data relay and collection. A primary mission of TIROS-N/NOAA is to monitor the 70 percent of the globe covered by water-where weather data is sparse and provide continuous data to the National Weather Service for use in numerical forecast modeling. Each TIROSN/NOAA carries six primary systems: The advanced very high resolution scanning radiometer (AVHRR) senses clouds over both ocean and land, using the visible and infrared parts of the spectrum. It stores measurements on tape, and later plays them back to NOAAs command and data acquisition stations. The satellites also broadcast in real time, and the broadcasts can be received around the world by anyone equipped with a direct readout receiving station.

The tiros operational vertical sounder (TOVS) is a 3-part TIROS system to measure: Temperature profile of the Earth's atmosphere from the surface to 10 millibars;

Water content of the Earth's atmosphere;

Total ozone content of the Earth's atmosphere;

The ARGOS Data Collection and Platform Location System (DCS) collects data from sensors placed on fixed and moving platforms, including ships, buoys, and weather balloons, and transmits data to a ground station antenna. Because ARGOS also determines the precise location of these moving sensors, it can serve wildlife managers by monitoring and tracking the transmitters placed on birds and animals.

The space environment monitor (SEM) measures energetic particles emitted by the sun over essentially the full range of energies and magnetic field variations in the Earth's near-space environment. Readings made by these instruments are invaluable in measuring the sun's radiation activity.

Search and Rescue Tracking (COSPAS/ SARSAT) equipment receives emergency signals from persons in distress. The satellites transmit the signals to ground receiving stations. The signals then are forwarded to rescue coordination centres. The rescue centres compute the location of the signals and provide the coordinates of the emergency site (usually within a few miles).

Earth Radiation Budget Experiment (ERBE) is a radiometer flown on NOAA 9 and 10, designed to measure all radiation striking and leaving the Earth. This enables scientists to measure the loss or gain of terrestrial energy to space. Shifts in this energy "budget" affect the Earth's average temperatures. Even slight changes can affect climatic patterns.

Tissue A group of similar cells that are organized into a structure with a specific purpose.

TM See thematic mapper.

TNL Thermal noise level.

Tolerance model of succession This model of succession suggests that the change in plant species dominance over time is caused by competition for resources. Later species are able to tolerate lower resource levels due to competition and can grow to maturity in the presence of early species, eventually out competing them.

Tolerance range Limits of tolerance a species has to an abiotic factor or condition in the environment.

Tombolo A coastal feature that forms when a belt sand and/or gravel is deposited between an island and the mainland. This feature is above sea-level for most of the time.

Topographic map Map that displays topography through the use of elevation contour lines. Base elevation on these maps is usually sea-level.

Topographic profile A two-dimensional diagram that describes the landscape in vertical cross-section.

Topography Configuration (relief) of the land surface; the graphic delineation or portrayal of that configuration in map form, as by contour lines; in oceanography the term is applied to a surface such as the sea bottom or surface of given characteristics within the water mass.

Topology Refers to properties of geometric forms that remain invariant when the forms are deformed or transformed by bending, stretching, and shrinking. Among the topological properties of concern in GIS are connectivity, order, and neighbourhood.

Topset bed Horizontal deltaic deposit composed of coarse alluvial sediment. Represents current or past surface of the delta.

Tornado A twisting, spinning funnel of low pressure air. The most unpredictable weather event, tornadoes are created during powerful thunderstorms. As a column of warm air rises, air rushes in at ground level and begins to spin. If the storm gathers energy a twisting, spinning funnel develops. Because of the funnel's cloud and rain composition and the dust, soil, and debris it draws up, the funnel appears blackish in colour. The most energetic storms result in the funnel touching the ground. In these tornadoes, the roaring winds in the funnel can reach 300 mph, the strongest winds on Earth. Funnels usually travel at 20 to 40 mph, moving toward the northeast. When tornadoes form over lakes or oceans they suck water into the funnel cloud and are called waterspouts.

Tornado alley A geographic corridor in the United States which stretches north from Texas to Nebraska and Iowa. In terms of sheer numbers, this section of the United States receives more tornadoes than any other.

Tornado warning A warning issued to the public that a tornado has been observed by an individual in a specified region. This warning can also be issued if meteorological information indicates a high probability that a tornado will develop in a specified region.

Tornado watch A forecast issued to the public that a tornado may occur in a specified region.

Total column ozone A measurement of ozone concentration in the atmosphere.

Total Ozone Mapping Spectrometer (TOMS) Flown on NASAs Nimbus-7 satellite, its primary goal is to continue the high-resolution global mapping of total ozone on a daily basis. The Nimbus-7 launch in 1978 enabled TOMS to begin delivering data in 1979 and continue providing information until 1993. TOMS has mapped the total amount of ozone between the ground and the top of the atmosphere, provided the first maps of the ozone hole, and continues to monitor this phenomenon.

Because of its longevity, TOMS also has obtained information on the more subtle trends in ozone outside the ozone hole region. This results from development of a powerful new calibration technique that removes the instrument measurement drift that developed over the years. With this technique applied to the TOMS 14.5-year data record, a global ozone decrease of 2.69 percent per decade was detected.

To ensure that ozone data will be available through the next decade, NASA will continue the TOMS program using U.S. and foreign launches. In 1991, the former Soviet Union launched a Meteor-3 satellite carrying a TOMS instrument provided by NASA. A third TOMS will be launched onboard a NASA Earth probe satellite in 1994, and the Japanese advanced earth observations satellite (ADEOS) will carry a fourth TOMS when it launches in 1996.

Towering cumulus Another name for cumulus congestus, it is a rapidly growing cumulus or an individual dome-shaped clouds whose height exceeds its width. Its distinctive cauliflower top often mean showers below, but lacking the characteristic anvil of a cumulonimbus, it is not a thunderstorm.

Trace Generally, an immeasurable or insignificant quantity. A precipitation amount of less than 0.005 inch.

Tracking and Data Relay Satellite System (TDRSS) An orbiting communications satellite, developed by NASA, used to relay data from satellite sensors to ground stations and to track the satellites in orbit.

Traction Erosional movement of particles by rolling, sliding and shuffling along the eroded surface. Occurs in all erosional mediums (air, water, and ice).

Trade winds Surface air from the horse latitudes that moves back toward the equator and is deflected by the Coriolis Force, causing the winds to blow from the Northeast in the Northern Hemisphere and from the Southeast in the Southern Hemisphere. These steady winds are called trade winds because they provided trade ships with an ocean route to the New World.

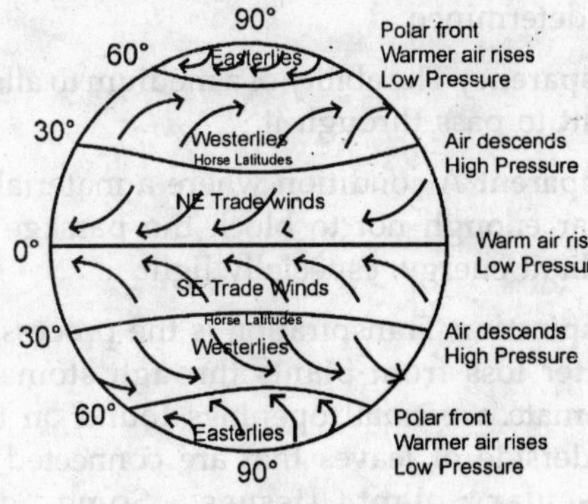

Fig. Trade winds

Trajectory The curve that a body, such as a celestial object, describes in space. This applies to air parcel movement also.

Transferability The extent to which a good or service can be moved from one location to another; the relative capacity for spatial interaction.

Transform fault Massive strike-slip fault continental in size. Examples of such faults occur along tectonic plate boundaries and at the mid-oceanic ridge.

Transhumance The seasonal movement of people and animals in search of pasture. Commonly, winters are spent in snow-free lowlands and summers in the cooler uplands.

Transit Precision surveying instrument; a theodolite in which the telescope can be reversed in direction by rotation about its horizontal axis.

Translation Is the process of converting data or commands from one computer format to another, or from one computer language to another.

Translucent Not transparent, but clear enough to allow light to pass through.

Transmissometer An electronic instrument system which provides a continuous record of the atmospheric transmission between two fixed points. By showing the transmissivity of light through the atmosphere, the horizontal visibility may be determined.

Transparency The ability of a medium to allow light to pass through it.

Transparent A condition where a material is clear enough not to block the passage of radiant energy, especially light.

Transpiration Transpiration is the process of water loss from plants through stomata. Stomata are small openings found on the underside of leaves that are connected to vascular plant tissues. Some dry environment plants do have the ability to open and close their stomata. Transpiration is a passive process largely controlled by the humidity of the atmospheric and the moisture content of the soil. Of the transpired water passing through a plant only 1 % is used in the growth process. Transpiration also transports nutrients from the soil into the roots and carries them to the various cells of the plant.

Transport One of three distinct processes involved in erosion. It is the movement of eroded material in the medium of air, water or ice.

Traverse Sequence of lengths and directions of lines connecting a series of stations, obtained from field measurements, and used in determining positions of the stations.

Tree A large woody plant that has a trunk which supports branches and leaves.

Tree line Either the latitudinal or elevational limit of normal tree growth. Beyond this, limit, closer to the poles or at higher or lower elevations, climatic conditions are too severe for such growth.

Triangulation Method of extending horizontal position on the surface of the Earth by measuring the angles of triangles and the included sides of selected triangles.

Triassic Geologic period that occurred roughly 208 to 245 million years ago. During this period, the first dinosaurs appeared.

Tributary A smaller branching stream channel that flows into a main stream channel. Opposite of distributary.

Trilateration Method of surveying wherein the lengths of the triangle sides are measured, usually by electronic methods, and the angles are computed from the measured lengths. Compare with triangulation.

Triple point The intersection between two atmospheric boundaries, such as a gust front and an outflow boundary, and often a focal point for thunderstorm development. It is also the condition of temperature and pressure under which the gaseous, liquid, and solid forms of a substance can exist in equilibrium.

Trophic level Level of organization in the grazing food chain.

Tropic of Cancer A line of latitude located at 23°30' north of the equator. The Sun is directly overhead the Tropic of Cancer on the summer solstice in the Northern Hemisphere (June 20 or 21). It marks the northernmost point of the tropics, which falls between the Tropic of Cancer and the Tropic of Capricorn.

Tropic of Capricorn A line of latitude located at 23°30' south. The Sun is directly overhead the Tropic of Capricorn on the summer solstice in the Southern Hemisphere (Dec. 20 or 21). It marks the southernmost point of the tropics.

Tropical air mass An air mass that forms in the tropics or subtropics over the low latitudes. Maritime tropical air is produced over oceans and is warm and humid, while continental tropical air is formed over arid regions and is very hot and dry.

Tropical cyclone The term used for cyclones that originate over tropical waters. It is a warm core low pressure system which is non-frontal and has organized circulation. Depending on sustained surface winds, the system is classified as a tropical disturbance, a tropical depression, a tropical storm, or a hurricane or typhoon.

Tropical depression A tropical cyclone in which the maximum sustained surface winds are 38 miles per hour (33 knots) or less. Characteristically having one or more closed isobars, it may form slowly from a tropical disturbance or an easterly wave which has continued to organize.

Tropical disturbance An area of organized convection, originating in the tropics and occasionally the subtropics, that maintains its identity for 24 hours or more. The cloudiness and showers may or may not be associated with a tropical wave, but the system is non-frontal in nature. It is often the first developmental stage of any subsequent tropical depression, tropical storm, or hurricane.

Tropical Ocean Global Atmosphere (TOGA) TOGA is a program jointly sponsored by the United Nations World Meteorological Organization (WMO); the International Council of Scientific Unions (ICSU); the United Nations Educational, Scientific, and Cultural Organization (UNESCO) Intergovernmental Oceanographic Com-mission (IOC); and the ICSU Scientific Committee on Oceanic Research (SCOR).

TOGA has four major objectives: To collect and catalog observations of the tropical atmosphere and ocean;

To assess the evolution of the tropical atmosphere/ocean system in real time;

To promote the development of short-term climate-prediction computer models for the tropics;

To study the influence of the tropical atmosphere/ocean system on the climate at higher latitudes.

Tropical rainforest Forested biome found near the equator and dominated by evergreen vegetation.

Tropical storm A tropical cyclone in which the maximum sustained surface winds are from 39 miles per hour (34 knots) to 73 miles per hour (63 knots). At this point, the system is given a name to identify and track it.

Tropical storm formation Tropical storms generally form in the eastern portion of tropical oceans and track westward. Hurricanes, typhoons, and willywuhes all start out as weak low pressure areas that form over warm tropical waters (e.g., surface water temperature of at least 80°F). Initially winds and cloud formations over the warm tropical waters are minimal. Both intensify with time. Formation of tropical storms also requires a significant Coriolis effect to induce proper spin in the wind formation. As the storm begins to organize itself into a coherent pattern, it will experience increased activity and intensity.

When a storm develops a clearly recognizable pattern, it is referred to as a tropical depression. When wind speeds reach 35 knots (40.3 mph), it is called a tropical storm and is given a name. When wind speed equals or exceeds 74 mph, the storm is called a hurricane. In the western Pacific, a hurricane is referred to as a typhoon. In waters around Australia it is called a cyclone or willy-willy.

Hurricanes intensify when moving over areas of increased water temperatures, and weaken over colder water surfaces. Upper atmosphere wind shear (different wind direction and speeds at different elevations) will frequently prevent or slow intensification of tropical storms by "spreading out" the storm horizontally and preventing the formation of strong updrafts of warm, humid air. Movement over a land-mass will weaken hurricane winds but will result in large-scale rain that can result in large-scale flooding. When encountering a strong frontal system (such as a polar front) the hurricane will curve and track along the leading edge of the front or become implanted in it.

Satellite infrared imagery can identify surface water temperatures that will foster tropical storm development.

Tropical wave Another name for an easterly wave, it is an area of relatively low pressure moving westward through the trade wind easterlies. Generally, it is associated with extensive cloudiness and showers, and may be associated with possible tropical cyclone development.

Tropics Technically, the area between the Tropic of Cancer (21-1/2 N latitude) and the Tropic of Capricorn (21-1/2 S latitude), characterized by the absence of a cold season. Often used to describe any area possessing what is considered to be a hot, humid climate.

Tropopause The boundary zone or transition layer between the troposphere and the stratosphere. This is characterized by little or no increase or decrease in temperature or change in lapse rate with increasing altitude.

Troposphere Layer in the atmosphere found from the surface to a height of between 8 to 16 kilometres of altitude (average height 11 kilometres). The troposphere is thinnest at poles and gradually increases in thickness as one approaches the equator. This atmospheric layer contains about 80 % of the total mass of the atmosphere. It is also the layer where the majority of our planet's weather occurs. Maximum air temperature occurs near the Earth's surface in this layer. With increasing altitude air temperature drops uniformly with increasing height at an average rate of 6.5° Celsius per 1000 meters (commonly called the Environmental Lapse Rate), until an average temperature of -56.5° Celsius is reached at the top of the troposphere.

Tropospheric emission spectrometer A high-resolution infrared spectrometer for monitoring the minor components of the lower atmosphere.

Trough An elongated area of low atmospheric pressure that is associated with an area of minimum cyclonic circulation. The opposite of a ridge.

True anomaly One of six Keplerian elements, it locates a satellite on an orbit. True anomaly is the true angular distance of a satellite (planet) from its perigee (perihelion) as seen from the centre of the Earth (sun).

True north Direction of the North Pole from an observer on the Earth.

True south Direction of the South Pole from an observer on the Earth.

Tsunami An ocean wave with a long period that is formed by an underwater earthquake or landslide, or volcanic eruption. It may travel unnoticed across the ocean for thousands of miles from its point of origin and builds up to great heights over shallower water. Also known as a seismic sea wave, and incorrectly, as a tidal wave.

Tule fog Ground fog in the central valley of California and the leading cause of weather-related casualties in that state. It forms at night and in the early morning when the ground cools, lowering the air temperature near the ground to or below its initial dew point. A form of radiation fog.

Tundra High latitude biome dominated by a few species of dwarf shrubs, a few grasses, sedges, lichens, and mosses. Productivity is low in this biome because of the extremes of climate.

Turbulence The irregular and instantaneous motions of air which is made up of a number of small of eddies that travel in the general air current. Atmospheric turbulence is caused by random fluctuations in the wind flow. It can be caused by thermal or convective currents, differences in terrain and wind speed, along a frontal zone, or variation in temperature and pressure.

Turbulent flow Movement of water within a stream that occurs as discrete eddies and vortices. Turbulent flow is caused by channel topography and friction.

Turnkey system Is á complete hardware and software system designed, built, installed, and tested by a vendor for a customer. Supposedly, the system will be fully operational once it is installed and tested. The term derives from the idea that the customer need only "turn the key," or flip on a switch, and the system will be ready to use. Turnkey systems are often preferred by customers who do not wish, or lack the expertise, to become encumbered in the complexities of system installation, configuration, and testing.

Twilight Often called dusk, it is the evening period of waning light from the time of sunset to dark. The time of increasing light in the morning is called dawn. Twilight ends in the evening or begins in the morning at a specific time and can be categorized into three areas of decreasing light. Civil twilight is the time in the evening when car headlights need to be turned on to be seen by other drivers. Nautical twilight is when the bright stars used by navigators have appeared and the horizon may still be seen. Astronomical twilight is when the sunlight is still shining on the higher levels of the atmosphere, yet it is dark enough for astronomical work to begin. During dawn, the reverse order occurs until full daylight.

Twister A slang term used in the United States for a tornado.

Two tailed statistical test Is an inferential statistical test where the values for which one can reject the null hypothesis are located either side of the centre of the probability distribution.

Typhoon The name for a tropical cyclone with sustained winds of 74 miles per hour (65 knots) or greater in the western North Pacific Ocean. This same tropical cyclone is known as a hurricane in the eastern North Pacific and North Atlantic Ocean, and as a cyclone in the Indian Ocean.

U

Ultisols Soil order (type) of the United States Natural Resources Conservation Service Soil Classification System. Tropical or subtropical soil that is in an advanced state of development. These soils have pronounced eluviation, clay accumulation in a subsurface layer, and are often poor in base cations.

Ultramafic Rock that is rich in magnesium and iron content.

Ultraviolet radiation The energy range just beyond the violet end of the visible spectrum. Although ultraviolet radiation constitutes only about 5 percent of the total energy emitted from the sun, it is the major energy source for the stratosphere and mesosphere, playing a dominant role in both energy balance and chemical composition.

Most ultraviolet radiation is blocked by Earth's atmosphere, but some solar ultraviolet penetrates and aids in plant photosynthesis and helps produce vitamin D in humans. Too much ultraviolet radiation can burn the skin, cause skin cancer and cataracts, and damage vegetation.

Unconfined aquifer Aquifer that is not restricted by impervious layers of rock.

Unconfined groundwater Groundwater that is not restricted by impervious layers of rock.

Unconformity A break in the sequence of sedimentary strata. Often the unconformity surface is the result of erosion.

Undercast In aviation, it is an opaque cloud layer viewed from an observation point above the layer. From the ground, it would be considered an overcast.

Undercut bank Steep bank found on the inside of stream meanders. Formed by the erosion that occurs when a stream channel moves horizontally.

Underemployment A condition among a labour force such that a portion of the labour force could be eliminated without reducing the total output. Some individuals are working less than they are able or want to, or they are engaged in tasks that are not entirely productive.

Underpopulation Economically, a situation in which an increase in the size of the labour force will result in an increase in per worker productivity.

Uniform region A territory with one or more features present throughout and absent or unimportant elsewhere.

Uniformitarianism Is a theory that rejects the idea that catastrophic forces were responsible for the current conditions on the Earth. The theory suggested instead, that continuing uniformity of existing processes were responsible for the present and past conditions of this planet.

Universal Time (UT) The mean solar time of the meridian at the Prime Meridian. Universal Time replaced the time standard known as Greenwich Mean Time in 1928. Universal Time is commonly used to denote solar time.

Universal time coordinate One of several names for the twenty-four hour time which is used throughout the scientific and military communities. Other names for this time measurement are Zulu (Z) or Greenwich mean time (GMT).

Universal Transverse Mercator (UTM) grid Military grid system based on the transverse Mercator projection, applied to maps of the Earth's surface extending from the Equator to 84 Degrees north and 80 degrees south latitudes.

Universe All of the observable phenomena in the celestial cosmos.

Unloading The releasing of downward pressure on rocks because of removal of overlying material by erosion. Unloading can cause the development of horizontal bedding in once solid rock.

Unstable atmosphere Condition in the atmosphere where isolated air parcels have a tendency to rise. The parcels of air tend to be warmer than the air that surrounds them.

Unstable equilibrium In an unstable equilibrium the system returns to a new equilibrium after disturbance.

Unstable/ instability Occurs when a rising air parcel becomes less dense than the surrounding air. Since its temperature will not cool as rapidly as the surrounding environment, it will continue to rise on its own. Refer to instability. Contrast with stable air.

Updraft A small scale current of air with vertical motion. If there is enough moisture, then it may condense, forming a cumulus cloud, the first step towards thunderstorm development. Contrast with a downdraft.

Upland Highland; ground elevation above the lowlands along rivers or between hills.

Upper air westerlies Consistent winds that exist in the upper troposphere that flow east to west from about 20° of latitude to the poles.

Upper air/upper level The portion of the atmosphere which is above the lower troposphere. It is generally applied to the levels above 850 millibars. Therefore, upper level lows and highs, troughs, winds, observations, and charts all apply to atmospheric phenomena above the surface.

Upper mantle Layer of the Earth's interior extending from the base of the crust to 670 kilometres below the surface. Part of the Earth's mantle layer. The upper mantle is composed of peridotite, an ultramafic magma primarily made up of the minerals olivine and pyroxene. The top layer of the upper mantle, 100-350 km below surface, is called the asthenosphere.

Upslope effect The cooling of an air flow as it ascends a hill or mountain slope. If there is enough moisture and the air is stable, stratiform clouds and precipitation may form. If the air is unstable, there might be an increased chance of thunderstorm development. Contrast with a downslope effect.

Upslope fog Fog that forms when warm, moist surface air is forced up a slope by the wind. It is adiabatically cooled to below its initial dew point, which means the air cools by expansion as it rises. It forms best where there is a gradual slope, and it can become quite deep, requiring considerable time to dissipate. An example is Cheyenne Fog.

Upwelling The process by which water rises from a lower to a higher depth, usually as a result of divergence and offshore currents. It influences climate by bringing colder, more nutrient-rich water to the surface. A vital factor of the El Nino event.

Upwind The direction from which the wind is blowing. Also the windward side of an object. The opposite of the downwind or leeward side.

Urban area Geographic area with a high density of people over a limited area. Homes and other types of buildings tend to be close together. Urban systems also tend to differentiate themselves spatially into particular types of human activities.

Urban heat island Observed condition that urban areas tend to be warmer than surrounding rural areas.

Urbanization Expansion of cities into rural regions because of population growth. In most cases, population growth is primarily due to the movement of rural based people to urban areas. This is especially true in Less Developed Countries.

UTM coordinate system Based on the universal transverse mercator map projection, is a planar locational reference system which provides positional descriptions accurate to 1 meter in 2,500 across the entire earth's surface except the poles. At the poles, the Universal Polar Stereographic projection is used. The UTM system divides the earth's surface into a grid in which each cell, excluding overlap with its neighbours, is 6 degrees east to west, and 8 degrees north to south (with the exception of the row from 72-84 degrees north latitude). For any position in the UTM grid, X-Y coordinates can be determined in eastings and northings. Eastings are in meters with respect to a central meridian drawn through the centre of each grid zone (and given an arbitrary easting of 500,000 meters). In the northern hemisphere, northings are read in meters from the equator (0 meters). In the southern hemisphere, the equator is given the false northing of 10 million meters.

UV Ultraviolet. See ultraviolet radiation.

V

Vacuum 1. Space devoid of atoms or molecules.

2. Emptying of air.

Valley A linear depression in the landscape that slopes down to a stream, lake or the ocean. Formed by water and/or ice erosion.

Valley breeze An anabatic wind, it is formed during the day by the heating of the valley floor. As the ground becomes warmer than the surrounding atmosphere, the lower levels of air heat and rise, flowing up the mountainsides. It blows in the opposite direction of a mountain breeze.

Valley fog Fog formed by the movement of cooler, more dense air from higher elevations to the warm valley bottom.

Valley train A linear accumulation of glaciofluvial outwash sediments found in a once glaciated valley.

Valley wall The side slope of a stream or glacial valley.

Van Allen belts Doughnut-shaped regions encircling Earth and containing high energy electrons and ions trapped in the Earth's magnetic field (the magnetic field has definite boundaries, and is distorted into a tear-drop shape by the solar wind). Explorer I, launched by NASA in 1958, discovered this intense radiation zone. These regions are called the inner and outer Van Allen radiation belts, named after the scientist who first observed them.

Vapour pressure The pressure exerted by the molecules of a given vapour. In meteorology, it is considered as the part of total atmospheric pressure due to the water vapour content. It is independent of other gases or vapours.

Vapour trail A cloudlike streamer or trail often seen behind aircraft flying in clear, cold, humid air. A vapour trail is created when the water vapour from the engine exhaust gases are added to the atmosphere. Also called a contrail, for condensation trail.

Variable ceiling Occurs when the height of a ceiling layer increases and decreases rapidly, The ascribed height is the average of all the varying values.

Variance A statistical measure of the dispersion of observation values in a data set. The variance of a sample is the sum of the square of each value in the data set subtracted from the mean divided by one less than the total number of observations in the data set.

Varve A thin yearly deposit of sediment found on the bottom of a lake. Within each yearly varve, there are variations in the colour and the texture of the material deposited. The thickness of the varve and its associated layers can be used to reconstruct past environmental conditions influencing the lake.

Vascular plant Plant that has vascular tissues to transport water, nutrients, and other metabolic products.

Vector Displays and databases build all geographic features from points, that is from discrete X-Y locations. Lines are constructed from strings of points, and polygons (regions) are built from lines which close. Vector methods are sometimes contrasted with raster techniques which record geographic features within a matrix of grid cells. The choice between vector and raster GIS has much to do with the application being considered since both methods have strengths and weaknesses. Many current GIS permit transformation between vector and raster input and output.

Veering A change in wind direction that shifts clockwise in the Northern Hemisphere at a certain location. In the Southern Hemisphere, it is counter-clockwise. This can either happen in the horizontal or the vertical (with height). For example, the wind shifts from the north to the northeast to the east. It is the opposite of backing.

Velocity The speed of movement of an object in one direction.

Ventifact A loose piece of rock that has been polished smooth by wind transported particles. Common in arid environments.

Venturi An increase in the velocity of a fluid or gas due to the constriction of flow.

Vernal equinox Taking place in the Northern Hemispheric spring, it is the point at which the ecliptic intersects the celestial equator. Days and nights are most nearly equal in duration. It falls on or about March 20 and is considered the beginning of spring in the Northern Hemisphere and autumn in the Southern Hemisphere. It is the astronomical opposite of the autumnal equinox.

Vertebrate Animal that does have a backbone.

Vertical aerial photograph Photograph taken from a overhead or near overhead angle from a platform in the atmosphere.

Vertical temperature profile A series of temperature measurements taken at various levels in the atmosphere that show the thermal structure of the atmosphere over a specific location. Obtained through a rawinsonde sounding or comparable method, and exhibited in a skew t-log p diagram.

Vertical visibility The distance an observer can see vertically into an undefined ceiling, or the height corresponding to the top of a ceiling light projector beam, or the height at which a ceiling balloon disappears during the presence of an indefinite ceiling.

Vertical wind profile A series of wind direction and wind speed measurements taken at various levels in the atmosphere that show the wind structure of the atmosphere over a specific location. Obtained through a rawinsonde sounding or comparable method, and exhibited in a skew t-log p diagram.

Vertisols 1. Soil order (type) of the United States Natural Resources Conservation Service Soil Classification System. Tropical or sub-tropical soil that has a high clay and base cation status.

2. Soil order (type) of the Canadian System of Soil Classification. This soil has a high clay content and exhibits the effects of extreme contraction-expansion due to temporal changes in soil water content.

Virga Streaks or wisps of precipitation, such as water or ice particles, that fall from clouds but evaporate before reaching the ground. From a distance, the event sometimes may be mistaken for a funnel cloud or tornado. Typically, it may fall from altocumulus, altostratus,or high based cumuonimbus.

Viscosity The amount of the resistance to flow in a fluid due to intermolecular friction.

Visibility A measure of the opacity of the atmosphere, and therefore, the greatest distance one can see prominent objects with normal eyesight. The National Weather Service has various terms for visibility. Surface visibility is the prevailing visibility determined from the usual point of observation. Prevailing visibility is considered representative of visibility conditions at the station. Sector visibility is the visibility in a specified direction that represents at least a 45 degree arc of the horizon circle. Tower visibility is the prevailing visibility determined from the airport traffic control tower (ATCT) at stations that also report surface visibility.

Visible That part of the electromagnetic spectrum to which the human eye is sensitive, between about 0.4 and 0.7 micrometers.

Visible light The portion of the electromagnetic spectrum that can be detected by the human eye. It travels at the same speed as all other radiation, that is at 186,000 mile per second. It has a wave length longer than ultraviolet light and shorter than x-rays. Refer to light waves.

Visual Flight Rules (VFR) Refers to the general weather conditions pilots can expect at the surface. VFR criteria means a ceiling greater than 3,000 feet and greater than 5 miles visibility. Contrast with MVFR and IFR.

Volatile Organic Compounds (VOCs) Organic molecules that are mainly composed of carbon and hydrogen atoms (hydrocarbons). The most common volatile organic compound release into the atmosphere is methane. Involved in the formation of photochemical smog.

Volcanic ash Small sized particles ejected from explosive volcanoes.

Volcanic pipe A dyke reaches the surface of the Earth. Also called volcanic neck.

Volcanic vent An opening on a volcano through which lava is released and rock fragments and ash are ejected.

Volcano A naturally occurring vent or fissure at the Earth's surface through which erupt molten, solid, and gaseous materials. Volcanic eruptions inject large quantities of dust, gas, and aerosols into the atmosphere. A major component of volcanic clouds is sulphur dioxide, a strong absorber of ultraviolet radiation. Chemical interactions between sulphur dioxide and water cause sulphuric acid aerosols which can scatter some of the incident solar radiation back to space, thus causing a global cooling effect. For example, Mt. Pinatubo in the Philippines erupted in June 1991, and in the following year the global surface temperature was observed to decrease by about 0.3o C.

Volt The unit of electromotive force, or difference of potential, which will cause a current of one ampere to flow through a resistance of one ohm. Named for Italian physicist Alessandro Volta (1745-1827).

Volume The occupation of space in three dimensions. Measured in cubic units.

Vortex A rapid spiralling motion of air or liquid around a centre of rotation.

Vorticity maximum A centre of vorticity, or the maximum of the vorticity field of a fluid.

Vorticity The measurement of the rotation of a small air parcel. It has vorticity when the parcel spins as it moves along its path. Although the axis of the rotation can extend in any direction, meteorologists are primarily concerned with the rotational motion about an axis that is perpendicular to the earth's surface. If it does not spin, it is said to have zero vorticity. In the Northern Hemisphere, the vorticity is positive when the parcel has a counterclockwise, or cyclonic, rotation. It is negative when the parcel has clockwise, or anticyclonic, rotation. An example is relative vorticity.

W

Wall cloud An abrupt lowering of a cloud from its parent cloud base, a cumulonimbus or supercell with no visible precipitation underneath. Forming in the area of a thunderstorm updraft, or inflow area, it exhibits rapid upward movement and cyclonic rotation. It usually develops before strong or violent tornadoes.

Warm To have or give out heat to a moderate or adequate degree. A subjective term for temperatures between cold and hot. In meteorology, an air parcel that is warm is only so in relation to another parcel.

Warm advection The horizontal movement of warmer air into a location. Contrast with cold advection.

Warm desert Desert found in the subtropics or interiors of continents at the middle latitudes where precipitation is low and surface air temperatures are high.

Warm front The leading edge of an advancing warm air mass that is replacing a retreating relatively colder air mass. Generally, with the passage of a warm front, the temperature and humidity increase, the pressure rises, and although the wind shifts (usually from the southwest to the northwest in the Northern Hemisphere), it is not as pronounced as with a cold frontal passage. Precipitation, in the form of rain, snow, or drizzle, is generally found ahead of the surface front, as well as convective showers and thunderstorms. Fog is common in the cold air ahead of the front. Although clearing usually occurs after passage, some conditions may produced fog in the warm air.

Warm high A high pressure system that has its warmest temperatures at or near the centre of circulation. Contrast with a cold high. Examples include a cut-off high and an omega block.

Warm low A low pressure system that has its coldest temperatures at or near the centre of circulation. Contrast with a cold low.

Warning A forecast issued when severe weather has developed, is already occurring and reported, or is detected on radar. Warnings state a particular hazard or imminent danger, such as tornadoes, severe thunderstorms, flash and river floods, winter storms, heavy snows, etc.

Warping Is a method which can overcome some of the problems of edge matching . It can also be used to resolve registration problems arising from digitizing source maps of different scales, different projections, and varying coordinate systems.

Wasatch winds Strong winds blowing easterly out of the Wasatch Mountains in Utah, sometimes reaching speeds greater than 75 miles per hour.

Wash 1. Coarse alluvial sediments.

2. The downslope movement of small particles of soil by overland flow. Also called sheetwash.

3. A term used in the United States for a shallow intermittent stream channel found in arid and semi-arid regions.

Watch A forecast issued well in advance of a severe weather event to alert the public of the possibility of a particular hazard, such as tornadoes, severe thunderstorms, flash and river floods, winter storms, or heavy snows.

Water consumption The complete removal of water from some type of source, like groundwater, for some use by humans. This water is not returned to the source. Compare with water withdrawal.

Water Refers to the chemical compound, H_2O, as well as its liquid form. At atmospheric temperatures and pressures, it can exist in all three phases: solid (ice), liquid (water), and gaseous (water vapour). It is a vital, life-sustaining part of life on earth.

Water cycle Also called the hydrologic cycle, it is the vertical and horizontal transport of water in all its states between the earth, the atmosphere, and the seas.

Water table Top surface of groundwater.

Water vapour (moisture) Water in a gaseous form.

Water vapour (H_2O) Water in gaseous form. It is one of the most import constituents of the atmosphere. Due to its molecular content, air containing water vapour is lighter than dry air. This contributes to the reason why moist air has a tendency to rise.

Water withdrawal The removal of water from some type of source, like groundwater, for some use by humans. The water is subsequently returned some period of time later after its is used. The quality of the returned water may not be the same as when it was originally removed. Compare with water consumption.

Waterfall 1. A location in the long profile of a stream where water flows vertically. A nickpoint.

2. Vertical drop in elevation that causes a stream's discharge to flow vertically.

Watershed Catchment area of a drainage basin.

Waterspout A small, weak tornado, which is not formed by a storm-scale rotation. It is generally weaker than a supercell tornado and is not associated with a wall cloud or mesocyclone. It may be observed beneath cumulonimbus or towering cumulus clouds and is the water equivalent of a landspout.

Watt A metric unit of measurement of the intensity of radiation in Watts over a square meter surface (W/m^2 or $W\ m^{-2}$).

Wave In general, any pattern with some roughly identifiable periodicity in time and/or space. It is also considered as a disturbance that moves through or over the surface of the medium with speed dependent on the properties of the medium. In meteorology, this applies to atmospheric waves, such as long waves, short waves, Rossby waves, and cyclonic waves. In oceanography, this applies to waves generated by mechanical means, such as currents, turbidity, and the wind.

Wave crest The curved tops or ridges of an oscillating wave.

Wave cut notch A rock recess at the foot of a sea cliff where the energy of water waves is concentrated.

Wave cut platform A flat or slightly sloping bedrock surface that forms in the tidal zone. Caused by wave erosion.

Wave cyclone A cyclone which forms and moves along a front. The circulation around the cyclone's centre produces a wavelike deformation on the front. May also be call a migratory cyclone or low.

Wave height Vertical distance between a wave's trough and crest.

Wave length The least distance between particles moving in the same phase of oscillation of a wave. In oceanography, it is the horizontal distance between the highest parts of two successive wave crests above the still water level, separated by a trough that is below the still water level, and it is measured in meters.

Wave period The time elapsed for a wave to travel the distance of one wavelength.

Wave refraction The re-orientation of a wave so that it approaches a shoreline at a more perpendicular angle. This process is caused by the differential reduction of water depth as a linear wave approaches a curved shoreline. A reduction in water depth causes a wave to slow down causing the waves approaching a nonlinear shoreline to curve with the shore's shape.

Wave trough Area in between wave crests.

Weather The state of the atmosphere at a specific time and with respect to its effect on life and human activities. It is the short term variations of the atmosphere, as opposed to the long term, or climatic, changes. It is often referred to in terms of brightness, cloudiness, humidity, precipitation, temperature, visibility, and wind.

Weather Facsimile (WEFAX) A system for transmitting visual reproductions of weather forecast maps, temperature summaries, cloud analyses, etc. via radio waves. WEFAX transmissions are relayed by NOAAs geostationary GOES spacecraft.

Weather map A weather map is a map that shows weather conditions for a time period. Weather maps show storms, fronts, temperatures, rain, snow, sleet, fog, etc.

Weather symbols Some commonly used symbols are illustrated in the chart on the right.

Weather terms Clear Sky cloud-free to 30 percent covered.

Sunny: Sunshine 70-100 percent of the day

Partly sunny and partly cloudy Both terms refer to 40 to 70 percent cloud cover. Partly sunny is used in the day; partly cloudy is used at night.

Fog: A cloud on the ground. Fog is composed of billions of tiny water droplets floating in the air.

Snow: Precipitation of ice crystals.

Snow flurries: Intermittent snowfall that may result in little accumulation.

Sleet: Pellets of ice that form when rain or melting snowflakes freeze while falling. (Occurs in cold weather; hail usually occurs in summer.)

Freezing rain: Rain that turns to ice on impact with the surface.

Rain: Extended period of precipitation. Associated with large storm systems rather than single clouds or thunder storms

Showers: Brief interval of rain that does not affect a large area

Squall: Fast-moving thunderstorm or line of thunderstorms that often can produce damaging winds, hail, and tornadoes.

Hail: Pieces of ice that fall from thunderstorms. Hail often is composed of concentric rings of ice that form as the particle moves through "wet" and "dry" areas of the thunderstorm.

Weather vane Originally used as a wind vane, it is an instrument that indicates the wind direction. The name developed based on observations on what kind of weather occurred with certain wind directions. Creative designs often adorn the tops of barns and houses.

Weather warning Statement that dangerous weather is likely or is occurring. Take action.

Weather watch Statement about a particularly dangerous weather system that may occur at some specified time in the future.

Weathering The decay and break-up of rocks on the earth's surface by natural chemical and mechanical processes. The mechanical action includes large changes of temperature, high temperatures, frost, or the impact of windborne sand or water. Chemical action includes the chemical reactions between atmospheric constituents in a moist environments or in rain water. Biological agents are mainly fungi which attack organic material.

Weathering landform Is a landform created by the physical or chemical decomposition of rock through weathering. Weathering produces landforms where rocks and sediments are decomposed and disintegrated. This includes landforms with some of the following geomorphic features: karst, patterned ground, and soil profiles.

Wedge Primarily refers to an elongated area of shallow high pressure at the earth's surface. It is generally associated with cold air east of the Rockies or Appalachians. It is another name for a ridge, ridge line, or ridge axis. Contrast with a trough. Wedge is also a slang term for a large, wide tornado with a wedge-like shape.

Westerlies Usually applied to the broad patterns of persistent winds with a westerly component. It is the dominant persistent atmospheric motion, centred over the midlatitudes of each hemisphere. Near the earth's surface, the westerlies extend from approximately 35 to 65 degrees latitude, while in the upper levels they extend further polarward and equatorward.

Western hemisphere The Western Hemisphere is another name for the Americas (or the New World).

Wet bulb depression Dependent on the temperature and the humidity of the air, it is the difference between the dry bulb and the wet bulb readings.

Wet bulb thermometer A thermometer used to measure the lowest temperature in the ambient atmosphere in its natural state by evaporating water from a wet muslin-covered bulb of a thermometer. The wet bulb temperature is used to compute dew point and relative humidity. One of the two thermometers that make up a psychrometer.

Wetland A wetland is an area of land that is often wet; the soil in wetlands are often low in oxygen. Wetland plants are adapted to life in wet soil. There are many types of wetlands, including: swamp, slough, fen, bog, marsh, moor, muskeg, peatland, bottomland, delmarva, mire, wet meadow, riparian, etc.

Wetting and drying Physical weathering process where rocks are mechanically disintegrated by the accumulation of successive layers of water molecules in between the mineral grains of a rock. Sometimes called slaking.

Whirlwind A small-scale, rapidly rotating column of wind, formed thermally and most likely to develop on clear, dry, hot afternoons. Often called a dust devil when visible by the dust, dirt or debris it picks up. Also slang for a landspout or a tornado.

Whiteout When visibility is near zero due to blizzard conditions or occurs on sunless days when clouds and surface snow seem to blend, erasing the horizon and creating a completely white vista.

Wien's law This radiation law suggests that the wavelength of maximum emission of any body is inversely proportional to its absolute temperature. The following equation mathematically describes this law: lmax = C/T

where lmax is the body's maximum emitted wavelength of radiation in micrometers (μm),

C is a constant equal to 0.2897,

and T is the temperature of the body in Kelvins.

Willy willy Australian term for tropical cyclone, hurricane.

Wilting point The point at which the rate of water leaving a plant's leaves is greater than the water uptake by the roots. At this point the plant will fail to recover its turgidity.

Wind A natural motion of the air; especially a noticeable current of air moving in the atmosphere parallel to the Earths' surface. Winds are caused by unequal heating and cooling of the Earth and atmosphere due to absorbed, incoming solar radiation and infrared radiation lost to space-as modified by such effects as the Coriolis force, the condensation of water vapour; the formation of clouds, the interaction of air masses and frontal systems, friction over land and water; etc. The chart above is an abbreviated version of the Beaufort Wind Scale, named for the British admiral who invented it in 1805.

Wind chill The wind can reduce significantly the amount of heat your body retains. The following wind chill chart does not take into account such variables as type of clothing worn, amount of exposed flesh, and physical condition, all of which would alter body heat.

simple wind-chill equation

Tw = TA - .5 x VA

Tw = wind chill

TA = air temperature

VA = wind speed

For example, if the temperature is 20o and the wind 20 mph: Tw 20 - 1.5×20

Tw = 20 - 30

Tw =-10°F .

Wind chill index The calculation of temperature that takes into consideration the effects of wind and temperature on the human body. Describes the average loss of body heat and how the temperature feels. This is not the actual air temperature. For an example, check out the wind chill chart.

Wind direction The direction from which the wind is blowing. For example, an easterly wind is blowing from the east, not toward the east. It is reported with reference to true north, or 360 degrees on the compass, and expressed to the nearest 10 degrees, or to one of the 16 points of the compass (N, NE, etc.).

Wind ripples Wind ripples are miniature sand dunes between 5 centimetres and 2 meters in length and 0.1 to 5 centimetres in height. They are created by saltation when the sand grains are of similar size and the wind has a constant speed. Also called sand ripples.

Wind shear The rate of wind speed or direction change with distance. Vertical wind shear is the rate of change of the wind with respect to altitude. Horizontal wind shear is the rate of change on a horizontal plane.

Wind shift The term applied to a change in wind direction of 45 degrees or more, which takes place in less than 15 minutes. It may the result of a frontal passage, from katabatic winds, sea breezes, or thunderstorms, and in some instances, the change may be gradual or abrupt.

Wind speed The rate of the motion of the air on a unit of time. It can be measured in a number of ways. In observing, it is measured in knots, or nautical miles per hour. The unit most often used in the United States is miles per hour.

Wind vane A mechanical device used to measure the direction of wind flow. Usually consists of a horizontal bar with a fin at one end and a aerodynamic pointer at the other end. The centre of horizontal is attached to a vertical spindle which is connected to a mechanical device that records direction.

Wind vector Arrow representing wind velocity. The arrow points in the direction of the wind. The length of the arrow is proportional to wind speed.

Wind velocity Vector term that includes both wind speed and wind direction.

Wind wave An ocean or lake wave resulting from the action of wind on the water's surface. After it leaves its fetch area, it is considered a swell.

Window Term used to denote a region of the electromagnetic spectrum where the atmosphere does not absorb radiation strongly.

Windward The direction from which the wind is blowing. Also the upwind side of an object. The opposite of the downwind or leeward side.

Winkel Tripel Projection A Winkel Tripel projection is a type of pseudocylindrical projection map in which both the lines of latitude and longitude are curved. The Winkel Tripel projection was adopted by the National Geographic Society in the late 1990s (replacing the Robinson projection).

Winter Astronomically, this is the period between the winter solstice and the vernal equinox. It is characterized as having the coldest temperatures of the year, when the sun is primarily over the opposite hemisphere. Customarily, this refers to the months of December, January, and February in the North Hemisphere, and the months of June, July, and August in the Southern Hemisphere.

Winter solstice Date when the declination of the sun is at 23.5° South of the equator. This date is usually December 21 or 22.

Winter storm Any one of several storm systems that develop during the late fall to early spring and deposit wintry precipitation, such as snow, freezing rain, or ice. Examples include blizzards, ice storms, and nor'easters.

World Meteorological Organization (WMO) From weather prediction to air pollution research, climate change related activities, ozone layer depletion studies and tropical storm forecasting, the World Meteorological Organization coordinates global scientific activity to allow increasingly prompt and accurate weather information and other services for public, private and commercial use, including international airline and shipping industries. Established by the United Nations in 1951, it is composed of 184 members.

World Ocean Circulation Experiment (WOCE) A study of the general global circulation of the oceans. It emphasizes the measurements and understanding needed to describe and understand the circulation, to simulate it, and to predict its changes in response to climatic changes.

WWW World weather watch.

X

X-axis Horizontal axis on a graph.

Xerophyte Plant that have adaptations to survive prolonged periods of soil drought.

X-ray radiation Form of electromagnetic radiation with a wavelength between 0.03 to 30 nanometres.

X-rays The portion of the electromagnetic spectrum that has a very short wave length. It has a wave length longer than gamma rays, yet shorter than visible light. X-rays can penetrate various thicknesses of all solids, and when absorbed by a gas, can result in ionization.

Y

Yagi A type of receiving antenna that has several rod elements mounted on a beam. Its directional pattern of sensitivity and ease of construction make it ideal for APT direct readout stations.

Yardang Rock that has developed a streamline form because of wind erosion. The long axis of these features is aligned with the dominant wind direction.

Y-axis Vertical axis on a graph.

Yazoo tributary Small tributary channel that is prevented from joining the main stream channel by the presence of levees. Yazoo tributaries tend to flow on the floodplain parallel to the main stream channel.

Year The interval required for the earth to complete one revolution around the sun. A sidereal year, which is the time it take for the earth to make one absolute revolution around the sun, is 365 days, 6 hours, 9 minutes, and 9.5 seconds. The calendar year begins at 12 o'clock midnight local time on the night of December 31st-January 1st. Currently, we operate under the Gregorian calendar of 365 days, with 366 days every four years, a leap year. The tropical year, also called the mean solar year, is dependent on the seasons. It is the interval between two consecutive returns of the sun to the vernal equinox. In 1900, that took 365 days, 5 hours, 48 minutes, and 46 seconds, and it is decreasing at the rate of 0.53 second per century.

Yellow snow Snow that is given golden, or yellow, appearance by the presence of pine or cypress pollen in it.

Younger-dryas A cold period during the generally mild Holocene epoch that occurred from about 10,000 - 8,500 BC. Scientists speculate that this cooling may have been caused by the release of fresh water trapped behind ice on North America into the North Atlantic Ocean.

Z

Zenith The point which is elevated 90 degrees from all points on a given observer's astronomical horizon. The point on any given observer's celestial sphere that lies directly above him. The opposite of nadir.

Zenith telescope Instrument for observing starts near the zenith (a point on the celestial sphere directly above the observer's position).

Zephyr A Mediterranean term for any soft, gentle breeze.

Zodiac The position of the sun during the course of the year as it appears to move though successive constellations. Also, the band where the ecliptic runs centrally through the celestial sphere and contains the sun, the moon, and all the planets except Venus and Pluto.

Zonal Movement of wind or ocean waters in a direction that is roughly parallel to the lines of latitude.

Zonal flow The flow of air along a latitudinal component of existing flow, normally from west to east. Contrast with meridional flow.

Zonal index The measure of the strength of the westerly winds of the middle latitudes. It is expressed as the horizontal pressure difference between 35 degrees and 55 degrees latitude, or as the corresponding geostrophic wind.

Zone of ablation Area of a glacier where losses of ice from melting, evaporation, and sublimation exceed additions of snow annually.

Zone of accumulation Area of a glacier where additions of snow exceed losses of ice from melting, evaporation, and sublimation.

Zone of aeration Horizontal zone that extends from the top of the water table to the ground surface. Soil and rock pore spaces in this zone may and may not have water.

Zone of saturation Groundwater zone within the Earth's bedrock where all available pores spaces are filled by water. Found beneath the water table.

Zooplankton Small heterotrophic organisms found inhabiting aquatic ecosystems.

Zulu time One of several names for the twenty-four hour time which is used throughout the scientific and military communities. Other names for this time measurement are Universal time coordinate (UTC) or Greenwich mean time (GMT).